A Textbook of
SOIL CHEMISTRY

The Author

Professor Saroj Kumar Sanyal has **brilliant academic career** with undergraduate, post-graduate, Ph.D. and Post-Doctoral studies at highly reputed Institutes, namely the **Presidency College**, affiliated to the **University of Calcutta; Indian Agricultural Research Institute (IARI), New Delhi; University of Cambridge, United Kingdom;** and the **International Rice Research Institute, Philippines.**

Professor Sanyal is endowed with strong teaching skills and enjoyed teaching soil science and physical chemistry spanning over **more than 35 years** at several Institutions. **Presently, Professor Sanyal has been selected as an Adjunct Professor in Soil Science & Agricultural Chemistry of IARI, New Delhi (2013-2018)** and **Bihar Agricultural University, Sabour, Bhagalpur, Bihar (2014-2019).** Professor Sanyal also discharged the **editorial responsibilities** for a number of journals that highlights his strong editorial skills.

As a researcher, Professor Sanyal has been successful in securing **funding support** from a large number of agencies, within and beyond the National Agricultural Research Systems (NARS). *Professor Sanyal's research work spanned in diverse, locally and globally relevant areas such as transport processes in soil-plant systems; chemistry and soil fertility relationships of soil potassium and phosphorus as plant nutrients, soil organic matter dynamics including metal-humic interactions, integrated nutrient and water management, farming system research, site-specific nutrient management, balanced fertilization and nutrient mining, dynamics of pollutants in soil-plant-water system along with mitigation of these pollutants through adoption of appropriate management practices and waste water recycling. Such research has led to several publications in national and international peer-reviewed journals.* During his research career, Professor Sanyal also acted as the Principal Investigator of multi-disciplinary and multi-institutional teams of researchers in addressing the researchable issues under funding from the ICAR, New Delhi, as well as other funding organizations. Professor Sanyal guided 26 research scholars to their Ph.D. degree from different Universities with funding support from several funding sources.

Presently, Professor Sanyal has been selected as the Chairman of the Research Advisory Committee, National Bureau of Soil Survey & Land Use Planning (ICAR), Nagpur (2013-2016) and also as the **Chairman** of the **Research Advisory Committee** of the **Central Research Institute for Jute & Allied Fibres, Barrackpore, West Bengal (2014-2017),** as well as the **Chairman, Research Advisory Committee** of the **Central Research Institute of Sericultural Research & Training Institute, Central Silk Board, Ministry of Textiles, Government of India, Berhampore, West Bengal (2014-2017)** and the **Chairman of the Scientific Advisory Committee, Tocklai Tea Research Association, Jorhat, Assam (2015-2019).**

Professor Sanyal was elected as the **President** of the **Section of Agriculture & Forestry Sciences in the 92nd Session (2004-2005) of the Indian Science Congress Association**. He also discharged the responsibility at the national level in reviewing the performance of a number of ICAR Institutes and also All India Coordinated Research Projects of ICAR as the Member/Chairman of the Quinquennial Review Teams (**QRTs**) appointed by the **ICAR**. These responsibilities highlight Professor Sanyal's strong academic standing among the scientific as well as the policy-making bodies of India. His scientific standing in the country was honored as the **Fellowship** of **several National Academies and Learned Societies,** such as the *National Academy of Agricultural Sciences, New Delhi; Indian Society of Soil Science, New Delhi; West Bengal Academy of Sciences & Technology, Kolkata.* Professor Sanyal delivered a large number of prestigious Lectures at different forum, as well as received **Awards, Laurels and professional recognition from the peers.**

As Professor Sanyal progressed through his professional career, he has been entrusted with higher responsibilities of managing the overall research programmes as the **Director of Research**, and finally the overall management as the **Vice-Chancellor** of **Bidhan Chandra Krishi Viswavidyalaya** (BCKV), the largest and the oldest State Agricultural University of the State of West Bengal, India.

A Textbook of
SOIL CHEMISTRY

Dr. Saroj Kumar Sanyal

M.Sc. (IARI), Ph.D. (Cambridge)
FNASS, FISSS, FAScT

Former Vice-Chancellor
Bidhan Chandra Krishi Viswavidyalaya, Mohanpur, West Bengal
Adjunct Professor in Soil Science and Agricultural Chemistry
Indian Agricultural Research Institute, New Delhi
Former Visiting Scientist
International Rice Research Institute, Philippines

2018

Daya Publishing House®

A Division of

Astral International Pvt. Ltd.

New Delhi – 110 002

Publisher's Note:

Every possible effort has been made to ensure that the information contained in this book is accurate at the time of going to press, and the publisher and author cannot accept responsibility for any errors or omissions, however caused. No responsibility for loss or damage occasioned to any person acting, or refraining from action, as a result of the material in this publication can be accepted by the editor, the publisher or the author. The Publisher is not associated with any product or vendor mentioned in the book. The contents of this work are intended to further general scientific research, understanding and discussion only. Readers should consult with a specialist where appropriate.

Every effort has been made to trace the owners of copyright material used in this book, if any. The author and the publisher will be grateful for any omission brought to their notice for acknowledgement in the future editions of the book.

Cataloging in Publication Data--DK
Courtesy: D.K. Agencies (P) Ltd. <docinfo@dkagencies.com>

Sanyal, Saroj Kumar, author.
A textbook of soil chemistry / Dr. Saroj Kumar Sanyal.
  pages cm
Includes bibliographical references and index.
ISBN 9789387057241 (International Edition)

1. Soil chemistry.   I. Title.

LCC S592.5.S26 2017 I DDC 631.41   23

Published by   : **Daya Publishing House®**
                         A Division of
                         Astral International Pvt. Ltd.
                         – ISO 9001:2015 Certified Company –
                         4736/23, Ansari Road, Darya Ganj
                         New Delhi-110 002
                         Ph. 011-43549197, 23278134
                         E-mail: info@astralint.com
                         Website: www.astralint.com

Dr. N. N. Goswami, *FNAAS, FISSS*
Former Vice-Chancellor,
C.S. Azad University of Agriculture & Technology,
Kanpur, Uttar Pradesh
Former Dean & Joint Director (Education),
Indian Agricultural Research Institute, New Delhi
Honorary Member & Past President,
Indian Society of Soil Science, New Delhi
President, Section of Agricultural Sciences,
Indian Science Congree Association (1985-86)
Chairman, Commission IV (Soil Fertility & Plant Nutrition),
International Society of Soil Science
(International Union of Soil Sciences) (1986-90)

Residence:
13/5, Primrose,
Gardenia Street,
Vatika City, Sohna Road,
Sector-49,
Gurgaon-122 018.
Haryana
Mobile: 9810799002
Email:
nagengoswami@gmail.com
argoswami@hotmail.com

Foreword

I take pleasure indeed to write the Foreword for the book, entitled **A Textbook of Soil Chemistry,** by **Dr. Saroj Kumar Sanyal.** It is well recognized that soil is a complex living natural body, and its physics, chemistry and biology has evolved over last one hundred years or so. The present book deals with the chemistry of soil which involves application of the basic concepts and principles of chemistry to a heterogeneous, complex, living system, affected not only by climate, crop and other vegetation, relief and parent material, but also by human interference over time.

The book elucidates and discusses about the earth's crust, rocks, minerals and soils, along with the weathering sequence and stability of the silicate minerals; soil colloids; ion-exchange processes; chemistry of soil acidity and salt-affected soils; soil organic matter including metal-humic interactions and clay-humus complex; chemistry of phosphorus and micronutrients in soil; chemistry of waterlogged soils; transport processes in soil covering isothermal and thermal diffusion of water and solutes in soil, and also the isothermal coupled transport of aqueous solutes in soils and clays; and environmental aspects of soil chemistry. A very significant feature of the book has been to relate the soil processes to the concepts in basic sciences.

As for the author, I have known Dr. Saroj Kumar Sanyal for more than last four decades. He has a brilliant academic record, with B.Sc. (Hons.) and M.Sc. degrees from highly reputed Institutes in India, *viz.* Presidency College, affiliated to the University of Calcutta, and Indian Agricultural Research Institute, New Delhi, respectively, and Ph.D. degree from Cambridge University, United Kingdom. Dr. Sanyal undertook Post-doctoral research at the International Rice Research Institute, Philippines. While at the Indian Agricultural Research Institute, New Delhi, Dr. Sanyal read with me soil chemistry. Dr. Sanyal has more than 35 years of teaching experience at graduate and post-graduate levels in different Universities in India and has established himself as an able research leader and a reputed Professor of Soil Science with major focus in soil chemistry and physical chemistrty of soil.

He has guided over two dozens of research scholars for their Ph.D. degree with funding support from diverse sources.

The book, being written by an outstanding researcher and teacher in soil chemistry, reads well with exemplary clarity. I do believe and hope that the book will provide an excellent text and source material for all engaged in research and teaching in soil science. To my knowledge, there have been very few books on soil chemistry, and this one by Saroj Kumar Sanyal will be a wonderful addition with novelty in its approach and treatment.

N.N. Goswami

Preface

The present book makes an attempt to discuss the basic tenets of the principles of soil chemistry, a field that has grown over the years. Wherever possible, the contents have been related to the underlying processes in soil. An endeavour is also made to relate the material presented to the concepts of basic sciences that provide the foundation of the subject of soil chemistry. Soil, by its very nature, is a complex heterogeneous system in which large number of live organisms exists. The latter profoundly influence the processes that take place in soil, with particular reference to the nutrient transformations. This book makes an effort to reckon this fact in chapters where it is of greatest relevance. The chapter on soil organic matter is an example in this regard.

The book intends to cater the needs of the post-graduate and Ph.D. students, *notwithstanding* the fact that it also addresses the need of the undergraduate students studying soil science as a part of their undergraduate course curricula. If the book succeeds to kindle the interest of the students in the field, in addition to satisfying the requirement of their curricula, the author would consider his effort meaningful.

I am indebted to my Professors who initiated me to the intricacies of the subject of soil science, in general, and soil chemistry as well as physical chemistry, in particular. I would also like to acknowledge the continued support that I received from my wife, Dr. (Mrs.) Jhuma Sanyal, Associate Professor of History at the Women's Christian College, Kolkata (Retired) during the preparation of the manuscript. Last but not the least, I thank my son, Mainak Sanyal, for his assistance in course of the composition of the manuscript. I would also like to thank the Publishers for their skill and care in bringing out this book.

Saroj Kumar Sanyal

Reviews

June 20, 2017

There has been a genuine need of a text book on Soil Chemistry to cater to the requirements of the students of Soil Science. "A Text Book on Soil Chemistry", authored by Prof. Saroj Kumar Sanyal, is an effort in that direction. "A Text Book of Soil Chemistry", sees the return of Prof. Saroj Kumar Sanyal as a teacher, as a writer. Prof. Saroj Kumar Sanyal, an academician *par excellence*, has been an exponent in the discipline of Soil Chemistry.

In the current book, Prof. S.K. Sanyal has put together the basic concepts of soil, *viz.* its formation, the rocks, parent materials, soil colloids, layer silicates, the ion exchange capacity, the process of nutrient movement and uptake, physical and chemical processes in soils, chemistry of acid soils, problem soils and their amelioration.

The book elaborates chemistry of major- and micro-nutrients and transformation of most reluctantly soluble soil phosphorus and nutrient kinetics. It deals with soil-fertilizer-plant continuum, depicting the soil and fertilizer-mediated environmental pollution.

Soil organic matter, which is a heterogeneous mixture of raw plants and animals debris, is a vital component of any living soil. Humus which is the ultimate decomposed organic material is perhaps the eye of a needle through which organic matter passes more than once. The humus, its complexes with metals, their behaviour, and biology are very elaborately discussed here.

Most relevant subjects, on which the book has dealt comprehensively, are environmental soil chemistry, heavy metals and metalloids, impact of solid waste on the environment and potential methods of their remediation. The aspects covered in this book relate to fundamentals and basics of soils, which is associated with a number of intrinsic processes in relation to nutrient transformation, availability and their kinetics.

The book is equally important to beginners, undergraduates, post-graduates, as well as Ph.D. students, so far as their course curricula are concerned.

Dr. D.D. Patra
Vice-Chancellor
Bidhan Chandra Krishi Viswavidyalaya
Mohanpur, West Bengal

July 26, 2017

Soil Chemistry-an important sub-discipline of Soil Science helps us to understand the intricacies associated with soil as a complex natural system, different soil processes and functions, and their relevance to the society. In fact, the need for a good textbook on this subject was felt since long, as only few books of foreign authors are available. In this backdrop, I find the present book an outstanding publication that meets very well the requirements and expectations of all concerned, especially the post-graduate students and teachers. Involvement of basic concepts of chemistry, and also of physics, in soil processes is explained in such a lucid manner that even the undergraduate students would find the book very useful to understand the basics of soil chemistry.

The book consists of eight chapters covering entire spectrum of soil chemistry,*viz.*, composition of earth's crust, rocks, minerals and soils; structure and properties of soil colloids; ion exchange; soil acidity and salt-affected soils; soil organic matter; chemistry of waterlogged soils, phosphorus and micronutrients; solute transport processes; and soil pollution and its amelioration/management. Inclusion of a chapter on environmental soil chemistry dealing with geogenic and anthropogenic pollution makes this book unique, as other available books hardly address this important area although the same has become so critical for crop production, animal/human health and agri-export in recent years. Moreover, the authored books have apparently an edge over the edited ones in terms of continuity of the subject across the chapters and overall comprehension.

The author, Prof. Saroj Kumar Sanyal, is an outstanding researcher and teacher. He is so well-recognized internationally for his contributions to the discipline of soil chemistry in general, and to soil pollution in particular. The book obviously embodies the subject matter knowledge and expertise of the brilliant author, as apparent from the clarity of expressions throughout the chapters. Nonetheless, the author may think of including (i) one chapter dedicated to important terms and laws pertaining to basic chemistry, and (ii) few pertinent questions at the end of each chapter, during subsequent revision of the book in future in order to further enhance its usefulness to the students and teachers. I am sure that the book will prove immensely useful resource to the students, teachers and researchers in the discipline of Soil Science, as also to the libraries of Academic Institutions in India and abroad.

Dr. B.S Dwivedi

Head, Division of Soil Science & Agricultural Chemistry

Indian Agricultural Research Institute, New Delhi

July 17, 2017

It gives me great pleasure in reviewing the book, entitled "A Textbook of Soil Chemistry", written by Prof. (Dr.) Saroj Kumar Sanyal, which is a significant contribution to aid soil science education and research.

The present book has dealt with the chemistry of soil which involves application of the basic concepts and principles of chemistry to the heterogeneous, complex and live soil system. Notwithstanding the fact that soil is a complex material, the present book has made a concerted effort with remarkable success to unravel the basic processes in soil, accompanying several important transformations, with direct bearing to its use for agricultural production. Professor Sanyal is particularly successful in relating the principles of basic chemistry to the intricate processes in soil, thereby leading to an in-depth understanding of the soil processes.

Prof. (Dr.) Saroj Kumar Sanyal, an academician *par excellence*, has been a leader in the subject of Soil Science, in particular of Soil Chemistry. In the present book, he exhibited his remarkable skill in preparing an accomplished knowledge base for the beginners, as well as the advanced learners in the field of study in this comprehensive text book.

The book, a legacy of Prof. Sanyal's outstanding teaching and research contributions in the field of soil chemistry, will certainly cater to the needs of the post-graduates, and the Ph.D. students in soil science, and will also serve as teaching material for teachers and scientists of the agricultural sciences.

Dr. K. Majumdar

Vice President, Asia and Africa Programs

International Plant Nutrition Institute (IPNI)

Gurgaon, Haryana

India

July 22, 2017

Dr. S. K. Sanyal has made great efforts in writing the book entitled, 'A Textbook of Soil Chemistry'. The book is well written and it covers all the important information related to soil chemistry with examples from Indian literature. The book layout is excellent and there are minimal errors. There is an urgent need for such book for the benefit of graduate students studying soil science in colleges and universities in India.

Dr. Yadvinder Singh

INSA Senior Scientist & Former Head

Department of Soil Science

Punjab Agricultural University, Ludhiana

June 29, 2017

The book entitled 'A Textbook of Soil Chemistry 'is the result of a very sincere and dedicated effort by an eminent Soil Scientist and well known academician in the field of Agricultural Sciences in India,Professor S.K.Sanyal. I am happy and pleased to have gone through the contents of the Book, which I find to be extremely interesting and valuable.

Soil Chemistry primarily deals with the chemical constitution, properties and reactions in soil. On a practical point of view, it provides the basis for solving problems,as diverse as soil fertility and plant nutrition,management of polluted and problematic soils,fixation,release,mobility of nutrients in soils,environmental degradation,*etc*. On this account,selection of the eight well-thought of Chapters by the author is appropriate and praiseworthy.

To sum up, Dr.Sanyal's effort to bring out this publication on 'A Textbook of Soil Chemistry' is praiseworthy. This book will cater to the needs of post-graduate students and research scholars of the State and the Central Agricultural Universities and Research Institutions in the disciplines of Soil Science and Agricultural Chemistry, Environmental Sciences,Horticulture,Agronomy,Agricultural Physics,*etc*.This book will serve as a research guide and teaching material for teachers and scientists of Agricultural Institutions in India and other developing countries.

I congratulate Dr.S.K.Sanyal for this valuable contribution and wish him all the success.

Dr. A.K. Sarkar

Retired Dean, College of Agriculture &

Former Chairman, Department of Soil Science & Agricultural Chemistry

Birsa Agricultural University, Kanke, Ranchi, Jharkhand

Contents

An Introduction to Earth's Crust, Rocks, Minerals and Soils

1. Earth's Crust

The chemical composition of the Earth's crust (Table 1.1) is dominated by handful of elements, namely oxygen, silicon, aluminium, iron, calcium, sodium, potassium, magnesium, titanium, phosphorus, manganese and fluorine. The less abundant elements include sulphur (0.052 per cent), chlorine (0.048 per cent), barium (0.043 per cent), carbon (0.032 per cent), rubidium (0.028 per cent), zirconium (0.022 per cent), chromium (0.0.020 per cent), strontium (0.015 per cent), vanadium (0.015 per cent), nickel (0.010 per cent), zinc (0.008 per cent), copper (0.007 per cent), cobalt (0.004 per cent), lithium (0.006 per cent), tin (0.004 per cent), lead (0.002 per cent), boron (0.001 per cent), and so on (Bear, 1976).

All these less abundant constituents except water occur only in very small quantities in the Earth's crust, and their total content is less than 1 per cent. The mean density of the upper crust ranges from 2.69 $g.cm^{-3}$ to 2.74 $g.cm^{-3}$, while for lower crust, it ranges over 3.00-3.25 $g.cm^{-3}$.

It is evident from Table 1.1 that oxygen accounts for nearly 50 per cent by weight of the composition of the Earth's crust, while by volume, its share is more than 90 per cent.

1.1. Soil

Soil is a complex natural formation having both weathered and unweathered materials as well as several altered products of both organic and inorganic

Table 1.1: Major Elements in the Earth's Crust

Element	Per cent by Weight	Per cent by Volume	Element	Per cent by Weight	Per cent by Volume
Oxygen	46.6	92.0	Silicon	27.7	0.80
Aluminum	8.13	0.77	Iron	5.00	0.68
Calcium	3.63	1.48	Sodium	2.83	1.60
Potassium	2.59	2.14	Magnesium	2.09	0.56
Titanium	0.440		Phosphorus	0.120	
Manganese	0.100		Fluorine	0.080	
Sulphur	0.052		Chlorine	0.048	
Barium	0.043		Carbon	0.032	
Rubidium	0.028		Zirconium	0.022	
Chromium	0.020		Strontium	0.015	
Vanadium	0.015		Nickel	0.010	
Zinc	0.008		Copper	0.007	
Cobalt	0.004		Lithium	0.006	
Tin	0.004		Lead	0.002	
Boron	0.001		Tungsten	0.0001	

Source: Adapted from Bear (1976).

origin, water regime, gaseous phase, as well as life forms–both plant and animal. Chemically, the soil solid phase is a mixture of both inorganic and organic materials. Those having more than 20 per cent organic matter are generally referred to as the organic soils, while ones with organic matter less than this limit are classified as mineral soils. The soils of the tropical and sub-tropical areas (such as ours) mostly fall in the latter category, that is, mineral soils.

The soil solid phase (< 2.0 mm size) is composed of three types of primary particles depending upon their sizes and hence the properties. These are sand–the coarsest particles (coarse sand: 2.0-0.2 mm; fine sand: 0.2-0.02 mm), silt (0.02-0.002 mm) and finally the finest and most reactive fraction, clay (< 0.002 mm). The distribution of these particle size-fractions in a soil determines its texture. The latter is a permanent property of the soil, and does *not* change when the soil is subjected to various activities including the agricultural production practices. The soil structure, on the other hand, refers to the arrangement of these primary particles through bonding, aided by several cementing materials-either natural or added by anthropogenic activities-in soil, into secondary aggregates. The latter are obviously amenable to changes by the agricultural activities involving soils. Such soil structure is vital in determining the physical *tilth* (condition or *health*) of the soil with direct bearing upon water and air movement in soil, crop root development, as well as the microbial activities in soil.

As stated above, the soil solid phase could be of inorganic or organic in nature.

1.1.1. Inorganic Phase

1.1.1.1. Rocks

The inorganic materials consist of both primary and secondary (or altered) minerals. These minerals originate from **rocks**, which could be *igneous, sedimentary or metamorphic* in nature. The igneous rocks are the oldest in origin, is formed through the cooling and solidification of molten magma or lava. Examples are granite, basalt, pumice, andesite, rhyolite, diorite, etc.

The igneous rocks may be of three types, namely intrusive (plutonic), extrusive (volcanic) or hypabyssal. The intrusive rocks are obtained from molten magma which cools and solidifies within the Earth's crust. Extrusive or volcanic igneous rocks are formed at the surface of the earth's crust from partial melting of rocks within the mantle and crust. Compared to intrusive igneous rocks, the extrusive igneous rocks cool and solidify more rapidly. The hypabyssal igneous rock, also known as subvolcanic rock, is formed at a depth in between the plutonic and volcanic rocks, *i.e.*, it is an intrusive igneous rock that is emplaced at medium to shallow depths within the crust, and has intermediate grain-size and often porphyritic texture between that of volcanic and plutonic rocks. Common examples of subvolcanic rocks are diabase, quartz-dolerite, micro-granite and diorite.

Sedimentary rocks are those which form through deposition and subsequent re-cementation of sediments generating from weathering materials from other rocks at the Earth's surface and within bodies of water, deposited over time, usually as layers at the bottom of lakes and oceans. Such sediment may cover minerals, plant pieces and other organic matter. The sediment is compressed for a prolonged period so as to consolidate into solid layers of rock. Even though sedimentary rocks cover the majority of the Earth's rocky surface, they account for only a small percentage of the Earth's crust, compared to the metamorphic (*see* below) and igneous types of rocks. Examples are provided by sandstone, shale, limestone, chalk (a soft, white form of limestone), dolomite, gypsum, conglomerate, coal, claystone, flint amber, etc. For instance, flint is a hard, sedimentary form of the mineral quartz in igneous rocks.

Metamorphic rocks are those formed over time from the igneous and the sedimentary rocks by extreme pressure and heat. These rocks can be formed by pressure deep under the Earth's surface, from the extreme heat caused by magma or by the intense collisions and friction of tectonic plates. Uplift and erosion help bring metamorphic rock to the Earth's surface. Examples include anthracite, quartzite, marble, slate, granulite, gneiss, schist, etc. Limestone forms the metamorphic rock marble, when subjected to extreme heat and pressure over time (metamorphism), while quartzite is obtained from the sedimentary rock sandstone and slate is a metamorphic rock that is formed from the sedimentary rock mudstone. Further, sandstone forms the metamorphic rock quartzite, whereas anthracite is a type of coal with a high carbon count, few impurities and with a high luster (meaning it looks shiny). Gneiss is a high-grade metamorphic rock, which has been subjected to higher temperatures and pressures than schist. It is formed by the metamorphosis of granite, or sedimentary rock. On the other hand, most schist is derived from clays and muds that have been subjected to metamorphism *via* the intermediate steps of production of shales, slates and Phyllite. Certain schists are also derived from fine-grained igneous rocks such as basalts and tuffs.

1. 1.1.2. Minerals

A mineral is an inorganic natural material with a definite chemical composition, having a crystalline structure, if formed under conducive environment, as well as defined physical properties (Biswas and Mukherjee, 1994).

Primary Minerals

These are those which, since their crystallization from the molten magma, have undergone no or small changes in their chemical composition. These primary minerals with magma or metamorphic origins pass into a soil from massive crystalline, and to a lesser extent, from a sedimentary rock (Zone, 1986). They generally occur in sand and silt fractions of soil.

Primary minerals cover the following:

Oxides cover *Quartz*, SiO_2; *Hematite*, Fe_2O_3, *Magnetite*, Fe_3O_4; *Rutile*, TiO_2; *Disthene*, Al_2SiO_3.

Silicates, namely Calcium, Magnesium and Ferrous Metasilicates, *e.g.*, $CaSiO_3$, $MgSiO_3$, $FeSiO_3$; Magnesium-Ferrous Orthosilicate, namely Mg_2SiO_4, $(Mg,Fe^{II})_2$, Fe_2SiO_4 which form the *Olivine* group of minerals.

Aluminosilicates include *Micas*, namely *Muscovite, Biotite, Phlogopite*, etc.; *Feldspars*, namely potash feldspars *Orthoclase*, $(KNa)AlSi_3O_8$; *Microline*, $KAlAlSi_3O_8$; *Plagioclase Feldspars*, namely *Albite*, $NaAlSi_3O_8$; *Anorthite*, $CaAlSi_2O_8$ as well as the corresponding intermediate products; *Pyroxenes* (*Single chains*) and *Amphiboles* (*Double chains*). These primary minerals will be discussed in more details in the companion **Chapter** on **Soil Colloids: Structural Aspects and Properties**.

Sulphides include iron sulphides, *Iron Pyrite*, FeS_2.

Phosphates include the most abundant phosphate mineral, namely *Apatites*. The latter cover the fluorapatites, $Ca_{10}(PO_4)_6F_2$, *Hydroxy Apatites*, $Ca_{10}(PO_4)_6 \cdot (OH)_2$ and *Carbonate Fluorapatite*, or *Francolites* $Ca_{10-a-b}Na_aMg_b(PO_4)_{6-x}(CO_3)_xF_{2+y}$ ($y = 0.4$ x, generally) (Sanyal and De Datta, 1991).

A gist of other important primary minerals will be given and discussed in the following **Chapter** on **Soil Colloids: Structural Aspects and Properties**.

Certain other minerals that occur in soils in much smaller quantities are zircon, $ZrSiO_4$, Ilmenite, $FeTiO_3$, Sphene, or Titanite, $CaTiSiO_3$, Fluorspar, CaF_2, and Garnets having the general composition, $X_3Y_2(SiO_4)_3$, where X=Ca, Mg, Fe^{2+}, Mn^{2+} and Y=Al, Fe^{3+}, Cr^{3+}.

Secondary Minerals

The disintegration of the primary minerals through break down and physical, chemical and biological weathering lead to the formation of the secondary minerals.

These minerals dominate the clay fraction. The presence of some of them is also noted in the fine silt fraction. The particle size of these fractions are small (as mentioned above), and as a result their specific surface area is large. Indeed the size of the clay fractions (< 0.002 mm) falls within the colloidal dimensions (~ 1 μm to 1 nm). These particles are charged having large specific surface charge density. Such colloidal fraction of natural moist soil is most reactive with large interfacial specific surface area and charge density (Sanyal, 2002a). Besides the crystalline clay minerals (the aluminosilicate layer-lattice clay minerals; *see* later), which are the dominant component, the inorganic soil colloidal fraction also includes amorphous silicates, sesquioxides (*e.g.*, oxides of iron, aluminium, manganese), as well as the clay-sized primary minerals such as hydrous oxides and hydroxides of iron, aluminium, manganese, silicon, etc. A detailed account of the soil colloidal fractions (that also includes the soil organic matter; *see* later) is given in the subsequent **Chapter** on **Soil Colloids: Structural Aspects and Properties.**

1.1.2. Soil Organic Matter

Soil organic matter (SOM) consists of a mixture of plant and animal residues in various stages of decomposition, substances synthesized microbiologically, and/or chemically, from the breakdown products, and the bodies of live and dead microorganisms and their decomposing remains (Schnitzer, 2000).

Although SOM is largely of plant origin (Stevenson, 1994), such organic matter is not recognizable under a light microscope as possessing the cellular organization of plant material, and is termed **Humus**. These humic substances (HS) are best described as a series of acidic yellow/brown to black polyelectrolytes with variable molecular weights. Thus the naturally occurring soil HS are *not* defined in terms of their chemical composition or typical functional group contents. They are defined instead operationally, in terms of solubility as well as the corresponding solubility behaviour in aqueous solution at different pH ranges (Schnitzer, 2000). Soil humic acids (HAs) are generally defined as being soluble in alkali and insoluble in acidic (pH = 1 to 2) solution. Such a definition lacks specificity since a large number of organic compounds, insoluble in acids, but soluble in alkali, is *not* necessarily HAs or soil HS. The fulvic acid (FA) fraction of SOM, on the other hand, is defined as that which is soluble in both acid and alkali, while the organic matter, that is not solubilized by alkali or acid, is referred to as the humin fraction, and usually makes up about 20 per cent of SOM (Sanyal, 2002b). Besides a potential source of confusion and uncertainty inherent in the above mentioned definitions, there is yet some degree of additional confusion as to the synonymy (or otherwise) of a number of terms often used interchangeably by the soil scientists, namely SOM, humus and HS. Stevenson (1994) identified SOM with humus, but according to Schnitzer (2000), total HS is synonymous with SOM and humus as long as losses occurring during the extraction and separation procedures are held to a minimum. Generally soil organic matter (SOM) or humic substances (HS) are defined as the sum of humic acid (HA), fulvic acid (FA) and humin fractions.

References

Bear, F. E. (Ed.) (1976). *Chemistry of the Soil*, Second Edition, Third Indian Reprint, Oxford and IBH Publishing Co., New Delhi, p. 322.

Biswas, T. D. and Mukherjee, S. K. (1994). *Textbook of Soil Science*, Second Edition, Tata McGraw-Hill Publishing Co. Ltd., New Delhi, p. 15.

Sanyal, S. K. (2002a). Soil Colloids. **In:** *"Fundamentals of Soil Science"* (G. S. Sekhon, P. K. Chhonkar, D. K. Das, N. N. Goswami, G. Narayanasamy, S. R. Poonia, R. K. Rattan and J. L. Sehgal, Eds.), Indian Society of Soil Science, New Delhi, pp. 229-259.

Sanyal, S. K. (2002b). Colloid chemical properties of soil humic substances: A Relook. **In:** *Bull. Indian Soc. Soil Sci.*, New Delhi, No. **21,** pp. 278-307.

Sanyal, S. K. and De Datta, S. K. (1991). Chemistry of phosphorus transformations in soil. *Adv. Soil Sci.*, **16,** 1-120.

Schnitzer, M. (2000). A lifetime perspective on the chemistry of soil organic matter. *Adv. Agron.*, **68,** 1-58.

Stevenson, F.J. (1994). *Humus Chemistry: Genesis, Composition, Reactions*, Second Edition, John Wiley and Sons, New York.

Zone, S. V. (1986). *Tropical and Subtropical Soil Science*, Mir Publishers, Moscow, p. 39-40.

Soil Colloids: Structural Aspects and Properties

2.1. Soil Colloids: Introduction

A colloidal system is a heterogeneous mixture of at least two phases (in Physical Chemistry, a *phase* is defined as any homogeneous and physically distinct part of a system which is separated from other such parts of the system by definite bounding surfaces) in which one phase is dispersed in a state of fine subdivision (particle size 1 µm to 1 nm) in another continuous medium. This continuous medium is termed as the dispersion medium in which the other phase, namely the disperse phase, is dispersed. The size limits of 1 µm to 1 nm are not very rigid and there may be variations on either side of the range. When the dispersed phase is a solid (*e.g.* soil colloidal particles) and the dispersion medium is water (*e.g.* soil water), the colloidal system is referred to as a sol. The properties peculiar to a colloidal system, which distinguish them from suspensions (particle size > 1µm), and true uni-phase solutions (particle size < 1 nm), are primarily those arising from the large interfacial area, which is charged, characterizing the colloidal system. In other words, most of the important properties of colloidal systems in soils may be attributed to the large specific surface (or interface) charge density. (Sanyal, 2002a).

The soil colloidal fraction is the site of important processes in soil, governing ion-exchange, nutrient availability and fixation, soil physical properties, especially soil structure, hydraulic conductivity, infiltration, and also soil management (Sanyal, 2002a). Soil colloids can be broadly classified into two types, depending on the nature of the linkages present and the types of compounds formed. These are:

1. Inorganic colloids, and

2. Organic colloids

The inorganic colloids cover aluminosilicate layer-lattice clay minerals; hydrous oxides and hydroxides of Fe, Al, Si and Mn in clay-size dimensions; amorphous silicates; and clay-sized primary minerals. The organic colloidal fraction in soil is constituted by soil humus.

2.1.1. Inorganic Soil Colloids

The aluminosilicate layer-lattice clays, the chief inorganic colloidal fraction in soil, are derived from the silicate minerals. These silicates are discussed below in brief first, before introducing the derived products, namely the aluminosilicate soil clays. Silicates are the most common minerals in the earth's crust, accounting for more than 90 per cent of the latter. Most of the soils are also dominated by these silicates which occur as primary minerals, inherited from igneous or metamorphic rocks, and as secondary minerals formed from the weathering products of primary minerals. These silicates consist of the SiO_4^{4-} - tetrahedron (*see* later) as the basic structural unit, present as either a single, or in several joint combinations. Depending on the arrangement of SiO_4^{4-} - tetrahedron in the structure, the silicate minerals are classified into different groups as demonstrated in Table 2.1.

2.1.1.1. Weathering and Stability of Silicate Minerals

The above noted classification of silicate minerals was found to be related to the sequence of crystallization under igneous conditions and also the sequence of resistance to decomposition by weathering conditions (Bear, 1976). Thus, the order of crystallization was found to be as follows: independent tetrahedrons $\longrightarrow$ single chains $\longrightarrow$ double chains $\longrightarrow$ sheets $\longrightarrow$ framework structures. Furthermore, the ratio of basic cations to Si^{4+} will be higher for a larger number of independent tetrahedrons linked in the structural composition, causing the corresponding mineral to be more basic. Similarly, the larger the number of alumina tetrahedrons, the more basic is the mineral. The most basic minerals crystallize first, followed by minerals with a decreasing degree of basicity (Bear, 1976). Additionally, the relative stability of the minerals to weathering also appears to be related to the degree of basicity, degree of linkage of the tetrahedrons, and other factors that induce a lowering of the basicity of the mineral and destruction of bonds linking the tetrahedrons (Bear, 1976). Thus, the weatherability of such silicates is known to be related to such basicity and it increases with the rise of the basicity.

Indeed, the ratio between the numerical value of the charge of the unit composition of a group of silicates to the number of silicon atoms (ions) present in the given unit composition (Table 2.1) is taken to provide a measure of the above mentioned basicity of the said group of silicates. Thus, the said basicity of the Nesosilicates $(SiO_4)^{4-}$ is 4 [(4/1) = 4]; for the Sorosilicates $(Si_2O_7)^{6-}$, it is 3 [(6/2) = 3]; for the Cyclosilicates $(Si_6O_{18})^{12-}$, it is 2; for the Inosilicates (single chains) $(SiO_3)^{2-}$, its

Table 2.1: Classification of Silicate Minerals

Silicate Class and Unit Composition Arrangement of SiO_4 Tetrahedra	Mineral	Ideal Formula
Nesosilicates $(SiO_4)^{4-}$ (singlet structure)	Olivine	$(Mg, Fe^{II})_2\ SiO_4$
	Forsterite	Mg_2SiO_4
	Fayalite	Fe_2SiO_4
	Zircon	$ZrSiO_4$
	Sphene	$CaTiO(SiO_4)$
	Topaz	$Al_2SiO_4(F, OH)_2$
	Garnets	$X_3Y_2(SiO_4)_3$, where X = Ca, Mg, Fe^{2+}, Mn^{2+} Y = Al, Fe^{3+}, Cr^{3+}
	Andalusite	Al_2SiO_5
	Sillimanite	Al_2SiO_5
	Kyanite	Al_2SiO_5
	Staurolite	$Fe_2Al_9O_6(SiO_4)_4(O, OH)_2$
Sorosilicates $(Si_2O_7)^{6-}$	Epidote	$Ca_2(Al,Fe)Al_2O(SiO_4)(Si_2O_7)(OH)$
Cyclosilicates $(Si_6O_{18})^{12-}$	Beryl	$Be_3Al_2(Si_6O_{18})$
		$(Na, Ca)(Li, Mg, Al)(Al, Fe, Mn)_6$-$(BO_3)_3(Si_6O_{18})(OH)_4$
Inosilicates (single chains) $(SiO_3)^{2-}$	Pyroxenes	
	Augite	$(Ca, Na)(Mg, Fe, Al)(Si, Al)_2O_6$
	Enstatite	$MgSiO_3$
	Hypersthene	$(Mg, Fe)SiO_3$
	Diopside	$CaMgSi_2O_6$
	Hedenbergite	$CaFeSi_2O_6$
	Pyroxenoids	
	Wollastonite	$CaSiO_3$
	Rhodonite	$MnSiO_3$
Inosilicates (double chains) $(Si_4O_{11})^{6-}$	Amphiboles	
	Hornblende	$(Ca, Na)_{2-3}(Mg, Fe, Al)_5Si_6$-$(Si, Al)_2O_{22}(OH)_2$
	Tremolite	$Ca_2Mg_5Si_8O_{22}(OH)_2$
	Actinolite	$Ca_2(Mg, Fe)_5Si_8O_{22}(OH)_2$
	Cummingtonite	$(Mg, Fe)_7Si_8O_{22}(OH)_2$
	Grunerite	$Fe_7Si_8O_{22}(OH)_2$

Contd...

Table 2.1–*Contd...*

Silicate Class and Unit Composition Arrangement of SiO_4 Tetrahedra	Mineral	Ideal Formula
Phyllosilicates $(Si_2O_5)^{2-}$ (sheet structure)	Micas	
	Muscovite	$KAl_2(AlSi_3O_{10})(OH)_2$
	Biotite	$K(Mg, Fe^{2+})_3(AlSi_3O_{10})(OH)_2$
	Phlogopite	$KMg_3(AlSi_3O_{10})(OH)_2$
	Chlorites	$(Mg, Fe)_3(Si, Al)_4O_{10}(OH)_2$- $(Mg, Fe)_3(OH)_6$
	Clay Minerals (selected)	
	Talc	$Mg_3Si_4O_{10}(OH)_2$
	Pyrophyllite	$Al_2Si_4O_{10}(OH)_2$
	Kaolinite	$Al_2Si_2O_5(OH)_4$
	Smectite	Variable
	Vermiculite	Variable
	Serpentines	
	Antigorite	$Mg_3Si_2O_5(OH)_4$
	Chrysotile	$Mg_3Si_2O_5(OH)_4$
Tectosilicates $(SiO_2)^0$ (framework structure)	Feldspars	
	Orthoclase	$KAlSi_3O_8$
	Albite	$NaAlSi_3O_8$
	Anorthite	$CaAl_2Si_2O_8$
	SiO_2 Group	
	Quartz	SiO_2
	Tridymite	SiO_2
	Cristobalite	SiO_2
	Zeolites	
	Analcime	$NaAlSi_2O_6.H_2O$
	Clinoptilolite	$(Na_3K_3)(Al_6Si_{30}O_{72}).24H_2O$
	Feldspathoids	
	Nephelene	$(Na, K)AlSiO_4$

Source: Adapted from Schulze (1989).

value is 2 and for the Inosilicates (double chains) $(Si_4O_{11})^{6-}$, the basicity is 1.67 [(6/4) = 1.67]; for the Phyllosilicates $(Si_2O_5)^{2-}$, it is 1, while for the Tectosilicates $(SiO_2)^0$, its value is the lowest, namely zero (0). As stated earlier, the weatherability of such silicates decreases with the fall of the ratio (basicity) defined above. Hence the ease of weatherability of the silicate minerals is the highest for the Nesosilicates and it falls in the following order:

Nesosilicates > Sorosilicates > Cyclosilicates > Inosilicates (single chains) > Inosilicates (double chains) > Phyllosilicates > Tectosilicates.

In other words, the crystallization of these minerals follows the order as given hereunder (Bear, 1976):

Olivine
↓
  Enstatite
↓
    Augite
↓
      Hornblende
↓
        Biotite

Bytownite
↓
  Labradorite
↓
    Andesine
↓
      Oligoclase
↓
        Albite

→ Quartz ←
Potash feldspar →
↓
Zeolite
↓
Solutions enriched in water

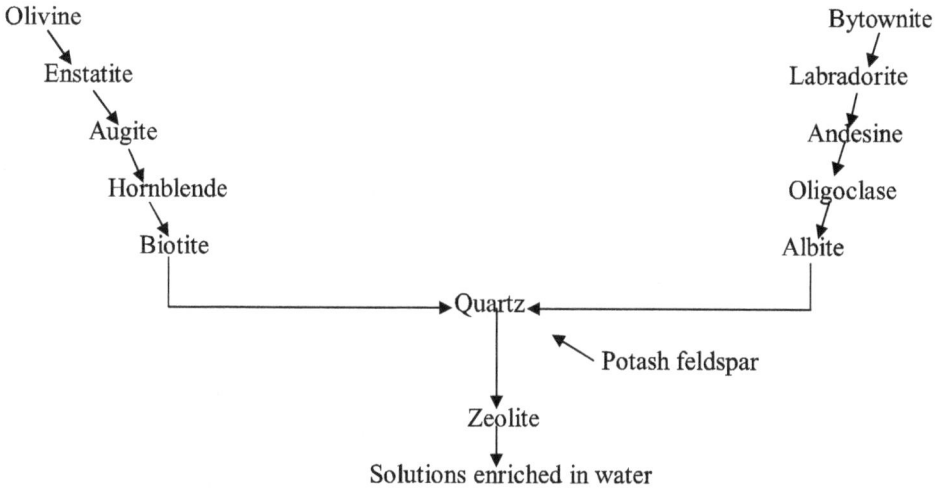

The weathering sequence of these minerals also closely follows such an order shown below (Bear, 1976).

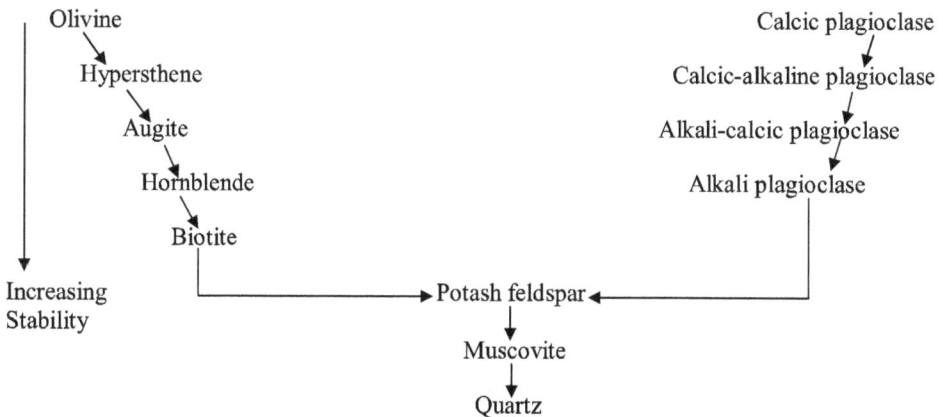

Olivine
↓
  Hypersthene
↓
    Augite
↓
      Hornblende
↓
        Biotite

Calcic plagioclase
↓
Calcic-alkaline plagioclase
↓
Alkali-calcic plagioclase
↓
Alkali plagioclase

Increasing Stability ↓

→ Potash feldspar ←
↓
Muscovite
↓
Quartz

2.2. Basics of Phyllosilicate Minerals

In common to colloids of diverse origin, the layer-lattice aluminosilicate clays (phyllosilicates) are characterized by large specific surface area and charge density. Because the crystal structure of phyllosilicates largely governs their properties, it is necessary to understand these structures. These structures have been extensively studied by employing various techniques, particularly since the discovery in the early 1920s that most of the clays are crystalline in nature, and subsequent application of the X-ray diffraction technique to elucidate the above mentioned crystal structures (Sanyal, 2002a).

2.2.1. Structural Units in Phyllosilicates

The convenient way of treating the atoms in the aluminosilicate structures as rigid spheres is oversimplified, but it provides a useful starting point.

Closest Packing of Spheres

The central concept of stability of crystal structure involves packing of anions around a central cation in such a way that each of these ionic spheres touches six of its nearest surrounding neighbours (Sanyal, 2002a). Such an arrangement is known as the hexagonal closest packing as shown in Figure 2.1. This forms a sheet leaving two kinds of central voids (A - type and B - type) between the spheres (Figure 2.1). Another sheet of such hexagonal packing, when stacked on the top of the first sheet, then a central cation placed in A-type of voids will have four surrounding spheres (such as anions) arranged in a tetrahedral manner around the cation, while a cation occupying a B-type of void will have six such surrounding spheres, above and below the central cation, pointing towards the apices of a regular octahedron (Figure 2.2).

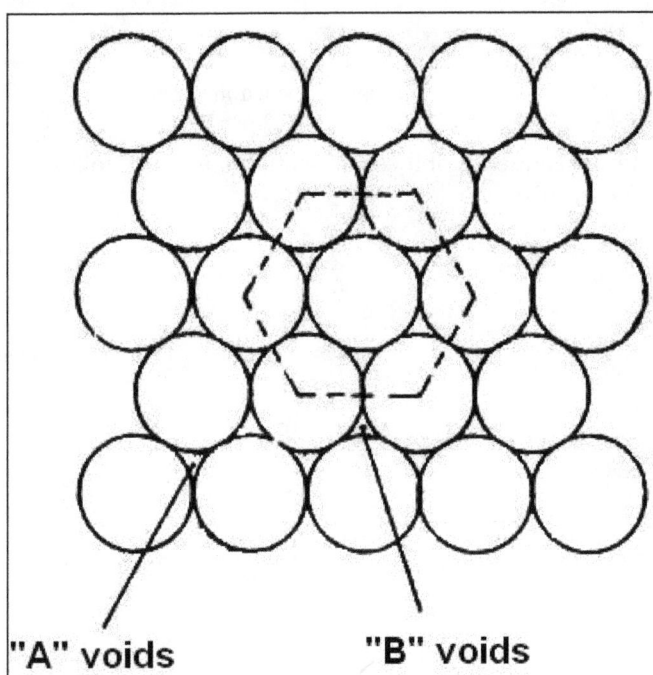

"A" voids   "B" voids

Figure 2.1: Hexagonal Closest Packing of Spheres in a Plane.
Source: Schulze (1989)

A cation occupying a tetrahedral site (the A-site) is said to be in a four-fold or tetrahedral coordination state (*see* later), being bonded to four oxide ions (O^{2-} ions) which form the apices of a regular tetrahedron (Figure 2.2). A cation occupying an octahedral site (B-site) is in six-fold (or octahedral) coordination state, being bonded to six hydroxyl (OH^-) or oxide ions, arranged in the shape of an octahedron (Figure 2.2).

Silica Tetrahedral and Alumina Octahedral Sheets

A silica tetrahedron (SiO_4^{4-}) and an aluminium octahedron [$Al(OH)_6^{3-}$] are schematically depicted in Figure 2.2. A silica tetrahedral sheet is formed by sharing

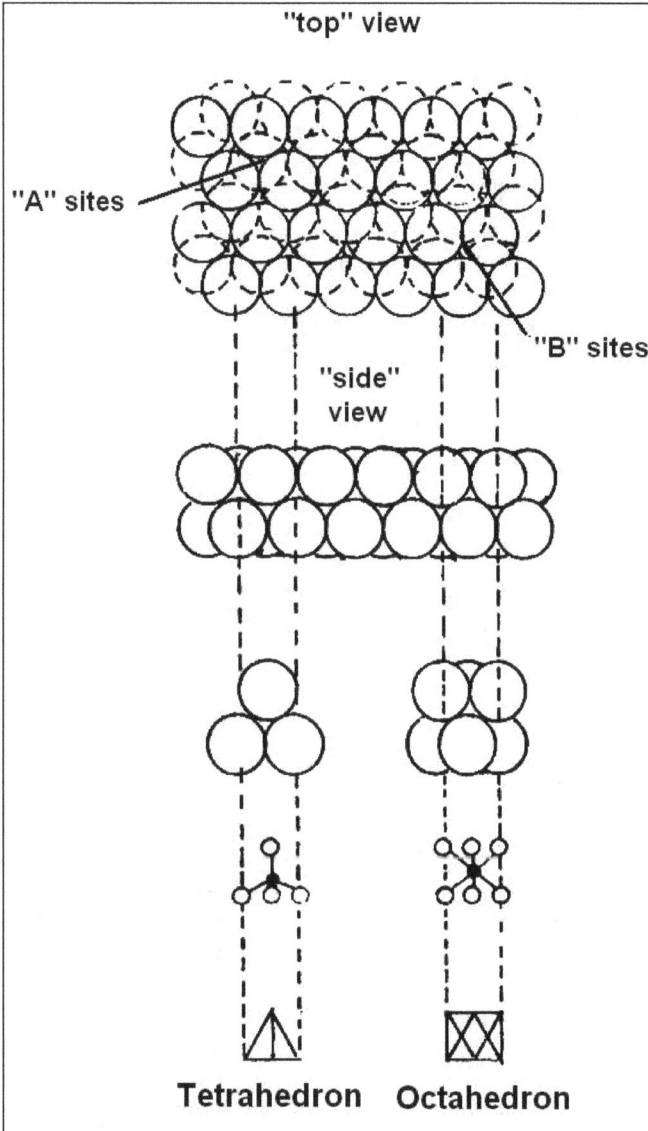

Figure 2.2: Tetrahedra and Octahedra as Consequence of Two Planes of Close-packed Spheres.
Source: Schulze (1989).

of three basal oxide ions of each (SiO_4^{4-})-tetrahedron with adjacent such tetrahedra in a and b directions (*i.e.* two dimensional) which will bring down the negative charge per unit tetrahedron (Figure 2.3). This leaves one unshared apical oxide ion in each tetrahedron, pointing upward from the basal plane accommodating the shared oxide ions. The apical oxide ion is free to form bond with other polyhedral atoms/ions. Indeed the term, 'Phyllosilicate' means "sheet silicates", which are thus minerals

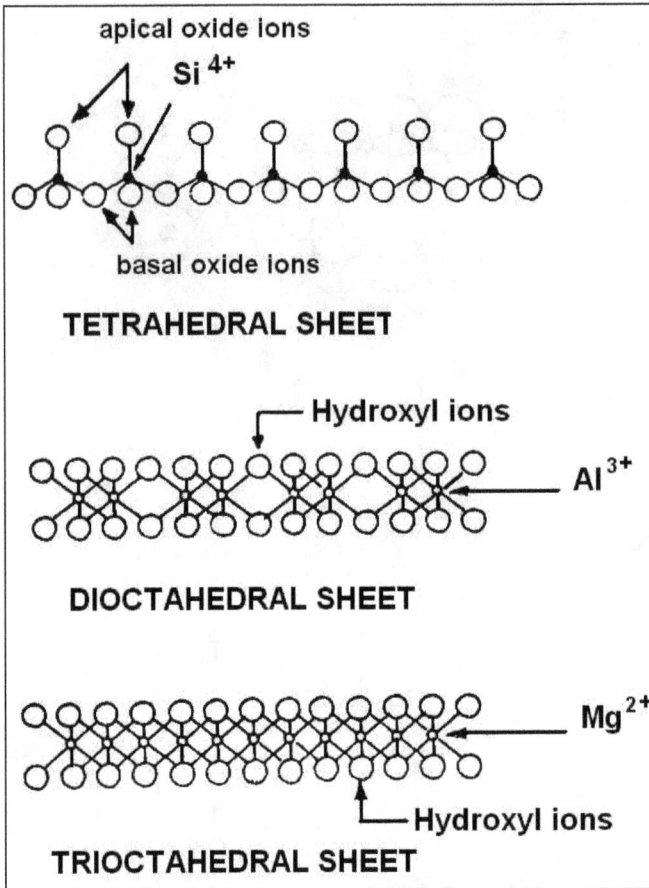

Figure 2.3: The Tetrahedral, Dioctahedral and Trioctahedral Sheets.
Source: **Schulze (1989).**

containing sheet-like arrangement of (SiO_4^{4-})-tetrahedra. There are two ways of filling the octahedral sites with a central cation (*e.g.* Al^{3+} or Mg^{2+}), depending on the valency of the cation. A divalent cation (such as Mg^{2+}) can be placed into each adjacent octahedral site of the neighbouring octahedra which are joined together by sharing of octahedral hydroxyl ions of a $[Mg(OH)_6^{4-}]$-octahedron that reduces the negative charge per unit Mg-octahedron. The network of shared octahedra, $Mg_3(OH)_6$, is such that each Mg^{2+} shares one-third of the uni-negative charge of each of the six octahedrally arranged hydroxyl ions surrounding it, because each hydroxyl ion is shared equally by three Mg^{2+} ions (Sanyal, 2002a). The resulting sheet is known as the trioctahedral sheet in which each octahedral site is occupied by a divalent cation (Figure 2.3).

For the dioctahedral sheet, however, only two out of each set of three octahedral positions are occupied by a trivalent cation (*e.g.* Al^{3+}) giving rise to the formula, $Al_2(OH)_6$ (Figure 2.3). It is thus evident that each of these tri-and di-octahedral sheets is electroneutral (Sanyal, 2002a).

Goldschmidt's Laws

The shape and size of a crystal structure is governed by the ratio of numbers of units (*e.g.*, cations and anions), ratio of ionic sizes and properties of polarization of its structural units. Besides, allowance must be made for repulsive forces in ionic compounds, especially when the central cation is small, *e.g.*, Li^+, and the surrounding anions have virtual contact with one another (Sanyal, 2002a).

The laws governing structural characteristics of the phyllosilicates are known as **Goldschmidt's Laws – Law I and Law II** (Sanyal, 2002a).

(i) Law I

In an ionic crystalline compound, isomorphous replacement of one cation by another, without incurring any change in the order of the crystal pattern, is permitted provided that the radii of the cation replaced and the cation substituting it agree within 10-15 per cent.

Such isomorphous substitutions may lead to charge imbalance in the crystal structure since the charge (valency) of the two cations concerned may differ. Thus substitution of Si^{4+} ion in a tetrahedral sheet (Figure 2.3) by Al^{3+} or that of Al^{3+} ion in dioctahedral sheet (Figure 2.3) by Mg^{2+} or Fe^{2+} ion will cause excess negative charge in sheet structures. This charge imbalance may be balanced by an appropriate and simultaneous substitution elsewhere, and/or by the retention of oppositely charged ions in the overall mineral structure (*see* later) so that the electroneutrality is maintained (Sanyal, 2002a).

(ii) Law II: Radius Ratio Law

When the number of anions surround a central cation (with the closest packing of spheres) such that they satisfy the charge (of the cation) completely, the number of anions that can be so accommodated around the cation depends on the ratio of the radius of the central cation to that of the surrounding anions. Thus, this ratio of ionic radii governs what is known as the coordination number of the central cation. The latter is thus basic to the shape of crystal forms (Sanyal, 2002a). The ionic radii of a number of cations and anions pertaining to phyllosilicates, the cationic to anionic radius ratio (r+: r−) [based on the radius of oxide ion, taken as the anionic radius (r−) of relevance to phyllosilicates], the range of the values of these ratios (r+: r−) corresponding to different coordination number of central cations, as well as the shape of the crystal structure or lattice are given in Table 2.2.

It can be shown by simple calculations that the tetrahedral site has a radius of 0.035 nm while an octahedral site possesses a radius of 0.050 nm. By considering these sizes and data presented in Table 2.2, as well as by taking due cognizance of the Goldschmidt's laws, one arrives at the following important conclusions (Sanyal, 2002a):

(i) A sphere of radius up to 0.41 times the radius of oxide ion can be accommodated in a tetrahedral site, while a sphere of radius up to 0.73 times the radius of oxide ion can be fitted into the octahedral site.

Table 2.2: Ionic Radius, Radius Ratio, Coordination Number of Central Cations of Phyllosilicate Structures and Shape of Crystal Lattice

Ion	Ionic Radius (nm)	Radius Ratio* (r⁺/r⁻)	Range of Radius Ratio for Coordination Number (given in Parentheses)	Coordination Number of Central Cation in Phyllosilicate Structure	Shape of Crystal Lattice
O^{2-}	0.140	–	< 0.15 (2)	–	Linear
F^-	0.133	–	0.15 – 0.22 (3)	–	Equilateral Triangle
Cl^-	0.181	–	–	–	–
Si^{4+}	0.039	0.278	0.22 – 0.41 (4)	4	Tetrahedral
Al^{3+}	0.051	0.364	0.41 – 0.73 (6)	4, 6	Tetrahedral Octahedral
Fe^{3+}	0.064	0.457	–	6	Octahedral
Mg^{2+}	0.066	0.471	–	6	–do–
Fe^{3+}	0.074	0.529	–	6	–do–
Mn^{3+}	0.080	0.571	–	6	–do–
Na^+	0.097	0.693	–	8	Cubic
Ca^{2+}	0.099	0.707	–	8	–do–
K^+	0.133	0.950	0.73 – 1.0 (8)	8, 12	Cubic Dodecahedral
Ba^{2+}	0.134	0.957	> 1.0 (12)	8, 12	– do-
Rb^+	0.149	1.050	–	8, 12	–do–

* Ratio of cation radius (r⁺) to O^{2-} ion radius (r⁻)

Source: (Sanyal, 2002a).

(ii) The Si^{4+} ion can remain in only four-fold coordination state in a tetrahedral sheet, while Al^{3+} ion can have both four- and six-fold coordination states in the tetrahedral and the octahedral sheets of phyllosilicates.

(iii) The cations such as Mg^{2+}, Fe^{3+} and Fe^{2+} ions can be accommodated only in the octahedral sites.

(iv) Potassium ion (K^+) can be accommodated in a dodecahedral arrangement (12-fold) of oxide ions, *i.e.*, in between two hexagonal array of oxide ions, above and below the K^+ ion.

2.2.2. Common Phyllosilicate Minerals in Soil

Among the common layer lattice minerals, the important ones are the 1:1 and 2:1 type minerals. These are described below.

The (1:1) layer structure is made up of one octahedral sheet and one tetrahedral sheet with the apical oxide ions (O^{2-} ions) of the latter being shared with the octahedral sheet as shown in Figure 2.4. This structure is characterized by the

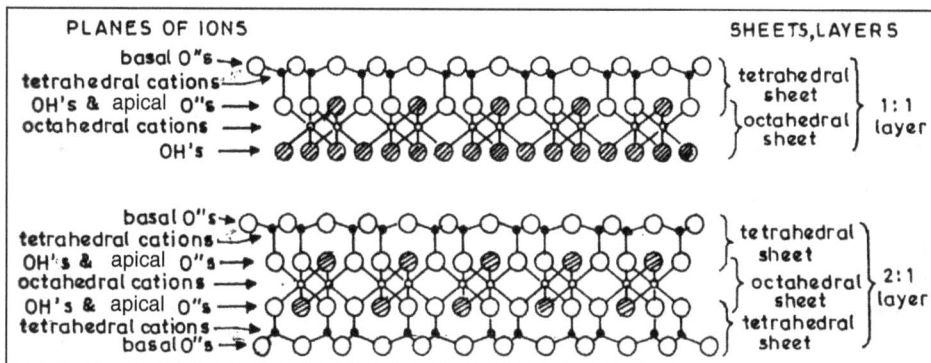

Figure 2.4: The 1:1 and 2:1 Phyllosilicates
Source: **Schulze (1989).**

presence of three planes of anions, namely (i) the basal O^{2-} ions of the tetrahedral sheet, (ii) the apical O^{2-} ions of the original tetrahedral sheet, common to it and the shared octahedral sheet, as well as the hydroxyl ions (OH^- ions) of the original octahedral sheet, and (iii) the OH^- ions of the octahedral sheet. The tetrahedral and the octahedral cations are, respectively, Si^{4+} and Al^{3+} ions. That is, here the original octahedral sheet is a dioctahedral sheet (Sanyal, 2002a). A similar (1:1) layer combination can be obtained between a tetrahedral sheet and a trioctahedral sheet with a divalent cation occupying all the octahedral positions in this sheet (Sanyal, 2002a).

A (2: 1) layer structure consists of an octahedral sheet sandwiched between two tetrahedral sheets. This is demonstrated for a combination of one dioctahedral sheet with two tetrahedral sheets, placed above and below it along the c-axis direction (*see* later) (Figure 2.4). There are thus four planes of anions, namely the two outer planes of basal O^{2-} ions of the two outer tetrahedral sheets and two inner planes, each of which is composed of the shared apical O^{2-} ions of an outer tetrahedral sheet, which are common to the corresponding tetrahedral sheet and the sandwiched octahedral sheet, as well as the remaining OH^- ions of the central octahedral sheet (Sanyal, 2002a). Such a (2:1) layer combination for a dioctahedral sheet in which a trivalent cation (*e.g.* Al^{3+}) occupies two-thirds of the octahedral positions in the corresponding octahedral sheet is depicted in Figure 2.4.

2.2.2a. Layer – Lattice* Minerals of 1:1 Type

Two (1:1) layer-lattice minerals- kaolinite and halloysite- are discussed below.

Kaolinite (Kaolin) Group

Among this group of phyllosilicates, kaolinite is the most dominant one. It is formed through (1:1) combination of tetrahedral and octahedral sheets, and these

---

* A space-lattice, such as the layer-lattice phyllosilicate clay minerals, refers to the regularity of internal structural units of crystalline substances (Sanyal, 2002a)

(1:1) layers are stacked one above the other along the c-axis (*see* later). Kaolinite is a dioctahedral mineral (Figure 2.5) with Al^{3+} being the octahedral cation. There is hardly any isomorphous substitution of either tetrahedral Si^{4+} or octahedral Al^{3+} ions in the (1:1) layers, which leads to virtual electroneutrality of these layers. Thus, kaolinite is characterized by a low cation-exchange capacity (C.E.C) (discussed in more details later, *see* Table 2.6). Further, the basal O^{2-} ions of any (1:1) layer are strongly held through electrostatic hydrogen bonding with the OH^{-} ions of octahedral sheet of the adjacent (1:1) layer stacked on to the former (Figure 2.5). As a consequence, the (1:1) layer does *not* expand along the c-axis in the presence of water and polar liquids such as glycerol, *nor* does kaolinite possess interlayer [between adjacent (1:1) layers] surfaces which are accessible to water molecules or ions, *e.g.* the cations (Sanyal, 2002a). Kaolinite is thus characterized by typically low specific surface area (discussed in more details later, *see* Table 2.5) as well as charge density. The basal spacing along the c-axis of kaolinite is 0.72 nm (*see* later for details, *see* Table 2.3).

Kaolinite tends to be abundant in the clay fraction of the soils in the advanced stages of weathering, such as Alfisols and Ultisols, which are also inherently less

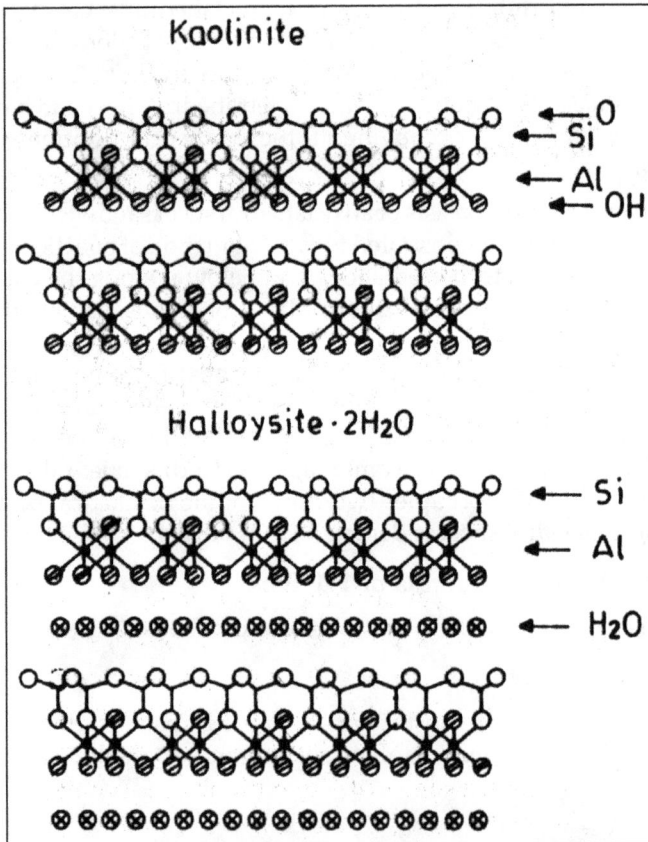

Figure 2.5: Structural Scheme of Kaolinite and Halloysite Clay Minerals.
Source: Schulze (1989).

fertile. Since the isomorphous substitutions in the (1: 1) layers of kaolinitic are negligible, kaolinite has the ideal formula of $Al_2Si_2O_5(OH)_4$. Primary minerals often weather to kaolinite. Thus, feldspars often weather to kaolinite in soils formed from igneous rocks.

Two less important minerals, dickite and nacrite, belonging to this group, possess the same composition as does kaolinite, although their structures are different. The following list gives some of the more important members of the Kaolin group (Sanyal, 2002a):

Mineral Type and Example	Composition
Dioctahedral	
Kaolinite	$Al_2Si_2O_5(OH)_4$
Trioctahedral	
Serpentine minerals	$(Mg_{3-x}.Al_x)(Si_{2-x}.Al._x)O_5(OH)_4$
Chrysotile	$Mg_3Si_2O_5(OH)_4$
Antigorite	$(Mg,Fe^{2+})_3Si_2O_5(OH)_4$
Amesite	$(Mg_2.Al)(Si.Al)O_5(OH)_4$
Greenalite	$(Fe^{2+}_{2\cdot2}.Fe^{3+}_{0.5}) Si_2O_5(OH)_4 [(Fe^{2+}, Fe^{3+})_{2-3}Si_2O_5(OH)_4]$

Halloysite

This mineral has a (1:1) layer structure similar to that of kaolinite, except that a layer of water molecules is hydrogen-bonded between two adjacent (1:1) layers (Figure 2.5). The fully hydrated form has the formula, $Al_2Si_2O_5(OH)_4.2H_2O$, and is also known as endelite which is characterized by the basal spacing along the c-axis of 1.01 nm. The fully hydrated form dehydrates irreversibly, on heating at low temperatures (70-80°C) or in vacuum at ordinary temperatures, to the anhydrous form, known as metahalloysite with the basal spacing of 0.72 nm (Sanyal, 2002a). The difference, (1.01 - 0.72 nm) or 0.29 nm, is about the thickness of a single molecular sheet of water molecules. In case the dehydration of the hydrated form of halloysite is not perfect, the resulting anhydrous form may have a basal spacing, going up to about 0.76 nm, due to entrapment of some residual water in between the collapsed adjacent (1:1) layers (Sanyal, 2002a).

The electron micrographs (*see* later) of the fully hydrated halloysite show it to be of elongated tubular shape, which might have arisen from the overlapping, curved sheets of the kaolinite type, with the basal spacing along the c-axis for any point on the tube nearly perpendicular to a plane tangent to the tube at that point. Kaolinite itself has a hexagonal morphology (Sanyal, 2002a).

Halloysite has very little or no isomorphous substitutions in the (1: 1) layers. This clay mineral generally occurs in soils of volcanic origin (*e.g.*, in the soils of Andept suborder). Halloysite forms in the early stages of weathering sequence, but is less stable than kaolinite to which it transforms as the degree of weathering progresses.

2.2.2b. Layer - Lattice Minerals of 2:1 Type

The (2:1) layer-lattice minerals are described under two groups - talc-pyrophyllite group and mica group (Sanyal, 2002a).

Talc-Pyrophyllite Group

Talc and pyrophyllite are the trioctahedral and dioctahedral members, respectively, of the simple form of (2:1) layer minerals, providing starting points for discussion of the structure of the clay minerals under this group. There is no isomorphous substitution in the tetrahedral and the octahedral sheets of these minerals, and they have the following ideal formulae:

Talc: $Mg_3Si_4O_{10}(OH)_2$ Pyrophyllite: $Al_2Si_4O_{10}(OH)_2$

Here, Mg^{2+} ions occupy all the octahedral positions in the octahedral sheet of talc, while Al^{3+} ions occupy two out of each three such octahedral positions in pyrophyllite. The (2:1) layers in these minerals, as a whole, are electrostatically neutral. The stacked (2:1) layers along the c-axis are held together by interlayer weak van der Waal's (induced dipole-induced dipole) attractions between the adjacent (2:1) layers, which are also sometimes supplemented by small ionic attractions as well, implying low degree of isomorphous substitutions in the tetrahedral and octahedral sheets of the natural specimen (Sanyal, 2002a).

Both these minerals occur only rarely in soils, characteristically when inherited from low-grade metamorphic rocks, frequently with minerals such as kyanite.

Mica Group

Mica, a primary mineral, forms a convenient starting material for discussing the secondary phyllosilicate clay minerals under this group. This is because of the abundance of micas in nature, (as in India) and their excellent crystalline structure, leading to extensive studies of these layer silicate minerals (Sanyal, 2002a).

The structure of mica consists of negatively charged (2:1) layers that are "keyed" together by large interlayer cations. In other words, micas have talc and pyrophyllite type of (2: 1) layer structure, but with two important differences as mentioned below:

(i) One- fourth of Si^{4+} ions in the tetrahedral sheet are replaced by Al^{3+} ions, leading to an excess negative charge of -1.0 per formula unit in the (2: 1) layer.

(ii) The negative charge of this layer is balanced by the interlayer monovalent cations, commonly K^+ ions (in true micas; but in brittle micas, e.g. margarite, divalent interlayer cations such as Ca^{2+} ions balance the layer charge) which reside in the interlayer sites located between two hexagonal oxide ion-arrangement of two adjacent (2:1) layers.

Thus the ideal formulae of for a dioctahedral mica mineral (muscovite) is $KA1_2(Al.Si_3)O_{10}(OH)_2$ (Figure 2.6), while for trioctahedral micas, it is $KMg_3(Al.Si_3)O_{10}(OH)_2$, (phlogopite) and $K(Mg.Fe^{2+})_3(Al.Si_3)O_{10}(OH)_2$ (biotite). In dioctahedral paragonite, Na^+ ion is the interlayer cation in place of K^+ ion, the formula being $NaAl_2(AlSi_3)O_{10}(OH)_2$. In margarite (brittle mica), Ca^{2+} ion is the interlayer cation, as mentioned above, and the mineral has the formula, $CaAl_2(Al_2Si_2)O_{10}(OH)_2$. In some dioctahedral micas, Al^{3+} ions in the octahedral sheet are partly substituted by divalent cations so that the excess negative (2:1) layer charge is distributed between the tetrahedral and the octahedral sheets, *e.g.* phengite with composition, $K(A_{1.5}.M^{II}_{0.5}).(Si_{3.5}.Al_{0.5})O_{10}(OH)_2$, where M^{II} is a divalent cation (Sanyal, 2002a).

Mica, generally inherited from the parent rocks, is found in soils formed on weathering of igneous and metamorphic rocks, and their sediments. The most common mica minerals in soils include muscovite, biotite and phlogopite. Micas generally weather to lead to (2:1) layer phyllosilicates, namely vermiculite and smectite with concomitant release of interlayer K^+ ions. These K^+ ions serve as important plant nutrient. In general, dioctahedral micas, such as muscovite, are more resistant to weathering than are the trioctahedral micas such as biotite and phlogopite, which are thus less found in soils of advanced stages of weathering. The term illite or hydrous mica has been proposed to denote the group name for the clay-sized micaceous components in soils and sediments. Illite is predominantly dioctahedral, but it differs from ideal muscovite in having, on an average, more Si^{4+} and Mg^{2+} ions and H_2O, but less interlayer K^+ ions. The main distinguishing features between mica and illite are the following (Sanyal, 2002a):

(i) Poorer crystallinity of illite.

(ii) Less substitution of tetrahedral Si^{4+} ions by Al^{3+} ions in illite (one-sixth of Si^{4+} ions are generally substituted),

(iii) Lower excess negative (2:1) layer charge in illite (-0.60 to -0.65 per formula unit),

(iv) More substitution of octahedral Al^{3+} ions in illite by Mg^{2+} or Fe^{2+} ions (one-eighth of octahedral Al^{3+} ions are generally substituted),

(v) The interlayer K^+ ions in illite are partially replaced by Ca^{2+}, Mg^{2+} and H^+ (H_3O^+) ions; and

(vi) Lower K-content and higher H_2O-content in illite.

There is a degree of randomness in the stacking of the (2:1) layers in the direction of c-axis in illite, and the size of the naturally occurring illite particles (which appear as poorly defined flakes under an electron microscope) is very small, of the order of 1 to 2 μm or even less. In micas as well as in illites, the adjacent (2:1) layers along the c-axis are strongly held by the interlayer K^+ ions which fit very tightly into the interlayer sites, which match the size of K^+ ions ($r_+ = 0.133$ nm, Table 2.2). The interlayer K^+ ions are present in a 12-fold (dodecahedral) coordination state, as explained earlier. As a result, these micaceous minerals do not expand along the c-axis in the presence of water and polar liquids, and are characterized by a basal spacing of 1.01 nm (*see* Table 2.3).

Natural glauconite is green in colour and is essentially iron-rich dioctahedral micas (iron potassium phyllosilicate of marine origin) with very low weathering resistance and is highly friable. It crystallizes with a monoclinic geometry. It is also called *green mica*. Its name is derived from the Greek *glaucos*, meaning 'blue', referring to the common blue-green color of the mineral; its sheen (mica glimmer) and blue-green color presumably relates to the surface of the sea. Its color ranges from olive green, black green to bluish green, and yellowish on exposed surfaces due to oxidation. Like illite, glauconite is K-deficient, and has a lower degree of tetrahedral substitution than the parent mica. Glauconite has the chemical formula given below:

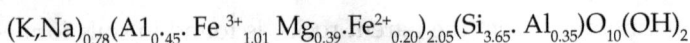

$$(K,Na)_{0.78}(Al_{0.45} \cdot Fe^{3+}_{1.01} \, Mg_{0.39} \cdot Fe^{2+}_{0.20})_{2.05}(Si_{3.65} \cdot Al_{0.35})O_{10}(OH)_2$$

In fact, both illite and glauconite are poorly defined terms, and may represent a mixture of minerals rather than distinct discrete phase (Sanyal, 2002a). Glauconite may be confused with chlorite (also of green color; *see later*) or with a clay mineral.

Vermiculite

The fine-grained vermiculites in soil may be tri- or dioctahedral in nature. Trioctahedral vermiculites are formed mainly by the alteration of trioctahedral micas (particularly phlogopite and biotite) and chlorites. Such weathering of mica is accompanied by the release of interlayer K^+ ion, and its replacement by a hydrated cation, generally Mg^{2+} ion. Dioctahedral vermiculites in soil clays (Figure 2.6) are formed primarily by alteration of dioctahedral illite (Sanyal, 2002a).

The chemistry of vermiculites is distinguished from that of micas (charge per formula unit: mica, -0.9 to -1; vermiculite, -0.6). It is due, chiefly, to the oxidation of octahedral Fe^{2+} ions to Fe^{3+} ions or hydroxylation of O^{2-} ions to OH^- ions, or both. Such a layer charge imparts to vermiculate a high CEC and high affinity for weakly hydrated cations such as K^+, NH_4^+ and Cs^+ [of rather low charge-to-radius ratio, that is, low ionic potential of feeble surface (positive) charge density]. Potassium fixation by vermiculite may also be considerable. A general formula for vermiculite may be given as (Sanyal, 2002a):

$$(Mg. Ca)_x (Mg. Fe^{2+})_3(Si_{4-x} Al_x)O_{10} (OH)_2 \cdot y \, H_2O \text{ with } x = 0.5\text{-}0.7 \text{ and } y \approx 4.$$

The hydrated interlayer cations (*e.g.* Mg^{2+} ions) strongly bind the adjacent (2:1) layers which are capable of limited expansion along the c-axis in the presence of water and polar liquids. The internal surface area is more in vermiculite than in illite. The vermiculites often occur as large crystals with platy morphology like that of micas, but are much softer (Sanyal, 2002a). The vermiculite is characterized by a basal spacing of 1.44 nm (*see* Table 2.3).

Smectites

The smectite group consists of clay minerals with (2:1) structure presented earlier for mica and vermiculite, but with a still lower excess negative layer charge (-0.60 to -0.25) per formula unit. Such excess negative layer charge is balanced by interlayer cations. The excess negative layer charge arises due to isomorphous substitution of octahedral and tetrahedral cations, depending on which different members of this group of minerals are obtained (Figure 2.6). Both dioctahedral and trioctahedral smectites are present in soil, the most dominant smectite being the dioctahedral members, namely montmorillonite, and the two other such members being beidellite and nontronite. Montmorillonites are distinguished from beidellite and nontronite on the basis of the location of the site of excess negative charge on the (2:1) layers. In montmorillonite, the layer charge arises from the substitution of

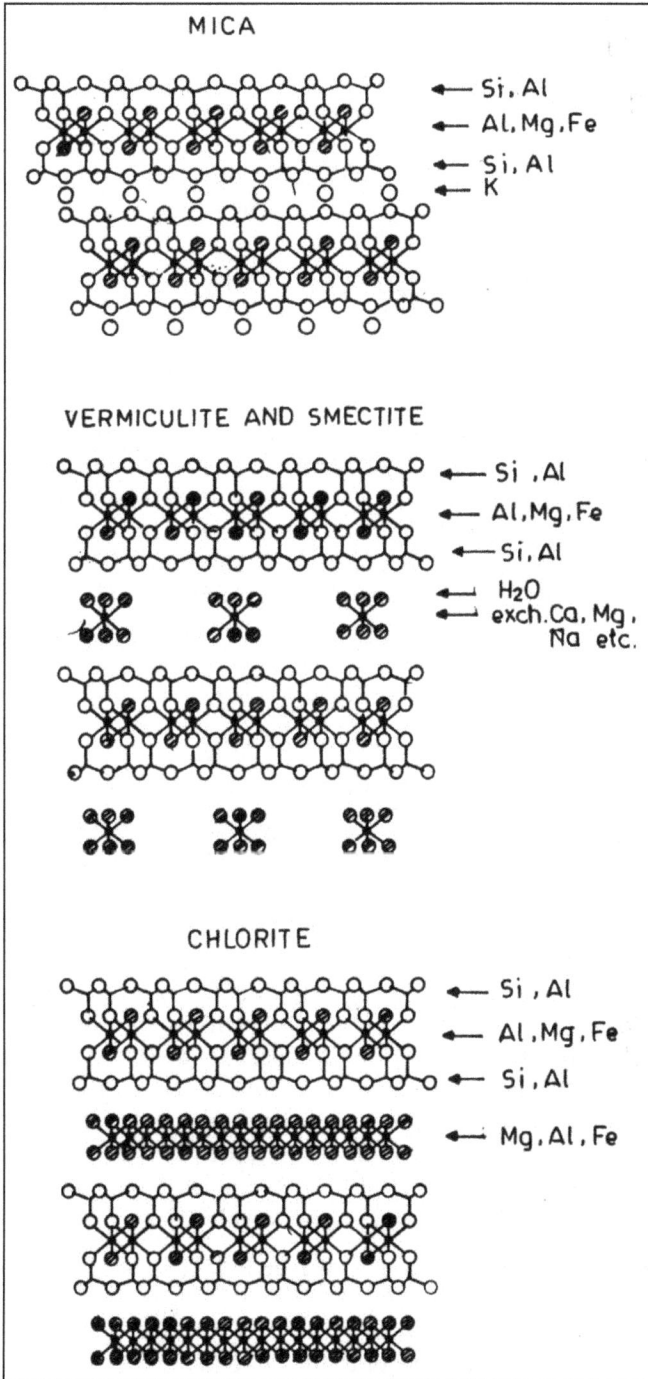

**Figure 2.6: Structural Scheme of Mica,
Vermiculite-Smectite and Chlorite Clay Minerals.**
Source: Schulze (1989).

octahedral Al^{3+} ion by Mg^{2+} ion, the structural formula being $(M^+_{0.43}.nH_2O)(Al_{1.57}.Mg_{0.43})Si_4O_{10}(OH)_2$, where $(M^+.nH_2O)$ is the monovalent hydrated interlayer cation. These cations, which are exchangeable, can have higher valences as well (e.g. Ca^{2+} ions). In beidellite and nontronite, the excess negative layer charge arises from the isomorphous substitution of Si^{4+} ions in the tetrahedral sheets by Al^{3+} ions. In nontronite, Fe^{3+} ions are the octahedral ions in place of Al^{3+} ions. This mineral is thus also referred to as iron-montmorillonite. The formulae of these clay minerals are the following (Sanyal, 2002a):

Beidellite: $(M_{0.43}.nH_2O)Al_2(Si_{3.57}. Al_{0.43})O_{10}(OH)_2$

Nontronite : $(M_{0.43}.nH_2O)Fe^{3+}_2(Si_{3.57}.Al_{0.43})O_{10}(OH)_2$

In montmorillonite, the adjacent (2:1) layers are held weakly along the direction of c-axis by the hydrated interlayer cations which are screened from the excess negative charge in the octahedral sheet by the intervening tetrahedral sheet, leading to weak Coulombic forces of attraction between the negatively charged sandwiched (between two tetrahedral layers) octahedral layer and the interlayer hydrated cations. As a result, the mineral expands along the direction of c-axis in the presence of water and polar liquids (such as glycerol), from about 1.4 nm to 1.7-1.8 nm. It also possesses large internal surface area (see Table 2.5). The important trioctahedral smectite minerals with the corresponding compositions are given below (Sanyal, 2002a):

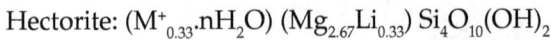

Hectorite: $(M^+_{0.33}.nH_2O) (Mg_{2.67}Li_{0.33}) Si_4O_{10}(OH)_2$

Saponite: $(M^+_{0.33}.nH_2O)Mg_3(Si_{3.67}Al_{0.33})O_{10}(OH)_2$

Where $(M^+.nH_2O)$ is the monovalent hydrated interlayer cation.

There are less common minerals of the smectite group, namely sauconite (Zn^{2+} ions in the octahedral layer of this trioctahedral smectite; $Na_{0.3}Zn_3(SiAl)_4O_{10}(OH)_2.4H_2O$), pimelite ($Ni^{2+}$ ions in the octahedral layer of this trioctahedral mineral; $Ni_3Si_4O_{10}(OH)_2.4H_2O$), and a vanadium-rich smectite. Smectites are important minerals, governing many important properties of soil due to their high specific surface area (see Table 2.5), and the consequent large sorption capacities. A remarkable property of smectites is their ability to swell when wet, and shrink on drying. Such swell-shrink behaviour is especially observed in Vertisols, and in the soils belonging to the Vertic subgroup. This causes large (two-dimensional) cracks in these soils, for instance, on wetting and subsequent drying. These cracks present a special problem in the management of the physical tilth of these soils for agricultural purposes, as well as engineering problems when houses, roads and other structures are built on smectitic soils (Sanyal, 2002a).

2.2.2c. Layer - Lattice Minerals of 2:1:1 or 2:2 Type

Chlorites

The chlorite structure consists of a mica-like (2:1) layer which carries excess negative charge because of isomorphous substitution in the layers. This negative charge, however, is balanced, in place of interlayer cations (such as interlayer K^+ in micas), by a positively charged interlayer of octahedrally coordinated hydroxide sheet. There are thus two octahedral sheets in the overall structure unit, one in the (2:1) layers, and another between two adjacent (2:1) layers. These interlayer octahedral sheets may be either trioctahedral [brucite-type, $Mg_3(OH)_6$], or dioctahedral [gibbsite-type, $Al_2(OH)_6$] which assume excess positive charge through appropriate substitutions forming, for instance, a layer of structural formula, $[Mg_2Al(OH)_6]^+$ (Figure 2.6). Either octahedral sheet in the structure [*i.e.*, that of the (2:1) layer and the interlayer hydroxide sheet] may be di- or trioctahedral, and may contain Mg^{2+}, Fe^{2+}, Mn^{2+}, Ni^{2+}, Al^{3+}, Fe^{3+} and Cr^{3+} ions. The tetrahedral cations are Si^{4+} and Al^{3+} ions. These minerals are often referred to as (2:1:1) or (2:2) layer-lattice minerals. The general formula of a trioctahedral chlorite is given as (Sanyal, 2002a):

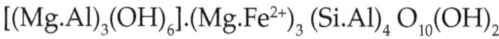

$$[(Mg.Al)_3(OH)_6].(Mg.Fe^{2+})_3 (Si.Al)_4 O_{10}(OH)_2$$

Trioctahedral chlorites are named according to the dominant divalent cation in the octahedral sheet of the (2:1) layer, *e.g.* clinochlore, chamosite, nimite and pennantite are the recommended names for Mg^{2+}-, Fe^{2+}-, Ni^{2+}-, and Mn^{2+}- dominant chlorites with ideal end-member formulae given as follows (Sanyal, 2002a):

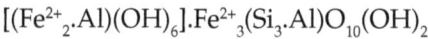

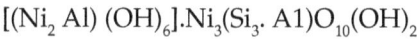

$$[(Mg_2Al)_3(OH)_6].Mg_3(Si_3Al)O_{10}(OH)_2$$
$$[(Fe^{2+}_2.Al)(OH)_6].Fe^{2+}_3(Si_3.Al)O_{10}(OH)_2$$
$$[(Ni_2 Al) (OH)_6].Ni_3(Si_3. Al)O_{10}(OH)_2$$
$$[(Mn^{2+}_2 Al)(OH)_6].Mn^{2+}_3(Si_3.Al)O_{10}(OH)_2$$

In addition to the strong electrostatic binding of the adjacent mica-like negatively charged (2: 1) layers in chlorite by the positively charged interlayer octahedral hydroxide sheet, the hydroxyl ions of each interlayer sheet maintain long hydrogen bond contacts with the basal oxide ions of the (2:1) layers above and below the interlayer sheet (Figure 2.6). As a consequence, chlorite minerals do not expand along the c-axis in the presence of water and the polar liquids, and are characterized by a basal spacing of about 1.43 nm (*see* Table 2.3).

Chlorite minerals in soils are generally primary minerals, inherited from either metamorphic rocks, or igneous rocks in which pyroxenes, amphiboles and biotite have undergone alteration to chlorite. Chlorites weather to vermiculite and smectite minerals (Sanyal, 2002a).

2.2.3. Interstratification in Phyllosilicates

In addition to the discrete particles of different layer-lattice aluminosilicates, discussed so far, the soil clays also consist of particles which are, in fact, a mixture of two and less commonly three (or even four) distinct aluminosilicate clays, stacked along the c-axis. Such interstratification leads to the formation of mixed-layer or interstratified minerals; some examples are mica-vermiculite, mica-smectite, mica-chlorite, kaolinite-smectite, chlorite-vermiculite, chlorite-smectite, etc. The layer sequence can be fully ordered, for example: ABABAB....ABBABBA.........., etc. or else partly ordered, or even fully random, *e.g.* ABABBABABBBAA...., etc. Such irregular interstratification is more commonly found in soil clays. The interstratified clay minerals are formed through several processes, namely (i) partial removal of interlayer K^+ ions from micas/illites or of interlayer hydroxide from chlorite interlayers, (ii) fixation of K^+ ion by some vermiculite layers to yield mica/illite, and (iii) the formation of hydroxide interlayers to form chlorite-like structures (Sanyal, 2002a).

2.2.3a. Palygorskite and Sepiolite

The structure of the fibrous minerals, namely palygorskite (formerly also known as attapulgite) and sepiolite, differs from the typical (1:1) and (2:1) layer structures in terms of lacking continuous octahedral sheets, but having instead continuous tetrahedral sheets. Both these structures may be described as ribbons of (2: 1) phyllosilicate structure, one ribbon being linked to the adjacent one by inversion of (SiO_4) tetrahedra along a set of Si-O-Si bonds. These sheets are held together by Al^{3+} and/or Mg^{2+} ion in octahedral coordination between the apical oxide ions of successive (SiO_4) sheets (Figure 2.7). The structurally important unit in these minerals is the amphibole double silica structure (Table 2.1) (Sanyal, 2002a).

2.2.4. Aluminosilicates with Short-Range Order: Allophane and Imogolite

Unlike (1:1) and (2: 1) structures, which exhibit crystalline structure, allophane and imogolite are characterized by short-range order, and are essentially "poorly

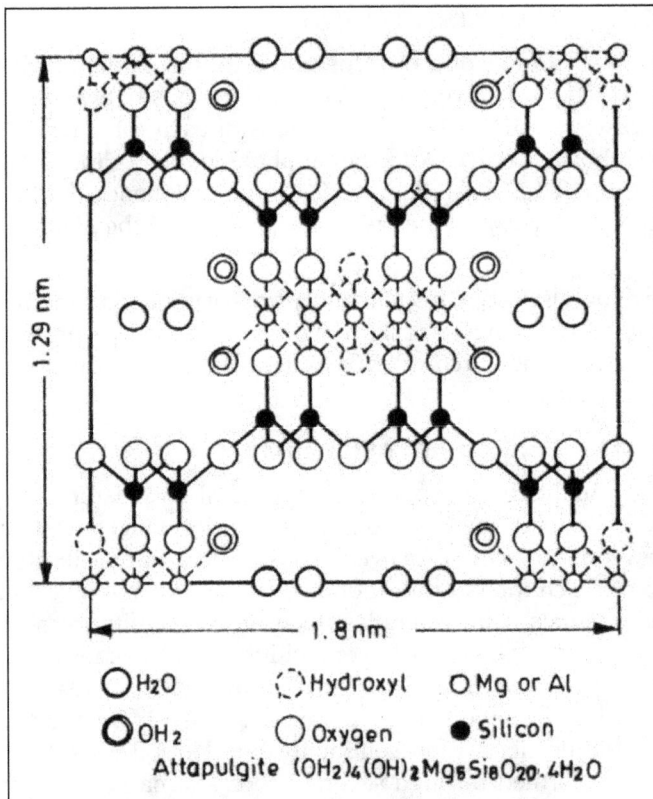

Figure 2.7: Structural Scheme of Attapulgite.

crystalline" or the structure is disordered. These aluminosilicates are derived from tropical weathering of basic igneous rocks, more acidic rocks during the development of Spodosols, and particularly, during weathering of volcanic ash. Indeed these are important minerals in the Andept sub-order. In allophane, the molar ratio of (SiO_2/Al_2O_3) is highly variable, ranging from about 0.84 (or even as low as 0.2 for silica-poor materials) to as high as 2.0 (the characteristic ratio for kaolinite). These sesquioxides in allophane are chemically combined, having Si-O-Al bonds (Sanyal, 2002a).

Allophane minerals generally consist of small spheres (of diameter 3.5 - 5.0 nm) of ill-defined structure. These spheres are often clumped together leading to irregular aggregates which frequently form coating on crystalline aluminosilicates. There is also some indication that there is a continuous chemical transition from allophane to halloysite to kaolinite, which can be viewed as a continuous process of dehydroxylation through condensation replacement with SiO_4, and this, in turn, leads to a higher degree of crystallinity of these minerals (Sanyal, 2002a).

Imogolite consists of a gibbsite sheet with isolated Si^{4+} ions bridging three O^{2-} ions over the vacant octahedral site; coordination of Si^{4+} ion is completed with OH^- ions directed away from the octahedral sheet. The short Si-O bonds impart curvature to the octahedral sheet, which forms a tubular structure with OH^- ion on the outside and (Si-OH) groups directed towards the centre of the tube. The tubes are arranged in bundles, 10 to 30 nm across and several micrometers long (Sanyal, 2002a).

Allophanic soils often also contain much organic matter, due possibly to adsorption on such amorphous surfaces of organic molecules, and are characterized by very low bulk density, high water holding capacities and other unusual physical properties. Allophane and imogolite have high specific surface areas, high phosphate fixing capacity, and also a considerable specific sorption capacity for many inorganic and organic compounds (Sanyal, 2002a).

2.2.5. Other Minerals in the Soil Clay Fraction

The soil clay contains relatively small amounts of hydrous oxides and hydroxides of Fe, Al, Si and Mn and clay-sized primary minerals. Calcite, dolomite, large flakes of mica, pyrite, feldspar, gibbsite, goethite (α-$Fe^{III}OOH$), hematite (α-Fe_2O_3) and other minerals are also found in some clay fractions of soil.

Goethite and hematite are the two most common iron oxide minerals, occurring in Alfisols and Ultisols. Gibbsite occurs in strongly weathered acidic soils such as Udults and Ustults. A sizable proportion of these oxides and hydrous oxides in the corresponding soil clay fractions may exist as discrete and very small particles of large specific surface area. The latter imparts to them a high phosphate and silica sorption capacities (Sanyal, 2002a).

2.3. Identification of Clays

2.3.1. X-ray Diffraction Analysis

Crystals are composed of an orderly three-dimensional structure with regular periodicity along the three crystallographic axes. The X-rays are electromagnetic radiation of very short wavelength (λ), of the order of 10 nm to less than 1 nm. Thus, they have very high penetrating energy as required by the Planck's equation, $E = h\nu = hc/\lambda$, where E is the energy of radiation of wavelength λ and frequency ν; c, is the velocity of light (*i.e.*, the velocity of the electromagnetic radiation concerned), and h is the Planck's constant.

When an X-ray beam of wavelength of about 0.1 nm (1 Å) is incident on a crystal, it causes electrons of the atoms of the crystal in its path to vibrate with a frequency of the incident X-ray. A part of the energy of the X-ray is absorbed by these electrons, which scatter (emit) this energy as the X-ray of the same λ. Since X-rays have high penetrating energy, each atomic layer in the crystal can be expected to scatter only a small part of the X-ray beam. If the crystal particles do not have a spacing of the same order of magnitude as the λ of the X-rays, simple reflections and scattering of the incident X-rays occur, and the simple scattered radiations interfere destructively. However, in some specific directions, they reinforce one another, producing a cooperative scattering effect, known as diffraction (Sanyal, 2002a). For such diffraction to occur, the interatomic distance in the crystal has got to be small enough (of the same order of magnitude as the incident X-ray wavelength) to enable the crystal lattice to act as a diffracting grating. For this, the difference between the path lengths traversed (Figure 2.8) must be equal to an integral multiple of the wave length (λ) of the incident X-ray beam (Sanyal, 2002a)

Let a single monochromatic X-ray beam be incident on aligned planes of lattice points, with separation of d, at angle θ, as shown in Figure 2.8. It is clear that points A and C are on one plane, while B is on the plane below. It is also apparent that the points ABCC' form a quadrilateral (Bragg and Bragg, 1913).

There will be a path difference between the ray that gets reflected along AC' and the ray that gets transmitted, then reflected, along AB and BC, respectively. This path difference is (AB + BC) - AC', with AB and BC being equal. The two separate waves will arrive at a point in the same phase, and hence undergo constructive

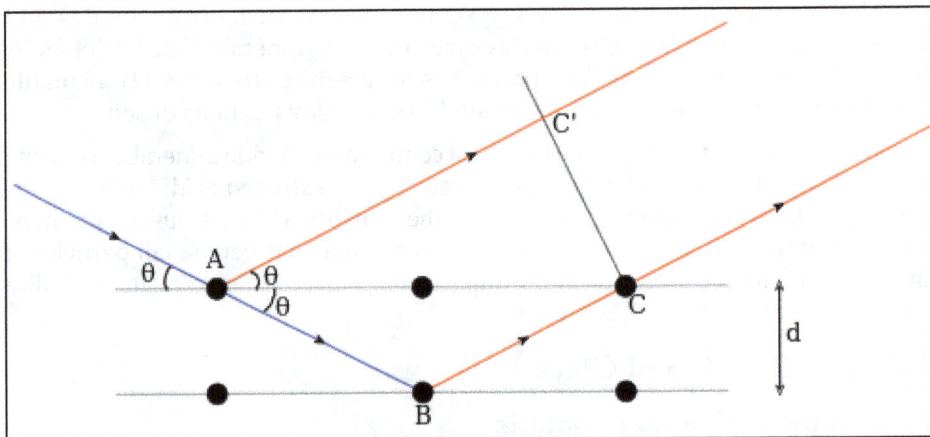

Figure 2.8. X-ray Reflection from Equidistant Planes of a Crystal.

interference, if and only if this path difference, as stated above, is equal to any integer value of the wavelength of the X-ray concerned, *i.e.*,

$(AB + BC) - AC' = n\lambda,$

Where n is a positive integer and λ is the wavelength of incident X-ray. It is to be noted here that moving particles, including electrons, protons and neutrons, have an associated wavelength called the *de Broglie wavelength*. A diffraction pattern is obtained by measuring the intensity of scattered waves as a function of the scattering angle. Very strong intensities, known as Bragg peaks, are obtained in the diffraction pattern at the points where the scattering angles satisfy Bragg condition (*see* later) (Bragg and Bragg, 1913).

Therefore, it follows from Figure 2.8

$AB = BC = (d/\sin \theta)$ while $AC = (2d/\tan \theta)$, from which it follows that

$AC' = AC. \cos \theta = (2d/\tan \theta) \cos \theta = (2d \cos \theta/\sin \theta) \cos \theta = (2 d/\sin \theta) \cos^2 \theta$

It therefore follows that for diffraction to occur,

$n\lambda = (AB + BC) - AC' = (2 d/\sin \theta) (1 - \cos^2 \theta) = (2 d/\sin \theta) \sin^2 \theta$

$[AB = BC = (d/\sin \theta)]$

Or, $n\lambda = 2d \sin \theta$ (2.1)

With n = 1, 2, 3,...., etc.

Equation 2.1 gives the Bragg's Law (Bragg and Bragg, 1913).

Here, n is the 'order of reflection' of the X-ray. As stated, Eq. 2.1 is the outcome of the Bragg's diffraction law. Thus, from the knowledge of λ of the incident X-ray and the angle (θ) of incidence on a given array of crystal planes, *e.g.*, the (ab) planes of a phyllosilicate crystal, stacked along the direction of c-axis, the spacing (d) between the identical crystal planes or units (*i.e.*, basal spacing along c-axis of a given phyllosilicate crystal) can be computed.

A re-arrangement of Eq. 2.1 leads to Eq. 2.2, *i.e.*,

$\lambda = 2(d/n) \sin \theta = 2d' \sin \theta$ (2.2)

Where, $d' = d/n$. That is, the higher-order reflections from planes stacked along a given axis (*e.g.* c-axis) can be imagined as though they were the first-order reflections from planes with spacing $d' = d/n$. Thus, a second-order reflection (n = 2) from a given array of planes will correspond to an interplane (interlayer) spacing of (d/2) where d is the corresponding spacing for the first-order (n = 1) reflection (Sanyal, 2002a).

For X-ray identification, the clay minerals are subjected to several chemical and thermal pretreatments, and subjected repeatedly to X-ray diffraction after each such treatment. For routine measurements, three diffractograms are recorded with the following pretreatments:

(i) Magnesium saturated clay, both as such, and after glycerol solvation

(ii) Potassium saturated clay, heated to 300°C, and

(iii) Potassium saturated clay, heated to 550°C.

The Mg-saturated clays are subjected to glycerol (a polar liquid) solvation to examine whether or not the ab planes of the layer-lattice aluminosilicate clays expand along the direction of c-axis due to accommodation of the polar molecules in the interlayer space separating two adjacent identical units.

Table 2.3 gives the diagnostic X-ray diffraction spacing (basal spacing) for a number of clay minerals of common occurrence in soil. Thus, kaolinite is characterized by strong reflections at 0.72 and 0.36 nm which correspond, respectively, to the first-order and the second-order diffractions from the basal (ab) planes of kaolinite, stacked along the direction of c-axis. These peaks disappear on heating the K-saturated clay samples at 550 °C due to lattice collapse of kaolinite.

Table 2.3: Diagnostic X-ray Spacing (nm) for Clay Minerals

Clay Mineral	Mg-Saturated Glycerol Solvated Specimen			K-Saturated Specimen Heated to 300°C			K-Saturated Specimen Heated to 550°C		
Kaolinite	0.715*	0.357	0.238	0.715	0.357	0.238	–	–	–
Halloysite	1.07	0.76	0.34	0.74	0.368		–	–	–
Mica	1.01	0.498	0.332	1.01	0.498	0.332	1.01	0.498	0.332
Vermiculite	1.44	0.718	0.479	1.01	0.498	0.332	1.01	0.498	0.332
Chlorite	1.43	0.718	0.479	1.43	0.718	0.479	1.43	0.718+	0.479
Smectite	1.78	0.885	0.59	1.01	0.498	0.332	1.01	0.498	0.332

* The underlined spacings are strong and generally used as diagnostic peaks.

+ In many soil chlorites, this peak may not appear at all.

Source: Raman and Ghosh (1974).

2.3.2. Electron Diffraction Studies

An electron (mass 9.1 x 10^{-31} kg), accelerated by 10,000 V, attains a velocity of 5.9 x 10^{7}m sec^{-1}, and the electron having wave character (as per what is known as the *de Broglie equation*, namely $\lambda = h/mv = h/p$, where p is the momentum), such velocity imparts to it a wavelength of 1.2×10^{-11} m or 1.2×10^{-2} nm. This is somewhat less than the usual interatomic distances in a molecule and an electron beam may, therefore, be diffracted by a crystal structure (acting as a diffraction grating) in much the same way as the X-rays (Sanyal, 2002a). Thus, in electron diffraction studies, in place of X-ray beam, an electron beam of fast electrons, but having much lower penetrating energy than the X-rays, is made incident on the crystal, and the desired basal spacings of the clay minerals are recorded. Unlike X-rays, the electron beams fail to penetrate beyond the surface of the solid and scan the same, even though the exposure time required may be only few seconds as compared to few hours in case of X-ray diffraction. The latter is the outcome of high speed of the incoming electron beam under an applied high voltage (Sanyal, 2002a).

2.3 3. Electron Microscopy

As stated above, the wavelength (λ) of an electron beam can be made extremely small by accelerating the electrons under an applied high potential difference.

Such a low wavelength may become comparable to that of X-rays, but the electron beams, due to their negative charge, have one advantage over X-rays as a means of investigating the shape and size of particles. That is, appropriate arrangements of electrostatic and magnetic fields can be designed to act as lenses for bringing the electrons into sharp focus (Sanyal, 2002a). This has led to the development of electron microscopes, which are capable of resolving images as small as 0.5 nm in diameter with the magnification of over 2, 00,000 times. This is not sufficiently fine so as to resolve the interatomic distances, but important information regarding shape and size of the soil clay particles (particle size l μm to 1 nm) can be obtained.

In this technique, a drop of dispersed clay suspension is taken on the specimen-support plastic films, and observed directly in the electron beam, or else the specimen-support film is mounted on specimen-support grid (made of copper or plated bronze), and is examined after shadow-casting to provide sharper contrast, leading to better three-dimensional information about the particle surface, as well as its thickness (Sanyal, 2002a). Such electron microscopic studies show the shape of kaolinite as thin, plate-like hexagonal crystals, and that of halloysite as elongated tubular particles. The electron micrographs of illites suggest small poorly defined flakes, grouped together in irregular aggregates, whereas that of montmorillonites reveals irregular flake-shaped aggregates.

2.3.4. Methods Based on Thermal Properties

2.3.4.1. Differential Thermal Analysis (DTA)

In this method, the sample and a thermally inert material (*e.g.* anhydrous Al_2O_3) are heated slowly and uniformly under the same environment and the difference in temperature (ΔT) between the sample and the inert material is monitored as a function of time (t) or temperature (T°C). The structural changes in the course of heating may be accompanied by either absorption or evolution of heat, and accordingly an exothermic peak or an endothermic trough may be obtained in the resulting Differential Thermal Analysis (DTA) curves. Endothermic troughs may be associated with decomposition processes, including loss of CO_2, water, etc., phase changes, and so on. Exothermic peaks are generally associated with burning of carbon, oxidation of sulphur, crystallization, etc. The clay minerals are identified by the characteristic peaks or troughs in the corresponding DTA curves (Sanyal, 2002a). Table 2.4 records the DTA characteristics of clay minerals of common occurrence in soil.

2.3.4.2. Other Thermal Methods

In thermogravimetric (TG) methods, the weight loss of clay minerals in the course of heating is recorded as a function of temperature. In yet another type of method, such weight loss per degree rise in temperature is shown as a function of the furnace temperature, leading to differential thermogravimetric (DTG) curves (Sanyal, 2002a). Some representative DTA, TG and DTG curves for clay minerals in soil are presented in Figure 2.9.

Table 2.4: Differential Thermal Peaks of Common Clay Minerals in Soil

Clay Mineral	Differential Thermal Peak (Endothermic), °C
Kaolinite	500 - 550 (symmetrical)
Halloysite	480 -530 (asymmetrical)
Illite	550 - 650
Montmorillonite	850 - 900
Beidellite	850 - 920
Vermiculite	700 - 850
Chlorite	750 - 800
Attapulgite	225 – 350, 400 -525

Source: Raman and Ghosh (1974).

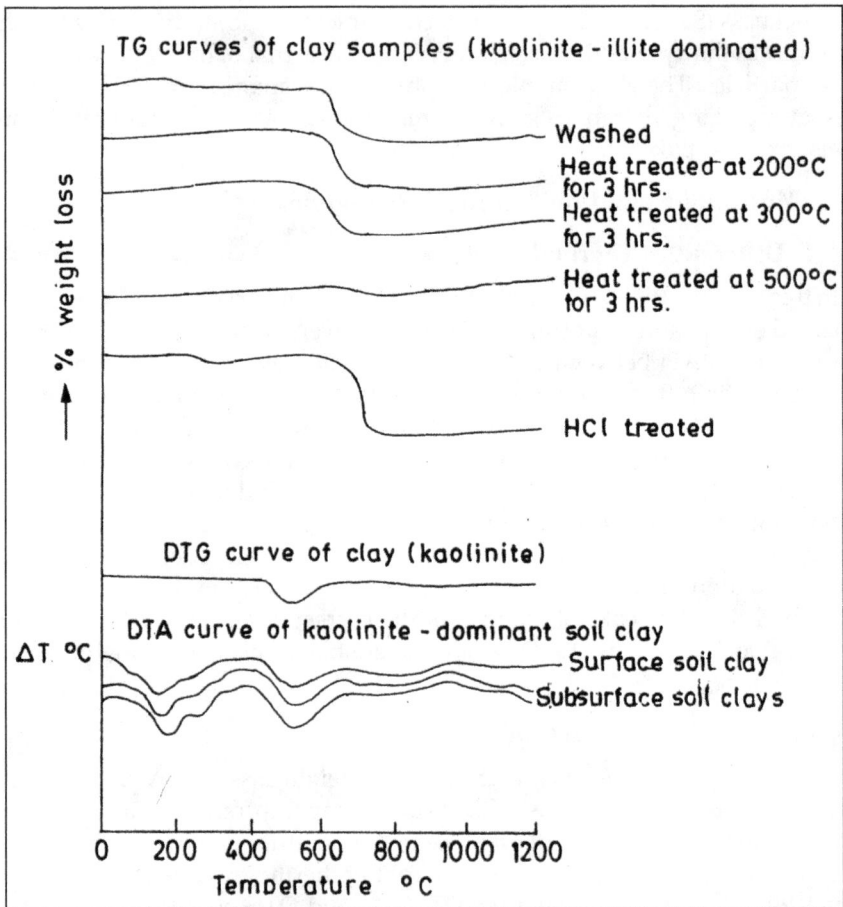

Figure 2.9: Thermogravimetric (TG), Differential Thermogravimetric (DTG) and Differential Thermal Analysis (DTA) Curves of Kaolinite-Illite/ Kaolinite-Dominated Clay Samples.

2.3.5. Surface Area Measurement

The soil clays are mixtures of clay minerals with one mineral being generally dominant. The clays being colloidal in nature, the most important feature, governing their properties, is their surface area (and also specific surface charge density). The specific surface area of a soil (or clay) is the function of the dominant mineral in the clay fraction of the soil. A comparison of the experimentally determined specific surface area of soils [by the method employing, for instance, the retention by soil surface of ethylene glycol (EG) or its monomethyl ether (EGME)] with the standard values for pure (individual) clay minerals often provides useful information regarding the dominant clay mineral present in a soil (Sanyal, 2002a). The specific surface areas of some clay minerals, as measured by the EG-retention method, are recorded in Table 2.5. It is clear that the total surface area of kaolinite is nearly equal to its external surface area, the corresponding internal surface being very low, as pointed out earlier (Section 2.2.2a) (Sanyal, 2002a).

Table 2.5: Specific Surface Area of some Clay Minerals (m².g⁻¹) as Determined by the Ethylene Glycol Retention Method

Clay Mineral	Range of Specific Surface Area		
	Internal	External	Total
Kaolinite	7-10	30-35	37-45
Illite	70-100	50-70	120-170
Muscovite	140-150	90-100	230-250
Montmorillonite	500-600	80-150	580-750
Chlorite	60-80	70-100	130-180
Vermiculite	700-800	80-100	780-900

Source: Raman and Ghosh (1974).

Montmorillonite, on the other hand, has a large internal surface area, it being a swelling type of clay. Illite has properties, intermediate between those of kaolinite and montmorillonite. Vermiculite possesses considerably high surface area, while chlorite has low internal surface area. Large surface areas of clays, in general, impart to them the properties of swelling, shrinkage, viscosity and sedimentation in water, as well as large adsorptive capacity for inorganic ions (*e.g.* plant nutrients) and also organic molecules such as humus (Sanyal, 2002a).

2.3.6. Ion-Exchange Properties

The cations satisfying the negative charge on clays undergo ion exchange reactions with externally added electrolytic cations. The cation-exchange capacity (CEC) of soils depends on the specific (negative) charge density of the clays (apart from that of the humic colloids), especially the dominant clay mineral in the soils. The representative values of CEC of some clay minerals of common occurrence in soils are recorded in Table 2.6. These values may prove useful in ascertaining the dominant clay mineral of a given soil by a process elaborated under surface area measurements (Sanyal, 2002a). As one would appreciate, there is a fairly large range of CEC values of illites and chlorites due, primarily, to variations in the degree of crystallinity, particle size, extent of isomorphous substitution of cations in the lattice, degree of interlayer K^+ and Mg^{2+} ion replacements, *etc.* (Sanyal, 2002a).

Table 2.6: Cation-exchange Capacity (CEC) [cmol (p$^+$) kg^{-1}]
of some Clay Minerals

Clay Mineral	CEC
Kaolinite	3-15
Halloysite (hydrated)	40-50
Montmorillonite	80-150
Nontronite	57-64
Illite	10-40
Vermiculite	100-150
Chlorite	10-40
Attapulgite	18-22

Source: Seal *et al.* (1974).

2.3.7. Y-Index of Martin and Russell

This index combines total K_2O content, ethylene glycol (EG)-retention by internal surface area and cation-exchange capacity (CEC) of the clay (Sanyal, 2002a). The Y-index is calculated by using Eq. 2.3:

$$Y = \frac{1}{3}\left[\frac{A \times 10}{6} + \frac{B \times 10}{20} + \frac{C \times 10}{25}\right]$$

$$(2.3)$$

Where A is the per cent K_2O in the material; B is the internal retention of EG (mg. g^{-1}); and C is the CEC [cmol (p$^+$) kg^{-1}]. The representative ranges of Y-index for different dominant clay minerals are as follows (Raman and Ghosh, 1974):

Y Index	Dominant Clay Mineral
0-5	Kaolinite type
5 -25	Illite type
25 and more	Montmorillonite type

2.3.8. Elemental Chemical Analysis

Chemical analysis has been an early essential step in ascertaining the nature of the dominant clay mineral in soils. The objective is to account for the total composition of the sample so that the total percentage of all elemental determinations (expressed as the corresponding per cent oxide present) would be approximately 100. For this purpose, it is necessary to carry out determinations in the solutions (prepared by fusing the sample with sodium carbonate) of Si^{4+}, Al^{3+}, Fe^{3+}, Fe^{2+}, Ti^{4+}, Mn (total Mn comprising various valence slates), Mg^{2+}, Ca^{2+}, Na^+ K^+, phosphate, and also water evolved below 105°C, as well as between 105°C and 1000°C. The findings suggest that the molar ratio, SiO_2/R_2O_3, (where R_2O_3 represents the sum of the sesquioxides, namely Fe_2O_3 and Al_2O_3), is generally more than 2.0 for montmorillonite-dominant soil clay, and less than 2.0 for kaolinite- dominant soil clay fraction. The micaceous clay minerals, *e.g.* illite, muscovite, biotite, is inferred from the K_2O-content, which for muscovite is 10.6 per cent and for biotite 9.4 per cent. For illites, it ranges between 6 and 10 per cent. Similarly, the chloritic and/or vermiculitic minerals are indicated by the relatively high MgO-content of the soil clay fraction (Sanyal, 2002a).

2.3.9. Other Methods

The dominant clay mineral in the clay fraction of soils may be inferred from the electrochemical properties of the soil clays, *e.g.* the nature of the pH-potentiometric titration curves and conductometric titration curves, either alone, or in conjunction with the viscosity properties. Several other methods involve measurements of sedimentation volume, selective dissolution analysis, optical microscopic study, staining test, infrared (IR) spectral analysis, etc. The soil clays being essentially a mixture of clays, a suitable combination of these methods is chosen for identifying, not only the dominant clay mineral, but also other minerals, less abundant in the clay fraction of soils (Sanyal, 2002a).

2.4. Organic Soil Colloids

These refer to the humic colloids, the main fraction of the soil organic matter, which are widely distributed all over the earth's surface. Nearly 70–80 per cent of the soil organic matter consists of humic substances, *viz.* humic acid, fulvic acid and humin. These are believed to be produced by continuous decomposition of plant, animal and microbial bodies, encompassing alteration of carbohydrates, proteins, fats, resins, wax, and so on, followed by synthesis of complex humic products. The humic colloids are dark in colour, partly aromatic and amorphous with very high specific surface area and charge density and are essentially hydrophilic in nature. These substances are characterized by flexible polyelectrolyte behaviour, having polyfunctional groups, and very high molecular weights, ranging from a few hundred to several thousands. A typical value of CEC of soil humic colloids (which is pH-dependent) may be of the order of 200–250 cmol (p^+) kg^{-1} or even higher. The typical functional groups of the organic colloids are carboxylic, phenolic hydroxyl, amino, amide, etc. and as a result, these act as proton donors or proton acceptors, depending on soil pH (Sanyal, 2002b).

The humic colloids in soil often impregnate the soil mineral matter, *e.g.*, the inorganic phyllosilicate clays (of soil) with which they remain in intimate association, forming clay-humic complexes.

2.5. Sources of Charges on Soil Colloids: Their Charge Characteristics

2.5.1. Isomorphous Substitution

Isomorphous substitution of cations by those of different charges (but having similar sizes within permissible limits) within the tetrahedral and the octahedral layer-lattice clay minerals leads to the development of *net* unbalanced charge within the lattice. Such a substitution takes place following the Goldschmidt's laws of crystal chemistry (*see* Section 2.2.1). Indeed an isomorphous substitution within Al-octahedral and Si-tetrahedral layers of higher-valent cations by lower-valent cations having similar sizes leads to permanent (pH-independent) negative charge. Examples are the substitution of Si^{4+} in tetrahedral layer by Al^{3+} in mica, vermiculite, illite, etc., and of Al^{3+} in the octahedral layer by Mg^{2+}/Fe^{2+} in montmorillonite. Positive charges on clay lattice may also arise through isomorphous substitution, *e.g.*, by substitution of a lower-valent cation by a higher-valent one, such as occurs in the trioctahedral sheet forming substituted $[Mg_2AlO_2(OH)_2^+]$ sheet. The latter is sometimes found in the trioctahedral sheet in some vermiculites and chlorites. The net permanent charge on the clay lattice obviously depends on the extent of these two types of isomorphous substitutions, although it is nearly always negative (Sanyal, 2002a).

The charge of soil clays arising out of isomorphous substitution remains constant and independent of soil pH.

2.5.2. pH-dependent Charge

There is yet another source of charge of soil colloids, especially for kaolinite, the aluminosilicate clay, and also for the hydrous oxides and hydroxides of Fe and Al in clay-size dimensions, allophones and non-crystalline clays, as well as the organic humic colloids of soil. The negative charge here arises through the dissociation of proton from the exposed hydroxyl group or bound water of constitution at the edge and surfaces of the inorganic colloids, and is naturally pH-dependent. The carboxylic and phenolic (OH) groups of soil humic substances also behave as weak acids, dissociation of which (thereby contributing towards soil acidity) would depend on the dissociation constant of the acid, and the pH of the surrounding medium. The carboxylic group ionizes at relatively lower pH (close to pH 5.5-6.0) than the phenolic (OH) group (which ionizes at pH 9.0), and contributes to the pool of pH-dependent negative charge. However, the substituted phenols, such as nitrophenols, are stronger acids than phenols (due to π-electron-withdrawing character of the nitro group) and ionize, contributing to negative charge of humic colloids, at intermediate pH between 6.0 and 9.0 (Sanyal, 2002a).

No soil organic fraction with a net positive charge has been reported so far at normal soil pH values. However, organic species (R, *see* later), bonded to (NH_2) functional groups, may protonate at low pH to yield positively charged $(R\text{-}NH_3^+)$ units, that hold anions, capable of anion exchange.

From the above discussion on pH-dependent charge of soil colloids, *e.g.* Fe and Al hydroxides, it is clear that the exposed edge/surface hydroxyl groups can act as both an acid (donating a proton to the surrounding OH group) or a base (accepting a proton from the surrounding soil solution) depending on the pH of the soil system, *i.e.*, relative preponderance of H^+/OH^- ions in the vicinity of exposed edge and surface OH group of soil colloid. Such amphoteric nature of these soil colloids warrants that these exposed hydroxyls would behave as an "uncharged" group (*neither* a proton donor *nor* a proton acceptor) at an intermediate pH. The pH at which the net surface charge of the soil colloids is zero is known as the zero-point of charge or ZPC (also known as point of zero-charge, PZC). The latter is thus a characteristic of soil colloids (such as in acid soils) which are mixtures of permanent and variable-charge minerals (Sanyal, 2002a). The concept of ZPC is illustrated in Scheme 1 with hydrous oxides of Fe and Al (in clay-size dimensions).

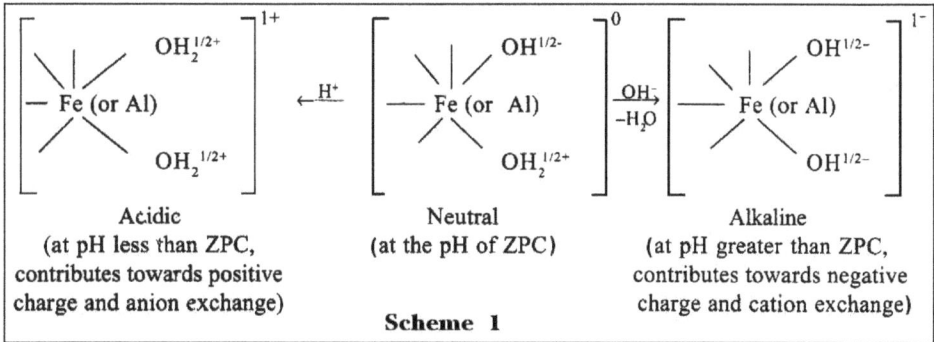

$$\left[\begin{array}{c} \diagdown\,|\diagup\quad OH_2^{1/2+} \\ — Fe\ (or\ Al) \\ \diagup\diagdown\quad OH_2^{1/2+} \end{array}\right]^{1+} \xleftarrow{H^+} \left[\begin{array}{c} \diagdown\,|\diagup\quad OH^{1/2-} \\ — Fe\ (or\ Al) \\ \diagup\diagdown\quad OH_2^{1/2+} \end{array}\right]^{0} \xrightarrow[-H_2O]{OH^-} \left[\begin{array}{c} \diagdown\,|\diagup\quad OH^{1/2-} \\ — Fe\ (or\ Al) \\ \diagup\diagdown\quad OH^{1/2-} \end{array}\right]^{1-}$$

Acidic	Neutral	Alkaline
(at pH less than ZPC, contributes towards positive charge and anion exchange)	(at the pH of ZPC)	(at pH greater than ZPC, contributes towards negative charge and cation exchange)

Scheme 1

The ZPC of a soil component depends upon its relative acidity with respect to water. Thus, ZPC of silicic acid is 2.0, that of goethite (α - $Fe^{III}OOH$) is 8.5 and that of gibbsite [Al $(OH)_3$] is more than 9.0. However, even at ZPC, the permanent charge of the soil colloid (such as the aluminosilicate clay) persists unchanged.

On the other hand, the modification of surface/edge charge characteristics of soil colloids (especially the inorganic colloids) leads to the modification of ZPC. Thus, phosphate-fixation by ligand-exchange (*e.g.* exchange of neutral OH_2 ligand to Fe or Al by the applied $H_2PO_4^-$ ion; *see* later) process raises the surface negative charge, and pushes the ZPC towards lower pH, *i.e.*, a stronger H^+ ion concentration is required in the surrounding soil solution to render the surface of the soil colloid neutral. Furthermore, intense weathering of a soil shifts its ZPC towards higher pH owing to greater accumulation of hydrous oxides/hydroxides of Fe and Al and also enrichment of soil layer-lattice clays with kaolinite (possessing mostly pH-dependent edge charge). Indeed, the ZPC of Ultisols or Alfisol is generally higher than that of an Inceptisol. Organic matter, on the other hand, pushes the ZPC towards lower pH (Sanyal, 2002a).

The development of charges on soil components is schematically shown in Scheme 2.

(i) **On clay minerals** (at edge/surface; pH-dependent negative charge)

(ii) **On clay minerals** (within lattice; permanent negative charge)

Mg^{2+}, Fe^{2+} replacing Al^{3+} in octahedral layer.

Al^{3+} replacing Si^{4+} in tetrahedral layer.

(iii) **On hydrous oxides of Fe and Al** (pH-dependent negative edge/surface charge)

Shown in Scheme 1.

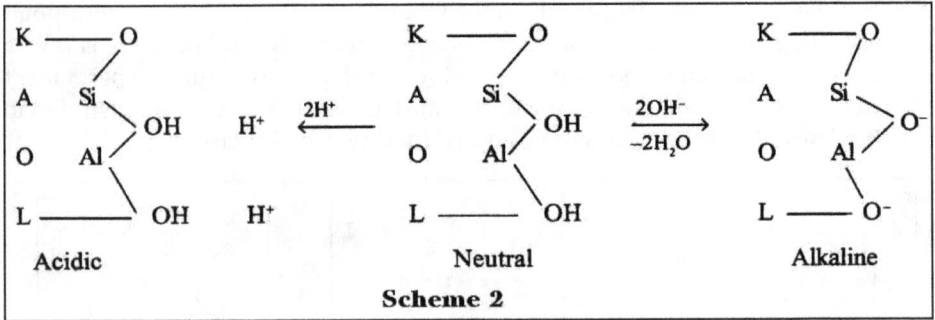

Scheme 2

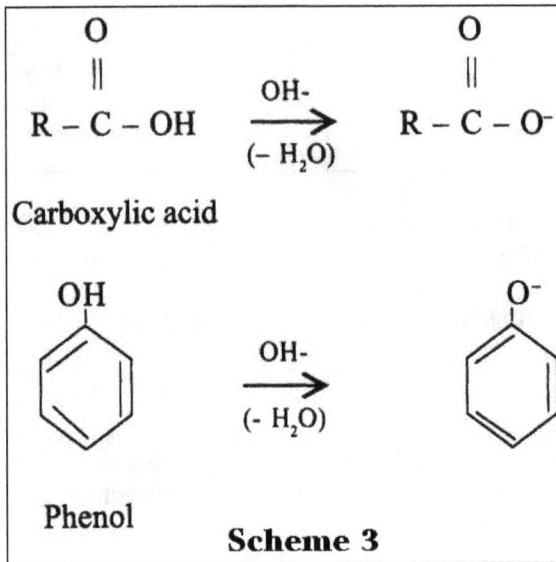

Scheme 3

(iv) **On soil organic matter** (pH-dependent negative charge)

Shown in Scheme 3.

2.6. Electrical Double Layer of Soil Colloids

The *net* charge (negative) on soil colloids is balanced by counter ions, such as interlayer cations of clay minerals, forming an electrical double layer. These counter ions, electrostatically attracted by the oppositely charged soil colloids, have a tendency (due to their thermal or kinetic energy) to diffuse away from the surface towards the bulk of soil solution where the counter ion concentration is lower. The equilibrium or balance between these two opposing (competitive) tendencies results in a distribution of the counter ions in such a manner that their concentration gradually decreases with increasing distance from the colloidal surface. In fact, it is postulated that the solution side of the aforesaid double layer consists of counter ions distributed into two broad regions, namely (i) a more or less fixed layer at a uni-molecular or uni-ionic (hydrated) distance from the colloidal particle surface (the so-called Helmholtz layer), trailed by (ii) a diffuse layer (the so-called Goüy-Chapman layer) of counter ions (in which some kinetic movement of the counter ions is possible), till the bulk solution is reached where the distribution of positive and negative ions (unaffected by the charge on the colloid which is far away) becomes uniform. Evidently, there is excess of cations and a deficit of anions in the diffuse double layer compared to their respective concentrations in the bulk solution (Sanyal, 2002a). The above model or description of the electrical double layer at the colloid/solution interface is known as the Stern model.

2.6.1. Stability of Soil Colloids and their Coagulation/Flocculation

The soil colloids are negatively charged, and as mentioned above, the counter ions (cations) are present in the fixed and the diffuse layer on the solution side of the electrical double layer at the soil colloid/soil solution interface. This causes a change of electrostatic potential from the solid surface (P) to the bulk soil solution (R) in such a manner that a sharp change in potential in the Helmholtz layer (PQ) is followed by a more gradual change in the Goüy-Chapman layer (QR). The potential difference, denoted by ξ (zeta), between the fixed part (Q) and the freely mobile portion (R) of the solution side of the double layer (Figure 2.10) governs the stability of the colloids. The higher the value of this potential (which is also known as the electrokinetic potential), the greater is the repulsion between two approaching soil colloidal particles, swarmed by the respective counter ions of like charge in the neighbouring soil solution (*e.g.*, the so-called 'clay micelles'), and hence less is the likelihood of these approaching particles to coalesce so that the larger will be the

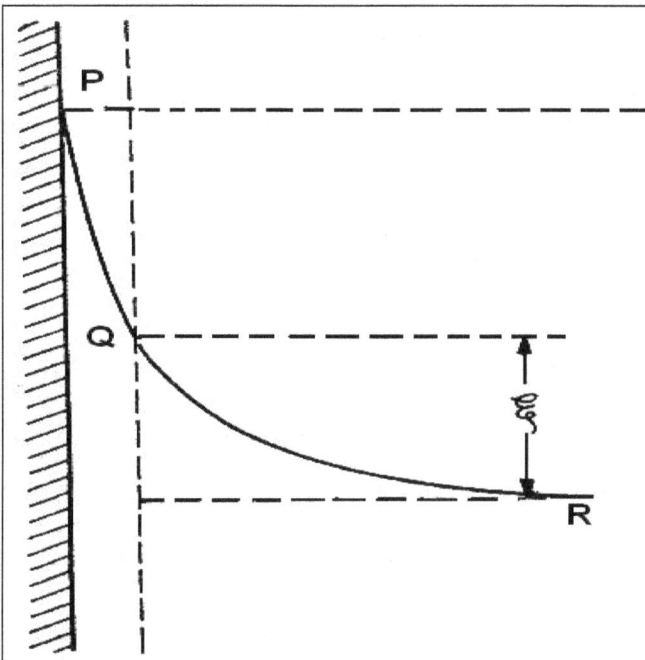

Figure 2.10: Electrokinetic (Zeta) Potential in an Electrical Double Layer.

buoyancy force to keep these colloidal particles afloat and dispersed. This will lead to the stability of the soil colloidal sol (Sanyal, 2002a).

By considering, as a first approximation, the electrical double layer at the soil colloid/soil solution interface to be equivalent to a parallel plate electrical condenser, with each plate carrying a charge of numerical value e per cm^2, the ξ potential can be approximated by the potential difference between the plates, at a distance d apart, as given by Eq. 2.4:

$$\xi = (4\pi ed)/D \tag{2.4}$$

Where, D is the dielectric constant of the medium (water). Such dielectric constant (D), which is the ratio of the static permittivity of the medium (i.e., water) to the permittivity of vacuum, characterizes primarily the solvent properties of liquid water, especially for the dissolution of the ionic and polar solutes.

On addition of an electrolyte to the soil colloidal system, the cations of the added electrolyte tend to accumulate on the solution side of the electrical double layer, in the vicinity of the plane PQ (Figure 2.10). Upon gradual increase in the concentration of the added electrolyte, this cationic accumulation may reach such a proportion that there will be a virtual neutralization of some of the charge on the soil colloidal surface through a phenomenon which is termed as 'ion-pair' formation. This is facilitated by an increase in the concentration of the added electrolyte, as well as an increase in the cationic charge (i.e., valency) of the added electrolyte due to a higher bonding energy of the added cation for the given colloidal surface (see later). This will bring down the value of 'e' (Equation 2.4), and also of 'd' owing to release of some of the counter ions from the diffuse portion of the double layer to the bulk solution. The latter arises from the fall in the effective charge on the solid (colloid) surface, and hence its electrostatic attraction for the counter ions. In other words, the electrical double layer will be *compressed* with reduced charge density. Both these factors will bring down the ξ potential (Equation 2.4), enabling the approaching colloidal particles to coalesce with a concomitant reduction in buoyancy. The particles will thus settle under gravity, leading to coagulation or flocculation of the disperse phase (i.e., soil colloids) out of the dispersion medium, namely soil solution (Sanyal, 2002a).

When the zeta (ξ) potential attains the value of zero, the given colloidal system is said to correspond to the *isoelectric state* of the system concerned. It is apparent that for the aluminosilicate layer-lattice clay minerals (and soil, in general), also possessing permanent (pH-independent) charge (in addition to the pH-dependent charge), the ξ potential is the combined function of the latter as well as the pH-dependent surface charge, while at the corresponding ZPC, only the *net* surface charge is zero, not *necessarily* the corresponding ξ potential. This underlines the difference between the ZPC and the *isoelectric point*. For the soil organic colloids and the hydrous oxides of iron and aluminium (possessing only pH-dependent charge), however, the two may be quite identical for obvious reasons (Sanyal, 2002a).

The flocculation-dispersion behaviour of soil colloids has a strong bearing on the maintenance of favourable soil structure, thereby aiding soil aeration and drainage.

2.7. Electrometric Properties of Soil Colloids

2.7.1. Potentiometric and Conductometric Titrations

It is advantageous to study the electrometric (or electrochemical) properties of soil colloids such as soil clays when the latter are rendered homoionic with respect to a single cation that satisfies the entire cation-exchange capacity of the given clay. An acid clay, for instance, may be prepared by repeated leaching of the clay with a dilute mineral acid (*e.g.* 0.1 N HCl), followed by electrodialysis of the resulting acid clay to remove the excess electrolytes. A better way of obtaining the H-clay is to pass the clay suspension finally through a column of H^+ ion-exchange resin. The titration of such clay acids with alkalis provides important information regarding not only the exchange capacity, but also the nature and properties of the acid clay.

It is well-known that the gradual neutralization of an aqueous acid (*e.g.*, acetic acid or HCl) by an alkali leads to a pH-metric (potentiometric) titration curve. An n-basic acid may show n-inflexions provided that the dissociation constants (K) of the $(n\text{--}1)^{th}$ and the nth stages of acidity are widely separated. Typically, the ratio (K_{n-1}/K_n) should be of the order of 10^4. In actual practice, however, the dissociation constants of a polybasic acid (such as orthophosphoric acid, H_3PO_4) are often closer together, and neutralization at one stage may overlap partially with the preceding or succeeding stages, giving rise to statistically monobasic acid behaviour with an average dissociation constant (Mukherjee, 1974). The total acidity of such an acid solution is independent of the alkali chosen to measure the former.

If, however, the anion of an acid is of colloidal dimension (such as the negatively charged clay lattice of an acid clay), the protons form altogether a different phase from the former, giving rise to a two-phase acid system. The H^+ ions (*i.e.*, the exchangeable cations) cannot be separated out into the intermicellary solution, for instance, by ultrafiltration. The clay acid behaves like a weakly dissociating acid (like aqueous acetic acid) towards a glass electrode measuring pH (Sanyal, 2002a).

On conducting titration with an aqueous alkali, the pH-metric titration curves not only show differences in their features, depending on the alkali used, but also register variations in total acidity. This is because the neutralization reaction here is a two-step process as shown below.

Step 1: Ion-exchange reaction between H^+ ion on the colloidal anion of the acid clay and the cation of the added alkali (Equation a):

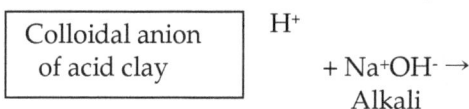

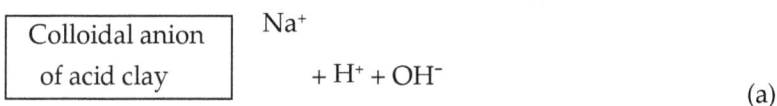

$$\boxed{\begin{array}{c}\text{Colloidal anion}\\\text{of acid clay}\end{array}}\ H^+$$

$$+ Na^+OH^- \rightarrow$$
$$\text{Alkali}$$

$$\boxed{\begin{array}{c}\text{Colloidal anion}\\\text{of acid clay}\end{array}}\ Na^+$$

$$+ H^+ + OH^- \tag{a}$$

Step 2: Neutralization of H^+ ion by OH^- ion in the solution phase (Equation b):

$$H^+ + OH^- \rightarrow H_2O \tag{b}$$

Some typical pH-metric titration curves of acid montmorillonite are shown in Figure 2.11.

The divalent Ca^{2+} ions, forming a stronger association with the colloidal anion of acid clay than the monovalent Na^+ ions, possess a greater H^+ ion replacing power (from the acid clay) than that of Na^+ ion. This leads to higher acidity (Figure 2.11) of the acid clay when titrated with $Ca(OH)_2$ than with NaOH. With $Ca(OH)_2$, the titrant, buffering (which amounts to flattening of the pH-metric titration curve; Figure 2.11) is observed at lower pH (Figure 2.11) which is the characteristic of a strong acid. With NaOH, such buffering is observed at relatively higher pH, a characteristic of a weak acid.

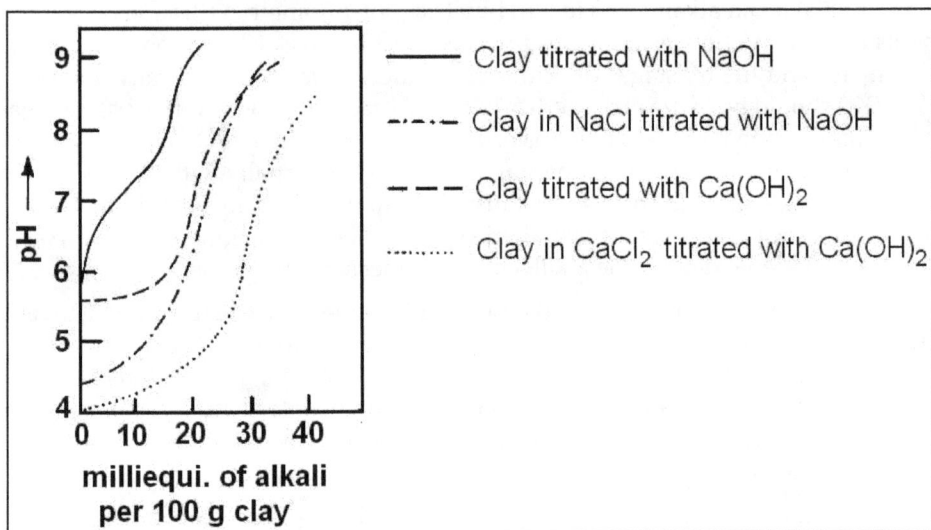

Figure 2.11: pH-metric Titration Curves of a Clay Separated from a Bengal Soil Showing the Effect of Alkali Used and Presence of Salts
Source: **Mitra and Kapoor (1969).**

Furthermore, the acidity of a colloidal acid increases (unlike a true acid) when titrated in the presence of neutral salts, so also the features of the titration curve change from those of a relatively weak acid to those of a stronger acid (Figure 2.11). This, again, results from an exchange of H^+ ion (also Al^{3+} ion, *see* later) for cations of the added salt preceding neutralization (Sanyal, 2002a).

2.7.2. Exchangeable Aluminium Ions in Acid Soils

An acid soil is actually a mixed H-Al system, *i.e.*, such a soil has both H^+ and Al^{3+} ions as exchangeable ions. As to the origin of exchangeable Al^{+3} ions, the H-soil or H-clay, when titrated immediately after preparation (on passage of the suspension through a column of H^+ ion-exchange resin), shows strong acid characteristics and a buffering at a low pH (Figure 2.12). The neutral salt extract contains little or no Al^{3+} ions. However, on aging (or on heating the suspension to, say, 95°C), the titration curve changes markedly to those of weak acids, buffering occurring at higher pH with the total acidity remaining unchanged. The neutral salt extract, however, shows the presence of an increasing amount of Al^{3+} ions with the time of aging (or heating) (Figure 2.12). The number of inflexions in the titration curve also increases on aging. Such a conversion of H-soil or H-clay to mixed H-Al system can be arrested by keeping the suspension in a non-aqueous solvent such as acetone or methanol (Mukherjee, 1974; Sanyal, 1995; Sanyal, 2002a).

The exchangeable Al^{3+} ions in H-clay or H-soil arise from the lateral mobilization of the octahedral Al^{3+} ions in the clay lattice (such as acid montmorillonite), effected

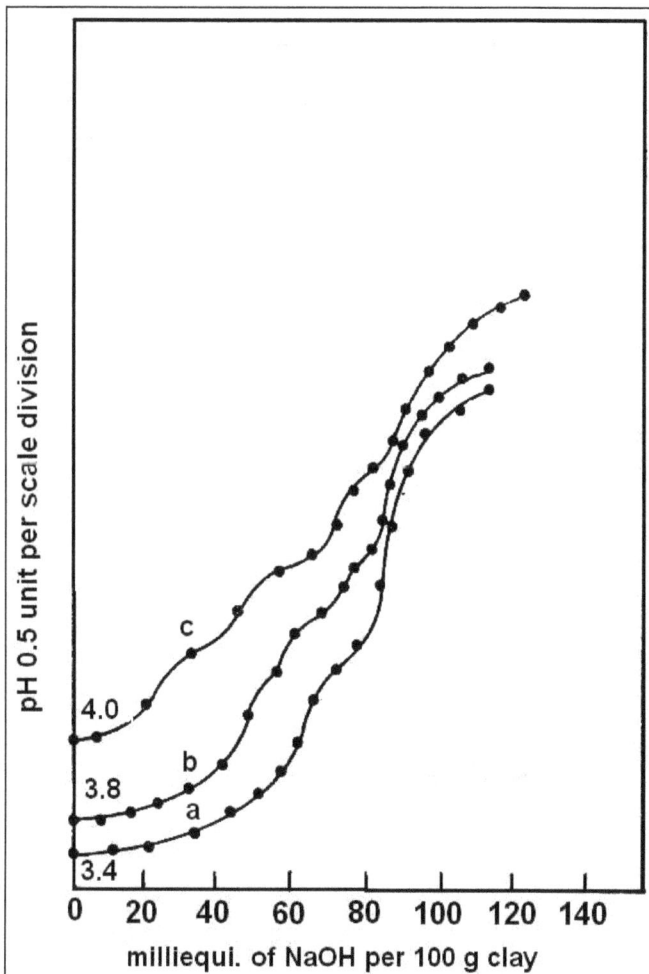

Figure 2.12: pH-metric Titration Curves of Acid Montmorillonite Clay: (a) Freshly prepared, (b) Aged for one week, and (c) Aged for nine weeks. *Source*: Mitra and Kapoor (1969).

by the exchangeable H^+ ions with which they exchange the position. This is a slow process, but it gets accelerated on heating. Such ionic migration is arrested in a non-aqueous solvent of relatively low dielectric constant (*e.g.* acetone/methanol). The exchangeable Al^{3+} ions being weaker acids (Lewis acid; *see* later) than H^+ ions, it is no wonder that the freshly prepared acid clays, showing strong acid behaviour, change-over to weaker acid type on aging (Figure 2.12), exhibiting a greater number of inflexions (and hence greater types of acidic groups) compared to the freshly prepared acid (*i.e.*, H-) clay. The total acidity, however, remains unaffected (Mukherjee, 1974; Sanyal, 1995; Sanyal, 2002a).

Homoionic acid clays (prepared in the usual way of acid leaching of the clay with dilute mineral acid, followed by electrodialysis), age considerably during preparation so as to contain appreciable amounts of mobilized Al^{3+} ions. If the aged

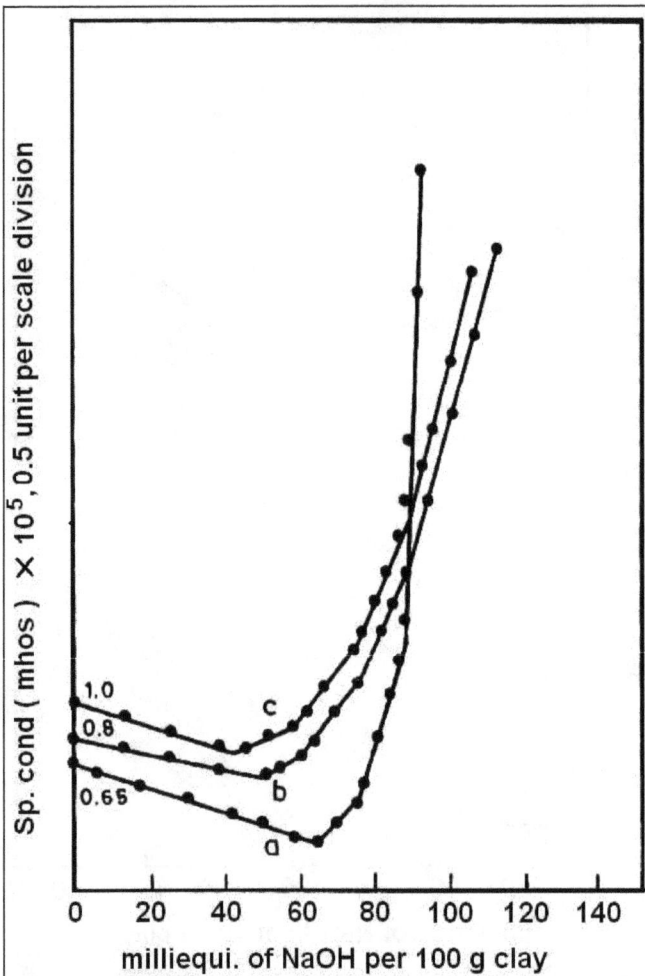

Figure 2.13: Conductometric Titration Curves of Acid Montmorillonite Clay: (a) Freshly prepared, (b) Aged for one week, and (c) Aged for four week. *Source*: Mitra and Kapoor (1969).

clay is passed through a column of H^+ ion-exchange resin, the exchangeable Al^{3+} ions are replaced by H^+ ions of the resin. The resulting H-clay, on aging, would again mobilize octahedral Al^{3+} ions. Thus, if the processes of aging and recovering the H-clay are repeated, the resulting H-clay would reach a stage when the clay would show signs of degradation owing to continued depletion of Al^{3+} ions from the clay structure (Mukherjee, 1974). Indeed, such degradation has been confirmed by means of electrometric titrations, X-ray and differential thermal analysis studies (Majumdar and Mukherjee, 1979).

The corresponding conductometric titration curves of acid montmorillonite also show stronger acid features for the freshly prepared H-clay, and a weaker (more polybasic) acid character on aging, the latter exhibiting a greater number of breaks in the conductometric titration curves. Figure 2.13 shows that the freshly prepared acid clay has fewer breaks (compared to aged acid clays), thereby signifying the appearance of new acidic species of weaker acidic strength in the aged acid clay (Sanyal, 2002a). The total acidity of the clay remains unchanged on aging (Figure 2.13).

The electrometric titration curves of soil humic colloids (humic acid, fulvic acid and hymatomelanic acid) show features characteristic of weak polybasic acid with the cation-exchange capacity being variable and a function of the alkali used for titration.

2.8. Electrokinetic Phenomena in Soil Colloids

It is well known that when an electric field is applied by sticking two water-filled glass tubes, encasing two platinum electrodes, into moist clay, and connecting these to an external source of electromotive force (E.M.F.) such as a battery, the liquid in the tube serving as a positive electrode (Figure 2.14a) becomes turbid while the water in the other tube remains clear. This indicates that clay particles move under the applied electric field towards the positive electrode. In another experiment, a clay plug is placed at the central part of a U-tube (Figure 2.14b) and is enclosed on both sides by sintered glass discs. On filling the U-tube with water, and applying an electric field through two electrodes inserted into water in the two arms of the U-tube, the water level in the arm having the negative electrode rises until the difference in water level between the two arms attains a constant value.

These two phenomena are known as electrophoresis and electro-osmosis, respectively, and are the direct consequences of the existence of electrical double layer at the clay/water interface, and slipping past of one layer of the said double layer over the other, the two layers possessing opposite (and equal) charges. In electrophoresis, the negatively charged clay particles are free to move, and therefore migrate towards the positive electrode, while in electro-osmosis, the clay particles being restricted from movement by the sintered glass discs, the positively charged solution side of the double layer would move under an electric field towards the negative electrode (Sanyal, 2002a).

A closer examination of electro-osmosis reveals that although the liquid flow is generally associated with the application of hydrostatic pressure difference (ΔP), the application of an electrical potential difference (*i.e.*, an electric field) is doing this

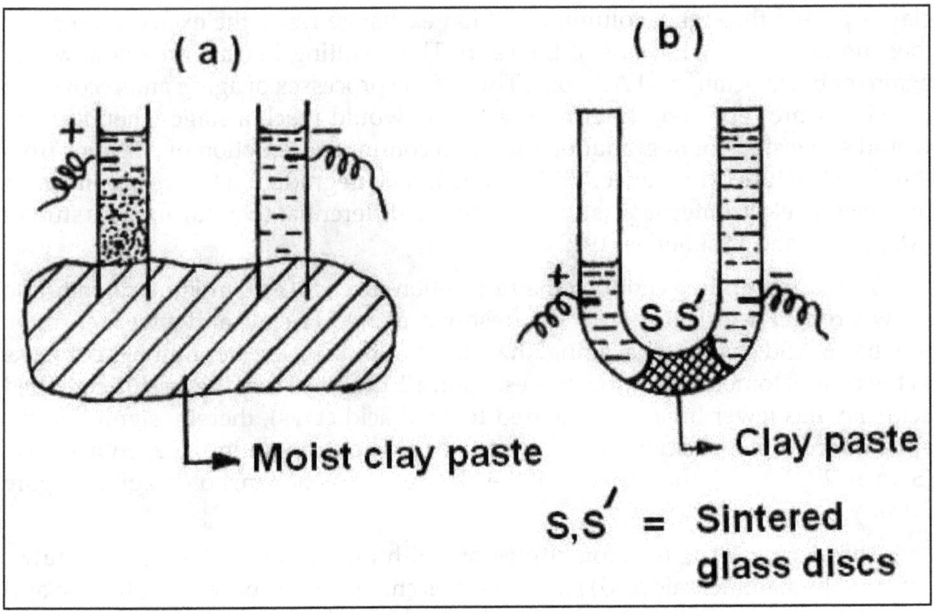

Figure 2.14: A Schematic Presentation of the Phenomenon of (a) Electrophoresis of clay, and (b) Electro-osmosis of clay.

job in electro-osmosis. By analogy, one would then expect a pressure difference to induce an electric current which is generally the result of application of an electric field. Indeed, this is found to be so when the clay plug (Figure 2.14b) is subjected to a pressure difference. The ensuing current is termed as the streaming current while the corresponding potential difference developed across the clay plug is known as the streaming potential. Here again, a mechanical separation of the solution side of the double layer (from the stationary clay particles) under the applied ΔP leads to development of streaming potential and the consequent current. It can be shown by the principles of irreversible thermodynamics, as applied to describe the coupled transport processes (Agar, 1963; Srivastava and Raj Pal, 1973; Sanyal and Mukherjee, 1988), by Eq. 2.5:

$$\left[\frac{v}{X}\right]_{\Delta P=0} = \left[\frac{i}{\Delta P}\right]_{X=0}$$

(2.5)

Where, X is the applied electric field across the clay plug (in electro-osmosis; Figure 14b), v is the corresponding velocity of electro-osmotic flow of water, ΔP is the pressure difference applied under zero (externally applied) electric field, and i is the corresponding streaming current. The term (v/X) represents the electro-osmotic mobility, while the term $(i/\Delta P)$ represents the streaming current density generated by a unit pressure difference. Both these quantities are characteristics of the electrical double layer at the colloid/solution interface (Sanyal, 2002a).

In the year 1878, Dorn discovered a related phenomenon (Shaw, 1970) in which colloidal particles under sedimentation in water lead to the development of an

E.M.F. between two electrodes inserted at different depths of the liquid column. This effect, opposite of electrophoresis, is known as Dorn effect, and the E.M.F. generated is known as sedimentation potential.

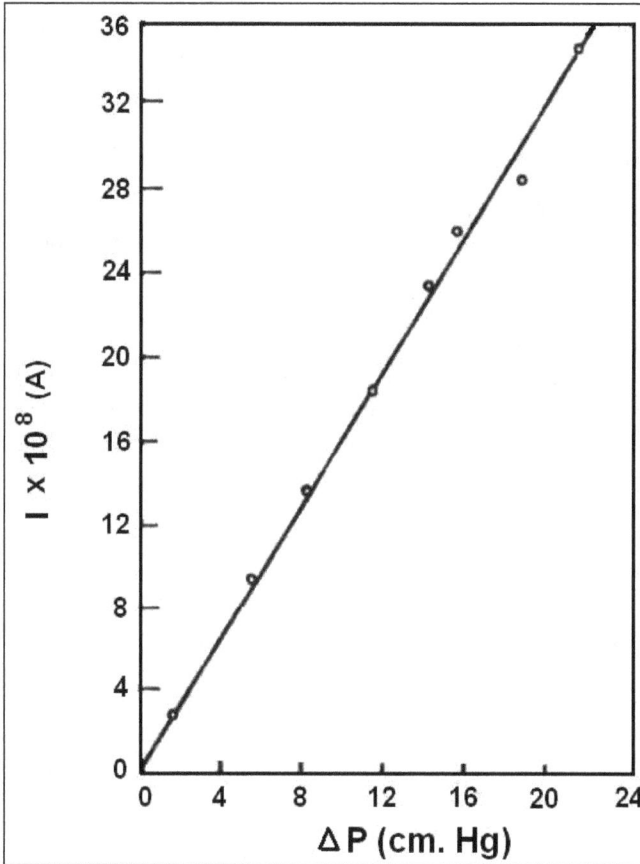

Figure 2.15: Relationship between Streaming Current (I) and Hydraulic Pressure difference (ΔP).
***Source*: Srivastava and Raj Pal (1973).**

The above four phenomena are known as electrokinetic phenomena for reasons explained above as regards the mechanism responsible for these processes. Figure 2.15 demonstrates the streaming current as a function of hydraulic pressure difference applied across kaolinite/water system.

2.9.   Visible and Infrared Spectroscopic Studies of Humic Colloids-Structural Elucidation

2.9.1. Basic Principles

The electronic transitions in humic molecules lead to characteristic spectral absorption bands in the ultraviolet (UV) and visible region (together covering the wavelengths from about 200 to 800 nm) of the electromagnetic spectrum. The electrons involved are those in σ-bond (formed through face-to-face overlapping of atomic orbitals, as found in single bonds between C and H in CH_4), π-bond (formed through lateral partial overlapping of atomic orbitals such as found in olefinic bonds; the electrons are more mobile than those in σ-bonds), and the non-bonding (n) electrons (on oxygen, sulphur, nitrogen and the halogens). The absorption of electromagnetic (*e.g.* light) energy (E) is quantized and is governed by the well-known Planck's equation (Eq. 2.6),

$$E = h\nu = hc/\lambda \tag{2.6}$$

Where, ν is the frequency; λ is the wavelength of the absorbed radiation; c, the velocity of light (*i.e.*, of the electromagnetic radiation concerned), and h is the Planck's constant.

The absorption of light energy in visible (and UV) regions by humic colloids is accompanied with promotion of the aforesaid electrons from the ground state to a higher energy state. It is frequently described as σ → σ*, n → σ*, n → π* and π → π* transitions. Here, σ* and π* denote the corresponding antibonding orbital of higher energy than those of the corresponding σ and π bonds, respectively. The π → π* transitions are associated with much lower energy than the σ → σ* transitions. Therefore, compounds having a large degree of unsaturation (π-bonds) will absorb energy at longer wavelength [Equation (2.6)], and hence the absorption bands will shift towards longer wavelengths. Extensive conjugation (*i.e.*, C-C bonds and C=C bonds occurring alternately in a molecule) causes further lowering of π → π* transition energy, owing to extensive delocalization of π-electrons. This leads to a greater shift of the absorption bands to longer wavelengths. This is particularly true in aromatic compounds, compared to their aliphatic analogues having the same number of C-atoms. Indeed, as the number of fused (co-planar) benzene rings increases, as in, say, anthracene, the extensive conjugative delocalization of π-electrons takes place, pushing the absorption band far in the visible spectrum of electromagnetic radiation. Such an absorption spectrum is given as a plot of absorbance (A) or optical density at various wavelengths (λ) of incident light as function of λ. The absorbance, following the Lambert-Beer's law, is given by Eq. 2.7:

$$A = \log (I_0/I) = abc \tag{2.7}$$

Where, I_0 and I are the intensities of the incident and the transmitted radiations, respectively, at λ corresponding to maximum absorption by a given substance, a is the molar extinction coefficient of the light-absorbing substance (*e.g.*, humic colloid), c, its concentration and b is the path length traversed by light through the given substance.

However, as compared to the absorption bands observed in the visible and UV regions (200-800 nm), the humic molecules (in common with organic molecules) show much larger number of peaks in the infrared (IR) spectrum (λ ranging from 2.5 to 15 μm, with the region 0.8 to 2.5 μm being known as near infrared and the one from 15 to 200 μm, known as far infrared region of electromagnetic spectrum). In the IR region, absorptions are often shown as function of reciprocal wavelength or what is known as wave number. The latter, for instance for the near IR region, ranges from 12,500 to 4000 cm^{-1}. The energy (E) of absorption in the IR region is smaller than that in UV or visible region, the corresponding wavelength being longer (Equation 2.6), and is associated with the vibration of molecules, including stretching (in which the distance between two bonding atoms changes, but the atoms remain in the same position) and bending (or deformation in which the position of the atoms changes relative to the original bond axis). These modes of vibrations are illustrated in Figure 2.16.

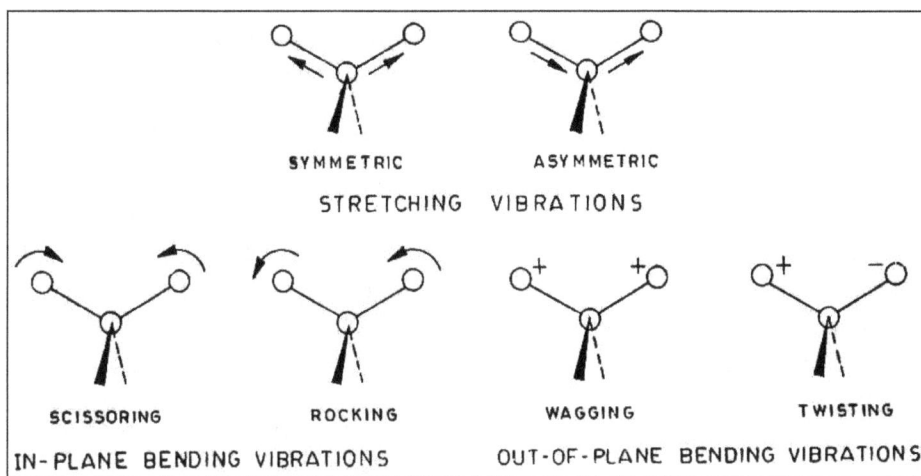

Figure 2.16: Vibrations of a Group of Atoms
(+ and – signify vibrations perpendicular to the plane of the paper).
***Source*: Dyer (1971)**

As before, the energy of the various stretching and bending vibrations of a bond is quantized, and incidence of IR radiation of identical frequency causes absorption of energy by the organic molecules, leading to an increase of amplitude of the given vibration. This, in turn, leads to an absorption peak in the IR spectrum. Perhaps the most significant application of the IR spectrum lies in the identification of the presence of a number of functional groups from the occurrence of the highly characteristic absorption peaks in the so-called "finger-print" region (7 to 11 μm) (Sanyal, 2002a).

2.9.2. Absorption of Energy in the Visible Region

The absorption of radiation energy in the visible region by soil humic and fulvic colloids often indicates the degree of aromatization and relative preponderance of aromatic and aliphatic nature of a colloid. Thus, the comparative absorption at 445 nm (denoted by E_4) and that at the longer wavelength of 665 nm (denoted by E_6) is taken to show the relative preponderance of aliphatic and aromatic groups, respectively, in the given humic molecule. The higher degree of aromatization (characterized by extensive delocalization of π-electrons) of an organic molecule causes stronger absorption at longer λ for reasons explained earlier. More important parameter in this context is neither E_4 nor E_6 values, each taken singly, but rather the ratio, (E_4/E_6), which provides a better index of the above-mentioned preponderance. The fulvic colloids (the lower polymer and an earlier fraction than humic colloid of the soil organic matter, the humic colloid being of a higher polycondensate nature; *see* later) possess typically higher (E_4/E_6) ratio than their corresponding humic fraction. This is obviously linked with a lower extent of aromatization of the former, and the presence of relatively larger proportion of aliphatic structures in the fulvic acids. Some representative values of the ratio, (E_4/E_6), for fulvic and humic colloids are given in Table 2.7 (Sanyal, 2002a).

Table 2.7: (E_4/E_6) Ratio for Humic and Fulvic Acids

Soil Location	Soil Order	(E_4/E_6) Ratio	
		Humic Acid	Fulvic Acid
Janji, Assam	Inceptisol	4.4	6.2
Kokrajan, Assam	Inceptisol	5.0	6.3
Raja Rammohanpur, West Bengal	Inceptisol	3.4	4.7
Golaghat, Assam	Alfisol	4.5	5.9
Deragaon, Assam	Alfisol	4.1	6.2
Synthetic humic acid (Synthesized from glycine and hydroquinone)	–	2.9	–

Source: Saha and Sanyal (1988); Sarmah and Bordoloi (1993); Lahiri and Chakravarti (1995).

2.9.3. Absorption of Energy in the Infrared Region

The functional group composition of humic colloids (*e.g.*, humic acid and fulvic acid) can be inferred from their infrared (IR) spectrum. Thus, the presence of phenolic (OH), amide, methyl, free-NH-, carboxylic, carbonyl and other important characteristics can be inferred from the finger-print region of the IR spectrum. In particular, the degree of aromatization can be ascertained from the presence of absorption bands/peaks in the region 1500 to 1600 cm^{-1} wave number. The different functional groups, present in the humic and fulvic acids of soil humic colloids, as revealed by a representative infrared spectral study of soil humic colloids, are summarized in Table 2.8.

Further, the IR spectrum of two synthetic humic acid samples (synthesized by alkaline oxidation of mixtures of the simple constituents of soil humic acids, namely the amino acids and phenolic compounds) is shown in Figure 2.17, in which per cent transmittance (bearing inverse relationship with absorption) is shown as a function of wave number.

To provide a comparison with soil humic colloids, the major IR bands of some representative clay minerals (the inorganic soil colloids) are listed in Table 2.9.

2.10. Coiling in Humic Substances: Hydrophobic Bond

The humic colloids exhibit hydrophobic and hydrophilic characters. While the various functional groups are generally hydrophilic, the (C-H) bonds are essentially hydrophobic. As a consequence, the humic colloids in aqueous medium (*e.g.*, soil

Table 2.8: Relative Intensity of Major IR Absorption Bands of Humic and Fulvic Acids of Soils

Frequency (cm⁻¹)	Assignment	Soils			
		S_1	S_2	S_3	S_4
Humic acids					
3450-3300	H-bonded OH, Free OH, Intermolecular bonded OH	B	B	B	B
2950 – 2850	Aliphatic C-H, C-H$_2$, C-H$_3$ stretching	S	S	S	S
1725 – 1640	C = O stretching of carboxylic acids, cyclic and acyclic aldehydes and ketones, quinones	Sh	W	M	W
1640 – 1585	C=C stretching vibration of double bonds of cyclic and acyclic compounds, benzene ring substitution	S	S	S	S
1540	NO$_2$ vibration of nitro groups	W	W	W	W
1515	C=C stretching vibration of benzene, pyridine, etc., benzene ring substitution, secondary amines	A	M	W	W
1470 – 1420	Aliphatic C-H deformation	W	W	S	S
1025	Si – O – Si vibration of silicates	S	S	S	S
Fulvic Acids					
3450-3300	H-bonded OH, Free OH, Intermolecular bonded OH	SB	SB	SB	SB
2950 – 2850	Aliphatic C-H, C-H$_2$, C-H$_3$ stretching	W	A	W	W
2850 – 2500	Carboxylate ion	Sh	Sh	A	A
1725 – 1640	C = O stretching of carboxylic acids, cyclic and acyclic aldehydes and ketones, quinones	M	M	M	M
1640 – 1585	C=C stretching vibration of double bonds of cyclic and acyclic compounds, benzene ring substitution	S	S	S	S
1515	C=C stretching vibration of bonzono, pyridine, etc., benzene ring substitution, secondary amines	W	W	W	W

A: Absent; B: Broad; M: Medium; S: Strong; Sh: Shoulder; W: Weak; S_1: Janji; S_2: Jorhat; S_3: Kakrajan; S_4: Golaghat.

Source: Sarmah and Bordoloi (1993).

solution) remain in a coiled state. Such coiling is, in general, less marked in fulvic colloids than in humic colloids; the latter possess a higher polycondensate nature and hence are more hydrophobic.

The interaction of the hydrophobic backbone of humic colloids with neighbouring water is (free) energetically (*i.e.*, thermodynamically) unfavourable (Sanyal, 1984). As a consequence, the water molecules are drawn together, as by surface tension, and the non-polar groups tend to aggregate (much like oil droplets in water), leading to coiling, thereby minimizing their contact surface area with water, and hence the mutual interaction. The term, 'hydrophobic bond', has been coined to represent the association (aggregation as referred above) of the hydrophobic parts, excluding neighbouring water. But the term 'bond' may well be considered a misnomer here in that there exists no particular bond localized between two atoms

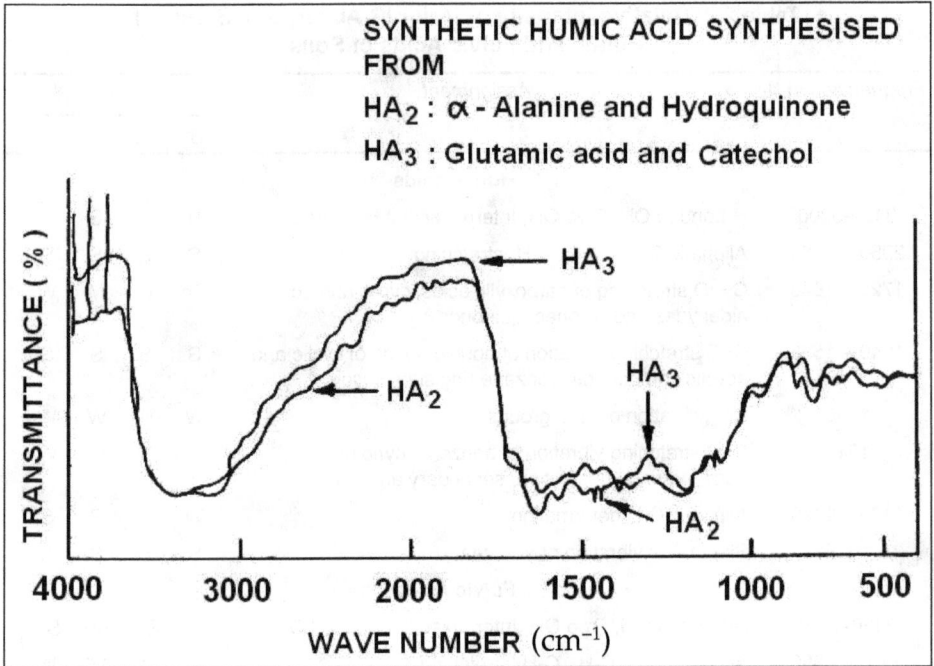

Figure 2.17: Infrared Spectra of Two Samples of Synthetic Humic Acids.
Source: Saha and Sanyal (1988).

of the hydrophobic residue, the association of which results under 'thermodynamic compulsion' (Sanyal, 1984).

2.11. Clay-Water Interactions

2.11.1. Rheological Properties of Soil Clays

The science of flow of fluids, or in general, the science of deformation processes, developing with time, is known as rheology. The rheological or flow behaviour of clay suspensions is largely governed by the mode and the extent of interaction between the clay particles. These interactions may be of three types: edge-to-edge (EE), edge-to-face (EF), and face-to-face (FF) associations. This presumes that a well-developed positively charged double layer exists at the edges of the clay particles, opposite in charge characteristics to that on the faces, which are predominantly negatively charged. As a result, there is an EF attraction causing "internal mutual flocculation", leading to high viscosity (*i.e.*, high resistance to flow). In fact, EF (and to an extent EE) association causes formation of card-house structure throughout the clay suspension. This imparts high viscosity to the system. On the other hand, FF association (through van der Waal's interaction, for instance) causes thickening of clay particles and an effective reduction of EF linkages. This leads to lowering of viscosity (Van Olphen, 1963; Sanyal, 2002a).

Table 2.9: The Major IR Absorption Bands of some Clay Minerals

Major IR Absorption Bands (cm⁻¹)		
Montmorillonite	*Illite*	*Kaolinite*
3620	3620	3700
3440	3460	3625
1650	2320	3490
1640 - 1630	1630	1430
1560	1460	1110
1505	1030 - 990	1025
1115	910	1000
1050	830	930
925	795	905
890	750	785
855	445	745
805	400	685
630	370	565
530		455
470		400
380		370
350		305

Source: Mondal *et al.* (1997).

In the presence of electrolytes, when both the edge and face double layers are compressed, the nature of viscosity variations of clay suspensions is determined by two factors: (i) the balance between the residual EF attraction and FF repulsion, and (ii) the enhanced van der Waal's FF and EE attractions. The viscosity of montmorillonite (in sodium form, for instance) suspension exhibits a high degree of sensitivity towards the presence of extraneous electrolytes. The illite and kaolinite suspensions show progressively less dependence of their viscosity on the concentration of added electrolytes. This provides a relatively simple method (employing the viscosity determinations) for identifying the presence of montmorillonite as the dominating clay in a soil clay fraction. It has been observed experimentally that the viscosity-buffer curves of montmorillonite in particular, and of clay mixtures, in general, in which montmorillonite is dominant, show a distinct peak/hump. This hump gradually becomes less marked and shifts towards lower degree of per cent base saturation as the proportion of montmorillonite in the clay mixtures decreases. This is illustrated by the viscosity-buffer curves of clays and clay mixtures in Figure 2.18.

The distinct behaviour of montmorillonite, as against that of kaolinite or mica, may be attributed to the nature of its clay lattice to expand in the polar liquids. It

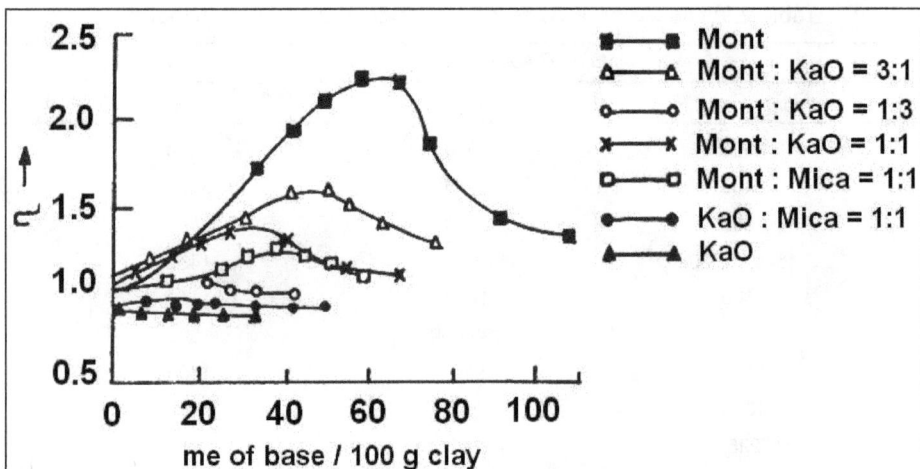

Figure 2.18: Viscosity Buffer-Curve of Clays and of their Mixtures.
Source: **Sahu and Das (1974).**

causes easy accommodation of the hydrated cation (*e.g.* Na$^+$) of the added base largely in the expanding interlayer space up to about 75-80 per cent base saturation, leading to a high value of the volume fraction (ϕ) of the dispersed clay particles (*i.e.*, fraction of total volume occupied by the clay particles). This imparts high viscosity according to the Einstein's equation (Eq. 2.8).

$$(\eta/\eta_0) = 1 + 2.5\phi \tag{2.8}$$

Where, η and η_0 are the viscosity coefficients of clay suspension and pure water, respectively, at the same temperature. The numerical factor 2.5 has been assigned to spherical suspended particles, and for other shapes, it is assigned different values. Hence, the value of η increases, relative to the value of η_0, with concomitant increase in the value of ϕ.

In mixtures of montmorillonite with non-expanding clays, *e.g.*, kaolinite, mica, etc., the volume-fraction at the same per cent base saturation will obviously be less, leading to a lower value of η.

Furthermore, this nature of montmorillonite to expand favours EF linkages, leading to high viscosity. However, at sufficiently high pH (*e.g.*, beyond 75-80 per cent base saturation in Figure 2.18), the positively charged double layer at the edges is significantly reduced so that the viscosity of the clay suspension falls due to snapping of EF linkages (Van Olphen, 1963; Sanyal, 2002a).

2.11.2. Clay-Sorbed Water

As mentioned earlier (in Section 2.6.1), one of the important properties of liquid water as a solvent is its dielectric behaviour, characterized primarily by its dielectric constant (D), which is the ratio of its static permittivity to the permittivity of vacuum, Such dielectric behaviour of water characterizes primarily the solvent properties of liquid water, especially for the dissolution of the ionic and polar solutes.

Measurement of such D values of adsorbed water on phyllosilicates may provide important information as to the nature of the corresponding surface chemical processes. The available information (Sposito, 1984) suggests that the average value of D of clay-sorbed water is about 20, in a range of values 2 to 50 at 25°C. One will recall that the value of D for bulk liquid water is about 80 at 25°C. Indeed the preferential orientation in clay-sorbed water under the ionic field of the charged clay and the exchangeable ions restricts the freedom of water molecules to assume their random orientation in bulk liquid water (through intermolecular hydrogen-bonding dipole-dipole interactions), which is determined by its kinetic energy at the given temperature. Such reduced orientation polarization of clay-sorbed water leads to an enhanced tendency of complex formation between the dissolved species as well as that between the exchangeable cations and clay minerals, thereby favouring the ion association. The latter (as explained earlier) leads to retardation of the development of the diffuse double layer near the clay colloid surface, and as a result, the electrical double layer will be compressed with reduced charge density.

Another important property of clay-associated water on its surface is its enhanced acidity, primarily for those water molecules in the solvation complex of the corresponding exchangeable cations for such clay minerals (*e.g.*, vermiculite and smectite surfaces). Thus,

$$M(H_2O)_n^{m+} + H_2O \rightarrow MOH(H_2O)_{n-1}^{(m-1)+} + H_3O^+$$

The equilibrium constant of such reactions has been measured for a number of the exchangeable cations to clay surfaces in aqueous medium, and has been shown to correlate positively to the corresponding ionic potential and the Lewis acid softness (Sposito, 1984). The latter is defined to be a molecular unit of relatively large size (as opposed to a hard Lewis acid), low oxidation state, low electronegativity, and high polarizability. As is well-known, the polarizability is a measure of the ease with which the electronic orbitals, basically the d-orbitals of a metal cation, can be deformed in an electric field. This being so, a higher Lewis acid softness would obviously tend to increase the covalent nature of the M-O bond in the solvated complex as mentioned, that, in turn, would facilitate the release of a proton from the solvated complex in aqueous clay system. As has been argued elsewhere (Sposito, 1984), with the increase in the ionic potential of an exchangeable cation, the intensity of the positive Coulombic field of the cation also increases, causing repulsion of a solvating water proton more likely, and hence the release of the latter is facilitated. Furthermore, the exchangeable cations tend to become more acidic as the amount of adsorbed water decreases. Indeed, the degree of dissociation of the sorbed water appears to be at least two-orders of magnitude higher than that in bulk liquid water at the same temperature. It has been also suggested that such increased acidity in the adsorbed water primarily arises from the presence of the exchangeable cations whose charge is screened by the solvating water molecules, and that it has very little to do with the charge distribution in the parent clay minerals (Sposito, 1984).

2.12. Surface Charge Characteristics of Soils/Clays

One of the important surface charge characteristics of soils dominated by variable-charge clays is the zero-point of charge (ZPC) (introduced earlier in Section 2.5.2), at which the value of the net charge of the variable-charge component is zero. Evaluation of the ZPC values of soils enables one to predict the response of the soil to the changes in surrounding conditions, *e.g.* fertilizer (particularly phosphate, sulphate, etc.) application in the field. In addition, the estimation of the soil net permanent charge (σ_p) is necessary for the location of ZPC.

The presence of variable-charge components in highly weathered tropical soils is known to control phosphate (P), as well as to an extent, of sulphate, borate and molybdate fixation and their mobility in these soils. Such P fixation leads to corresponding variation of surface charge density of the soil, knowledge of which is important in predicting the P availability to plants, as well as the P holding capacity of the given soil. The surface charge parameters cover, among others, surface potential (Ψ_0), total net surface charge density (σ_t), permanent charge density (σ_p), variable surface charge density (σ_v), ZPC and its shift on phosphate treatment of the given soil.

The above mentioned parameters, following the Goüy-Chapman model of Diffuse Double Layer, are given by the following equations (Sposito, 1984):

$$\sigma_t = \sigma_p + \sigma_v \tag{2.9}$$

$$\Psi_0 = (RT/F) [\ln (H^+)/(H^+)_{ZPC}] \tag{2.10}$$

$$= 0.059 \, (ZPC- pH) \text{ at } 25°C \tag{2.10a}$$

$$\sigma_p = [2C\varepsilon RT/\pi]^{1/2} \sinh [zF\Psi_0/2RT] \tag{2.11}$$

$$\sigma_v = [2C\varepsilon RT/\pi]^{1/2} \sinh 1.15 \, z \, (ZPC-pH) \tag{2.12}$$

where,

σ_t = Total net surface charge density (Coulomb.m^{-2})

σ_p = Permanent surface charge density (Coulomb.m^{-2})

σ_v = Variable surface charge density (Coulomb.m^{-2})

C = Ionic concentration in the bulk soil solution (mol.m^{-3})

ε = Static permittivity of aqueous medium = $\varepsilon_0 D$

Where ε_0 is the permittivity in vacuum (8.85×10^{-12} F. m^{-1} and D is the dielectric constant of the medium (78.5 at 298 K for water).

z = Valence of the counter ion

Ψ_0 = Surface potential (V)

R = Universal gas constant (8.317 J/mol K)

F = Faraday Constant (96487 Coulomb/mol)

T = Absolute temperature (K)

(H^+) = Hydrogen ion activity in soil solution

(H^+)$_{ZPC}$ = Hydrogen ion activity at the zero-point of charge (ZPC)

As mentioned earlier (in Section 2.5.2), the modification of surface/edge charge characteristics of soil colloids (especially the inorganic colloids) leads to the modification of ZPC. Thus, phosphate-fixation by ligand-exchange (*e.g.*, exchange of neutral OH_2 ligand to Fe or Al by the applied $H_2PO_4^-$ ion; *see* later) process raises the surface negative charge density (σ_v, for instance, and hence σ_t), and pushes the ZPC towards lower pH, *i.e.*, a stronger H^+ ion concentration is required in the surrounding soil solution to render the surface of soil colloid neutral. As a result, the surface potential (Ψ_0) is also rendered more negative. Such phosphate fixation thus raises the pH-dependent CEC, and a reduction in the leaching of the cations in highly weathered soils, characterized by low CEC. Furthermore, intense weathering of a soil shifts its ZPC towards higher pH owing to greater accumulation of hydrous oxides/hydroxides of Fe and Al and also enrichment of soil layer-lattice clays with kaolinite (possessing mostly pH-dependent edge charge). Indeed, the ZPC of Ultisols or Alfisol is generally higher than that of an Inceptisol. Organic matter, on the other hand, pushes the ZPC towards lower pH. Thus the stage of the pedogenic development of a soil may be assessed from the corresponding surface charge characteristics (Sanyal, 2002a).

2.13. Humic-Mineral Interactions: Clay-Humus Complex

Studies on naturally occurring clay-humus complexes have been fewer than those conducted with synthetic complexes. One reason, among others, for this trend could be related to the difficulty of the extraction and fractionation of these naturally occurring complexes *without* introducing artifacts (Sanyal, 2002b). Morra *et al.* (1991) suggested ultrasonic dispersion technique with application of no more than 3 to 5 kJ of the sonification energy to 10 g soil in 50 mL of aqueous suspension. Use of sedimentation and ultra-centrifugation techniques has also been resorted to for the purpose. Manoj Kumar (1995) reported from thermal analysis data that soil clay-organic complexes were associated with three-stage exothermic peaks, around 340°, 435° and 510°C, which was reported to ensue from oxidation of organic matter, complexed to clay mineral surface, sesquioxides surface and allophane, respectively.

Ghosh and Ghosh (1998) have reviewed extensively the organo-mineral interactions in soil. The extent of the interactions between soil minerals, including clays, and humic acid (HA) have been found to depend on the nature of the mineral, pH and, in some cases, the concentration of the background electrolyte (Yuan *et al.*, 2000; Sanyal, 2002b). The said interaction influences the solubility of the HA as well as the surface charge density and electrophoretic mobility of the mineral. While examining the role of soil mineral composition and cation suite in rendering the SOM of the tropical soils much younger than that of the temperate zones, it was further reported that kaolinitic was much less active in sorption reaction than were the expanding clays, dominant in soils of temperate zones (Sanyal, 2002b). The activity of the oxides in SOM stabilization was, on the other hand, determined not by their total amounts, but by their degree of crystallinity, particle-size, and association with kaolinite and mica. This greatly reduced the amount of oxides effectively participating in SOM stabilization (Sanyal, 2002b).

The functional groups in humic substances involved in binding metal ions are quite similar to those involved in clay-humic acid (HA)/fulvic acid (FA) associations. These include enolate (-OH), amino (-NH$_2$), imino (= NH), azo (-N=N-), heterocyclic N (aromatic ring N), carboxyl/carboxylate (-COOH/COO$^-$), phenolic (OH), ether group (-O-), sulfhydryl (-SH), carbonyl groups (C=O) (Stevenson, 1994).

Stable soil aggregates are formed by reactions between clay minerals and humic substances. Clay-humus complex formation retards the microbial degradation of humus. In mineral soils, about entire humus occurs in combination with clay. However, the mechanisms of clay-humus complexation process are far from being well understood (Sanyal, 2002b).

In chemical association between clay and humus, the metal ion tends to form a bridge, linking the two components, *e.g.*, Clay-Ca^{2+}-HA. In montmorillonite, bonding takes place at the basal surface of the mineral, while both basal surface and edges of illite plates are involved in complexation. For kaolinite, humus is bonded to the edges of the mineral.

Varadachari *et al.* (1997) observed that hematite, among goethite, hematite, gibbsite and boehmite, showed the greatest HA fixation at different oxide: HA ratio at all pH higher than 7.0. A gradual reduction in HA/FA fixation from pH 2.0 to 10.0 was reported for all the aforesaid minerals, except gibbsite which exhibited a sharp fall at pH > 7.0, and a maximum at pH 5.0. The extent of fixation was independent of specific surface area or ZPC of the given oxide minerals. This study was followed by an examination of the said humic-fixation/complexation process from the viewpoint of theoretical analysis of crystal surface structure of the oxide minerals (Varadachari *et al.*, 2000). For this purpose, the intrinsic charges on the surfaces and the edges of iron and aluminium oxides and sites for the coordinate bond formation were derived. These authors (2000) inferred that the HA bonding would be strongest in hematite, in agreement with earlier experimental observation (Varadachari *et al.*, 1997). The IR spectra rendered further evidence for the participation of (OH) groups of boehmite in strong HA linkages as opposed to gibbsite, involving very weak or no (OH) group involvement. The analysis of crystal structures at broken surfaces and edges of oxide minerals is thought to provide information on the factors primarily controlling oxide-HA complexation (Varadachari *et al.*, 2000).

The influence of crystal edges of the common aluminosilicate clays (kaolinite, illitic and montmorillonite) on clay-humus complexation was examined (Varadachari *et al.*, 1995) by way of blocking these edges with hexametaphosphate (HMP) and triethanolamine (TEA). The latter led to reduction of HA retention in the dried complexes for all except HMP-Ca-montmorillonite complex. These authors (1995) suggested that (i) edge bonding *via* exchangeable cations is the primary mode of interaction of kaolinite with HA, (ii) cations at the cleavage planes provide the major bonding sites for HA on illite, with crystal edges playing only a minor role, and (iii) increased availability of basal surfaces in montmorillonite by dispersion or swelling has strong influence on montmorillonite-HA complexation. Interaction of HA with montmorillonite causes disruption of its stacking arrangement due to prying open of the interlayer of HA molecules (Varadachari *et al.*, 1991). Indeed, HA fixation by montmorillonite with reduced charge was decreased substantially owing to loss of swelling capacity of the clay, while the exchangeable cations on the illite surface were found to form stronger bonds, compared with montmorillonite or kaolinite (Varadachari *et al.*, 1994).

There have been a number of other studies as well reported in the literature on reactions of crystalline clays with humic substances (Theng, 1979), but relatively little is known about the reactivity towards humic substances of allophane, a clay-sized alumino-silicate mineral, characterized by short-range order, occurring widely in Spodosols and Andosols (Sanyal, 2002b).

Yuan *et al.* (2000) investigated the interaction of allophane with HA and cations. The findings were interpreted in terms of ligand exchange between the carboxylate groups of HA and (OH) groups, associated with allophanic aluminium exposed on defect sites at the surface of allophane particles. This reaction generated a surface negative charge, requiring the co-sorption or chemical binding of extraneous cation of the background electrolytes used, namely aqueous $CaCl_2$ or NaCl. Much more HA was sorbed in presence of $CaCl_2$ than NaCl of identical ionic strength, due, presumably, to the possibility that Ca^{2+} ion, besides compensating the negative charge generated (like Na^+ ion also did), may have been bound specifically to the surface complex (Sanyal, 2002b), and was also more effective (being bivalent) in screening the negative charge on HA than was Na^+ ion.

The generation of surface negative charge on allophane through interaction with humic substance could enhance the capacity of allophanic soils to retain and immobilize positively charged species (as well as organic contaminants) such as nutrient cations and also heavy metal ions over the pH range of most soils (Yuan *et al.*, 2000). Liming of such soils, along with organic manure incorporation, would facilitate the humic-mineral matter interaction (through the added Ca^{2+} ions), and thus help stabilize organic matter in allophanic soils. The fixation of fulvic acid (FA) by clay minerals (at pH 2.0) was found to fall in the order: illite > kaolinite > montmorillonite (Varadachari *et al.*, 1994) The more numerous negative charges of FA (than on the corresponding HA fraction) and its more hydrophilic character were cited by Varadachari *et al.* (1994) to account for the poorer complexation of FA than that of HA by the given clay minerals.

It is fairly well-known that abiotic catalysts play a vital role in the transformation of phenolic compounds to humic substances (Sanyal, 2002b). These catalysts include primary (soil) minerals, layer-lattice aluminosilicates, metal oxides, hydroxides, and oxyhydroxide, as well as poorly crystalline aluminosilicates. Huang and coworkers have studied, since the early 1980s, the sequence of catalytic power of layer-lattice silicates and their reaction sites in the polymerisation of phenolic compounds, leading to the formation of the humic substances. Wang and Huang (1994) reported the Fe (III) in the octahedral sheet of nontronite to serve as a Lewis acid (*see* later) while catalyzing the oxidative polymerisation of hydroquinone, pyrogallol and catechol, by way of cleaving the corresponding aromatic ring structure to liberate CO_2. This is illustrated in Table 2.10 (Bollag *et al.*, 1998). The ease of cleavage decreased in the order: pyrogallol > catechol > hydroquinone; the proximity of phenolic (OH) groups apparently favouring the said cleavage (Sanyal, 2002b).

Table 2.10: Release of Carbon Dioxide in the Nontronite-polyphenol Systems at the End of a 90-hour Reaction Period

Reaction Condition		CO_2 Release
Nontronite	Polyphenol	(mmol[a])
+[b]	Pyrogallol	263
−[c]	Pyrogallol	54
+	Catechol	88
−	Catechol	34
+	Hydroquinone	49
−	Hydroquinone	21

a: Amount of CO_2 released in the system containing 1 g of Ca-nontronite (0.2 - 2 mm), 5 mmol of pyrogallol, catechol, or hydroquinone in 30 mL of aqueous solution, adjusted to pH 6.0; b: In the presence; c: In the absence.

Source: Bollage *et al.* (1998).

References

Agar, J. N. (1963). Thermogalvanic cells. **In**: *Advances in Electrochemistry and Electrochemical Engineering* (P. Delahey and C. W. Tobias, Eds.), Interscience, New York, Vol. **3**, pp. 31-121.

Bear, F. E. (Ed.) (1976). *Chemistry of the Soil*, Second Edition, Third Indian Reprint, Oxford and IBH Publishing Co., New Delhi, pp. 30-32.

Bollag, J.M., Dec J. and Huang, P.M. (1998). Formation mechanisms of complex organic structures in soil habitats. *Adv. Agron.*, **63**, 237-266.

Bragg, W. H. and Bragg, W. L. (1913). The reflexion of X-rays by crystals. *Proc Roy. Soc. Lond.*, Series *A.*, **88** (605), 428–438.

Dyer, J. R. (1971). *Application of Absorption Spectroscopy of Organic Compounds*. Prentice Hall of India Private Limited, New Delhi, p. 23.

Ghosh, S.K. and Ghosh, K. (1998). Organo-mineral complexation in soils. *Bull. Indian Soc. Soil Sci.*, New Delhi, No. **19**, pp. 68-79.

Lahiri, T. C. and Chakravarti, S. K. (1995). Distribution and nature of organic matter in some hill soils of West Bengal at various altitudes in the Eastern Himalayan region. *J. Indian Soc. Soil Sci.*, **43**, 464-466.

Majumder, R. N. and Mukherjee, S. K. (1979). Degradation characteristics of hydrogen montmorillonites. *J. Indian Soc. Soil Sci.*, **27**, 26-37.

Manoj Kumar (1995). Ph.D. Thesis, IARI, New Delhi; cited in Ghosh and Ghosh (1998).

Mitra, R. P. and Kapoor, B. S. (1969). Acid character of montmorillonite: Titration curves in water and some non-aqueous solvents. *Soil Sci.*, **108**, 11-23.

Mondal, A. H., Nayak, D. C., Varadachari, C. and Ghosh, K. (1997). Spectroscopic, thermal and electron microscopic investigations on clay-humus complexes. *J. Indian Soc. Soil Sci.*, **45**, 239-245.

Morra, M.J., Blank, R.R., Freeborn, L.L. and Shafili, B. (1991). Size fraction of soil organo-mineral complexes using ultrasonic dispersions. *Soil Sci.*, **152**, 294-303.

Mukherjee, S. K. (1974). Electrochemistry of clays and clay minerals. In :*Mineralogy of Soil Clays and Clay Minerals*. Bulletin No. **9** (S. K. Mukherjee and T. D. Biswas, Eds), Indian Society of Soil Science, New Delhi, pp. 87-102.

Raman, K. V. and Ghosh, S. K. (1974). Identification and quantification of minerals in clays. In :*Mineralogy of Soil Clays and Clay Minerals*. Bulletin No. 9 (S. K. Mukherjee and T. D. Biswas, Eds), Indian Society of Soil Science, New Delhi, pp. 117-142.

Saha, P. B. and Sanyal, S. K. (1988). Synthetic humic acids– Their constitution, shape and dimension, *J. Indian Soc. Soil Sci.*, **36**, 35-42.

Sahu, S. S. and Das, S. C. (1974). Study of electrochemical and physicochemical properties of clay minerals and their mixtures and of some soil clays to characterize their clay mineralogy. *Proc. Indian Natl. Sci. Acad.*, **40B**, 235-248.

Sanyal, S. K. (1984). Structure of water in solution of organics-Hydrophobic hydration. *Chem. Edu.* (UGC), **1** (No.2), 14-18.

Sanyal, S. K. (1995). Ionic environment of acid soils. In: *Acid Soil Management* (M. A. Mohsin, A. K. Sarkar and B. S. Mathur, Eds.), Kalyani Publishers, Ludhiana, New Delhi, pp. 31-47.

Sanyal, S. K. (2002a). Soil Colloids. In: *"Fundamentals of Soil Science"* (G. S. Sekhon, P. K. Chhonkar, D. K. Das, N. N. Goswami, G. Narayanasamy, S. R. Poonia, R. K. Rattan and J. L. Sehgal, Eds.), Indian Society of Soil Science, New Delhi, pp. 229-259.

Sanyal, S. K. (2002b). Colloid chemical properties of soil humic substances: A Relook. In: *Bull. Indian Soc. Soil Sci.*, New Delhi, No. **21**, pp. 278-307.

Sanyal, S. K. and Mukherjee, A. K. (1988). Heat of transport and heat capacity of transport of some aqueous electrolytes. *Can. J. Chem.*, **66**, 435-438.

Sarmah, A. C. and Bordoloi, P. K. (1993). Characterization of humic and fulvic acids extracted from two major soil groups of Assam. *J. Indian Soc. Soil Sci.*, **41**, 642-648.

Schulze, D. G. (1989). An introduction to soil mineralogy. In: *Minerals in Soil Environment* (J. B. Dixon and S. B. Weed, Eds). Soil Science Society of America, Madison, Wisconsin, U.S.A., pp. 1-34.

Seal, B. K., Lahiri, M. M. and Layek, D. (1974). Ion exchange in clay minerals and soils. In :*Mineralogy of Soil Clays and Clay Minerals*. Bulletin No. **9** (S. K. Mukherjee and T. D. Biswas, Eds), Indian Society of Soil Science, New Delhi, pp. 65-86.

Shaw, D. J. (1970). *Introduction to Colloid and Surface Chemistry*. Butterworths, London, pp 147-166.

Sposito, G. (1984). *The Surface Chemistry of Soils*. Clarendon Press, Oxford, UK.

Srivastava, R. C. and Pal, Raj (1973). *Non-Equilibrium Thermodynamics in Soil Physics*. Oxford and IBH Publishing Company, New Delhi, p. 180.

Stevenson, E.J. (1994). *Humus Chemistry: Genesis, Composition, Reactions*, Second edition, John Wiley & Sons, New York.

Theng, B.K.G. (1979). *Formation and Properties of Clay-polymer Complexes*, Elsevier, Amsterdam, The Netherlands.

Van Olphen, H. (1963). *An Introduction of Clay Colloid Chemistry*, Intersci., New York.

Varadachari, C, Chattopadhyay, T and Ghosh, K (2000). The crystallo-chemistry of oxide-humus complexes. *Aust. J. Soil Res.*, **38**, 789-806.

Varadachari, C., Chattopadhyay, T. and Ghosh, K. (1997). Complexation of humic substances with oxides of iron and aluminium. *Soil Sci.*, **162**, 28-34.

Varadachari, C., Mandal, A. H. and Ghosh, K. (1991). Some aspects of clay-humus complexation: effect of exchangeable cation and lattice charge. *Soil Sci.*, **151**, 220-227.

Varadachari, C., Mondal, A. H. and Ghosh, K. (1995). The influence of crystal edges on clay-humus complexation. *Soil Sci.*, **159**, 185-190.

Varadachari, C., Mondal, A.H., Nayak, D.C. and Ghosh, K. (1994). Clay-humus complexation: Effect of pH and the nature of bonding. *Soil Biol. Biochem.*, **26**, 1145-1149.

Wang, M.C. and Huang, P.M. (1994). Structural role of polyphenols in influencing the ring cleavage and related chemical reactions as catalyzed by nontronite. **In:** *Humic Substances in the Global Environment and Implications in Human Health* (Senesi, N., Miano, T.M., Eds.). Elsevier, Amsterdam, The Netherlands, pp. 173–180.

Yuan, G., Theng, B.K.G., Parfitt, R.L. and Percival. H. J. (2000). Interactions of allophane with humic acid and cations. *Eur. J. Soil Sci.*, **51**, 35-41.

Ion Exchange Processes in Soil

3.1. Ion Exchange in Soil

Ion exchange in soil system refers to exchange of equivalent amounts of ions between two phases in equilibrium in contact in a reversible process. When cations are involved, the process is termed cation exchange, while for anions, it is referred to as anion exchange. Such exchange may take place between the soil solid (soil colloid or exchange) phase and the soil solution phase, or less commonly though, between the soil solid phases in contact (*e.g.* the Inverse Ratio Law; *see* later in Section 3.4), or soil solid phase and growing plant in contact (contact exchange) (Sanyal *et al.*, 2009).

The experimental evidence of cation exchange was provided around 1850 when Thomson, followed by J. Thomas Way leached soil with ammonium sulphate solution, and upon filtration, calcium, and to a lesser extent magnesium, potassium ions were detected in the filtrate, while less of ammonium ions were recovered compared to what was used for leaching. Initially, this phenomenon was described as Base Exchange. But on recognition of the fact that hydrogen ions can also be so exchanged, the wider term, namely cation exchange was used (Sanyal *et al.*, 2009).

The composition of soil solution is known to change as a result of (1) dilution of soil solution (*e.g.* during rain or application of irrigation), (2) ion uptake by plants and leaching, (3) fertilizer application, (4) use of brackish groundwater for irrigation, (5) reclamation of salt affected soils by leaching or use of chemical amendments, (6) sewage and sludge farming and use of industrial effluents for irrigation, etc. This leads to migration of, for instance, cations from soil colloidal phase to soil solution, and/or *vice versa* (cation release/fixation). The process involved is obviously ion exchange (Poonia, 2002).

3.1.1. Cation Exchange Capacity

As stated above, interchange of a cation in soil solution phase with another on the surface of any surface-active material such as soil clay or organic matter in equivalent proportions in a reversible process is generally termed as cation exchange.

The cation exchange capacity (CEC) is defined as the capacity of the solid (*e.g.* soil) to adsorb and exchange cations. In other words, it is the sum total of exchangeable cations that can be adsorbed by a soil colloidal surface. It is expressed in the unit of cmol (p^+) kg^{-1} (in S.I. system) or meq per 100 g in the classical (cgs) unit. Here (p^+) denotes proton (Sanyal *et al.*, 2009).

Similarly, the anion exchange capacity (AEC) of soil colloid can also be defined. This is expressed (in S.I. system) in the units of cmol (e^-) kg^{-1}, (e^-) representing electron.

3.1.2. Bonding Energy of Exchangeable Cations

Considering the electrical double layer theory as applied to soil colloid system (see Section 3.2.3 later and Section 2.6.1 presented earlier), the solid surface in soil, being predominantly negatively charged, the counter cations on the solution side of the double layer, especially those in the diffuse double layer (Goüy-Chapman layer; Region QR of Figure 3.1) are the exchangeable cations, amenable to exchange with the externally added electrolytic cations (*e.g.* of the added fertilizer/amendments/ irrigation water) (Figure 3.1).

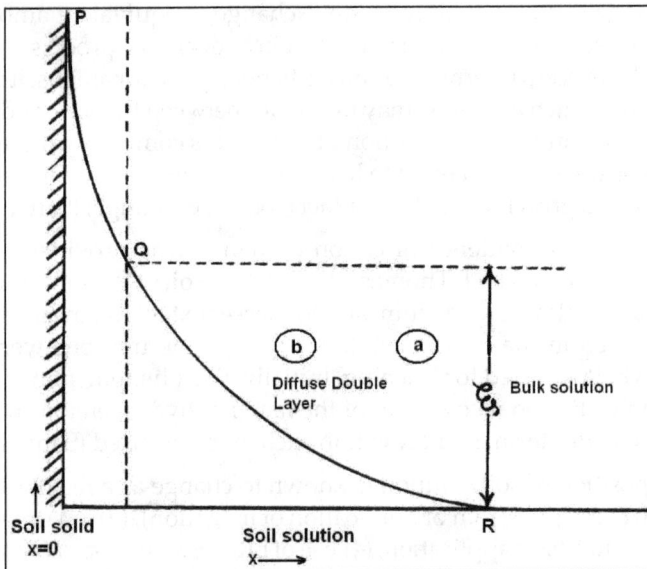

Figure 3.1: Electrical Double Layer at Soil-Soil Solution Interface.
Source: Sanyal *et al.* (2009)

The ionic strength (I) of a solution containing ions, which provides a measure of the intensity of the electric field due to such space charge in the solution, is given as

$$I = \frac{1}{2}\sum_i m_i Z_i^2$$

$$(3.1)$$

Where m_i is the molality and Z_i the valence (including sign) of the ion 'i' in the solution phase, with the summation (Σ) being extended to cover all the ionic species in the solution.

The activity (a_i) of an ion in a solution is related to its molality (m_i) by the equation,

$$a_i = \gamma_i m_i \qquad (3.2)$$

Where γ_i is the activity coefficient of the ion 'i' in the solution phase while the activity, a_i, is its effective concentration. At infinite dilution, $\gamma_i \to 1$, and $a_i \to m_i$

The activity coefficient (γ_i) of an ion in a solution is given by the Debye-Hückel Extended Law,

$$-\log\gamma_i = \frac{AZ_i^2\sqrt{I}}{1+Bd\sqrt{I}}$$

$$(3.3)$$

Where A and B are constants at a given temperature for the solvent (*e.g.* water for aqueous soil solution), d is the diameter of the hydrated ion, and I, the ionic strength of the solution.

For soil solution, which is generally dilute (except for the salt affected soil), the Debye-Hückel Extended Law is approximated by the Debye-Hückel Limiting Law, namely

$$-\log \gamma_i \approx AZ_i^2\sqrt{I} \qquad (3.4)$$

So that $\qquad -\log \gamma_i \alpha \sqrt{I} \qquad (3.5)$

for a given ion at a given temperature in aqueous solution (Sanyal *et al.*, 2009).

Considering the position of the same exchangeable cation at either 'a' or 'b' in the diffuse double layer (Figure 3.1), the bonding energy (B.E.) of the given ion for the present colloidal surface will obviously increase with the fall of distance (*i.e.*, x) of the ion from the colloidal surface (at x = 0 in Figure 3.1), according to the Coulomb's Inverse Square Law. That is, the B.E. at 'a' will be less than that at 'b'. However, the ionic strength (I) will increase as one moves closer to the surface (*i.e.*, with fall in x) with the concomitant increase in the ionic charge density (*i.e.*, Eq. 3.1). In other words, both B.E. and I would increase with the fall in x. That is, the bonding energy (B.E.) of the counter ions in the diffuse double layer is directly related to the corresponding ionic strength (I), *i.e.*,

$$\text{B.E. } \alpha I \qquad (3.6)$$

But from Eq. 3.5 $\qquad -\log \gamma_i \alpha \sqrt{I}$

Hence, $\qquad (\text{B.E.})_i \alpha - \log \gamma_i \qquad (3.7)$

Thus, the negative magnitude of the logarithm of the activity coefficient of a counter (exchangeable) ion provides a measure of the corresponding bonding energy of the counter ion for the soil solid surface (Sanyal *et al.*, 2009).

The adsorbability of cations to a charged surface (*e.g.* soil colloid) had often been discussed in terms of the *Hofmeister series* or *Lyotropic series*. Thus, the monovalent cations can be put in the following sequence according to their adsorbability.

$Li^+ < Na^+ < K^+ < Rb^+ < Cs^+$

Bivalent cations follow the sequence given below.

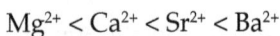

$Mg^{2+} < Ca^{2+} < Sr^{2+} < Ba^{2+}$

For the cations of varying valences, the *Lyotropic series* in respect of adsorbability is given as:

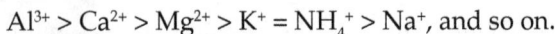

$Al^{3+} > Ca^{2+} > Mg^{2+} > K^+ = NH_4^+ > Na^+$, and so on.

Obviously the guiding factor for such orders of cation binding by a charged surface may be related to the concept of bonding energy that has been derived above (*e.g.*, Eqs. 3.3 and 3.7) in that the *net* bonding energy (or adsorbability) of the counter ions on a colloidal surface is the reflection of the relative weightage of the valency and the size of the hydrated cations concerned (Sanyal *et al.*, 2009).

3.2. Theories of Ion Exchange

3.2.1. Adsorption Isotherms

The adsorption process deals with the adsorption of molecules or ions from the soil solution on the surfaces of soil solids. The material adsorbed is called the **adsorbate**, while the material on which adsorption occurs is called the **adsorbent**. In some cases, the initial adsorption may be accompanied (and followed) by penetration of the adsorbed ions by diffusion into the adsorbent body, leading to further **absorption** of the adsorbed species. The general term *sorption* is sometimes used to denote both these processes, taking place simultaneously. Desorption is the reverse phenomenon of adsorption. It is of interest to note that the first quantitative studies on cation adsorption and release in soils were connected with NH_4^+ ion.

For instance, phosphate adsorption by soils is a process which is mainly responsible for rendering the soluble phosphates (*e.g.* applied phosphatic fertiliser such as single superphosphate or diammonium phosphate) in soil solution unavailable to plants. Desorption of once-sorbed plant nutrients from soils and clays is often slow and irreversible, leading to a large hysteresis effect (*see* later). Adsorbed ammonium ions are also less susceptible to nitrification in soil. The amount of material adsorbed per unit mass of the adsorbent depends largely upon the nature of the adsorbate and the adsorbent; it also depends on the concentration of the adsorbate at equilibrium. Mathematically, it is expressed by Eq. 3.8:

$x/m = f$ (Nature of adsorbate and adsorbent, C, T)   (3.8)

Where, x is the amount of adsorbate adsorbed on mass m of the adsorbent, C is the equilibrium concentration of the adsorbate and T is the temperature.

For a given mass and surface area of the adsorbent, the amount adsorbed increases with increase in the concentration of the adsorbate. When an adsorbent is placed in contact with a solution, the amount adsorbed gradually increases, and the concentration of the surrounding ions or molecules (adsorbate) decreases. This leads to the release of adsorbate from the adsorbent towards solution. When the rates of such desorption becomes equal to the rate of adsorption, equilibrium is established. When the concentration of adsorbate in solution increases (or decreases), the adsorbed substance also increases (or decreases) till a new equilibrium is established (Sanyal *et al.*, 2009).

Adsorption data at equilibrium are reported in the form of adsorption isotherms. An adsorption isotherm is a plot between the amount adsorbed per unit mass of the adsorbent and the equilibrium concentration of the adsorbent, at a constant temperature.

Different equations and models have been developed to predict precisely the shapes of the experimental adsorption isotherms between different adsorbates and adsorbents. The most commonly used ones are due to Langmuir (1918), Freundlich (1926), Tempkin (1940) and Brunauer *et al.* (1938), which are described below.

3.2.1.1. Langmuir Adsorption Theory

The basic theory of adsorption of gases on solids is due to Langmuir (1918) and it is based on the following assumptions (Poonia, 2002):

(1) The surface of solid is made up of adsorption sites, each of which can adsorb one gas molecule. If another molecule hits the adsorbed molecule, it is deflected back into the gas.

(2) All the adsorption sites are identical in their affinity for that gas molecule.

(3) The presence of a gas molecule on one site does not affect the properties of the neighbouring sites.

(4) The rate of adsorption is proportional to the number of adsorption sites and the pressure of gas. The rate of desorption is proportional to the number of occupied sites.

The system is considered to be at equilibrium when the rate of evaporation of the adsorbed gas is equal to the rate of its condensation.

The linear form of the Langmuir adsorption is given as in Eq. 3.9:

$$\frac{C}{x} = \frac{C}{x_m} + \frac{1}{x_m K}$$

$$(3.9)$$

Where, C is the concentration of the adsorbate at equilibrium, x is the amount adsorbed per unit mass of adsorbent at any instant and x_m is the maximum amount adsorbed per unit mass of adsorbent forming a monolayer on the adsorbent surface, and K is the equilibrium constant related to bonding energy. Experimental adsorption data, when plotted as (C/x) *versus* C, should yield a straight line with the slope $(1/x_m)$ and the intercept $(1/x_m K)$ on the ordinate (Y-axis). It may, however, be remembered that a linear plot between C/x and C is not adequate to prove that the given adsorption data follow the Langmuir equation. A linear plot of the data is a necessary, but *not* a sufficient condition of the Langmuir equation being obeyed (Sanyal and De Datta, 1991).

Khasawneh and Copeland (1973) introduced the concept of Supply Parameter (SP) involving the coefficients and constants from the Langmuir equitation (Eq. 3.9). Such SP is taken to provide the buffering capacity of the soil for the adsorbed nutrient due to change of concentration of the nutrient concerned. The SP is given by Eq. 3.10:

$$SP = \sqrt{(qC/x_m K)} \tag{3.10}$$

Where q is identical with x (specific adsorption of the nutrient concerned) of Eq. 3.9, while x_m and K are the Langmuir constants as given by Eq. 3.9 (Sanyal et al., 2009).

The adsorption maximum (x_m) of soils for various ions and molecules, such as phosphate, borate, sulphate, K, herbicides, etc., can be estimated from the slope of the linear plot between (C/x) and C. From such value of x_m, the surface area (A) of soils and/or clays (the adsorbent) can be estimated using the relationship (Eq. 3.11):

$$A = \frac{x_m \,(g)\, N \,(\text{number of molecules mol}^{-1})\, \pi r^2}{M \,(g\, \text{mol}^{-1})} \tag{3.11}$$

Where, r is the radius of the adsorbate molecule, N is the Avogadro's number and M is the molecular weight expressed in g of the substance. If the adsorbate happens to be a gas, then x_m and M in Equation (3.11) should, respectively, be replaced by the volume of the gas required to form a monolayer on the surface (v_m, cm^3), and volume occupied by a mole (molecular weight expressed in g) of the gas, i.e., 22.4 liter at 273 K, so that Eq. 3.11 becomes Eq. 3.12 (Poonia, 2002):

$$A = \frac{v_m \,(cm)^3 \, N \,(\text{number of molecules mol}^{-1})\, \pi r^2}{22400 \,(cm^3 \, \text{mol}^{-1})} \tag{3.12}$$

Generally, the Langmuir equation is *not* used to calculate the surface area of solids because it does *not* define total adsorption process, but covers only a certain fraction of adsorption sites.

It is important to emphasize here that soils are mixtures of inorganic and organic substances, consisting of montmorillonite, kaolinite, oxides of iron and aluminium, and decomposed plant material, etc., with each substance reacting towards competing cations differently. In a sense, soil exchangers are essentially polyfunctional ion exchangers. The condition of identical exchange sites (assumption 2), (identical bonding energy) and effectively no change of it during adsorption (assumption 3) thus become too restrictive for application of the Langmuir equation to a soil system. The experimental exchange isotherm obtained for a soil system is a weighted sum of isotherms for different types/classes of exchange sites rather than for identical exchange sites (Sanyal et al., 2009).

Use of Langmuir equation becomes further restrictive for cation exchange equilibria, as it does *not* take into account the competition between the cationic species. According to this equation, the amount of a cation on the adsorbed phase is governed only by the concentration of that cation, independent of the concentration of the competing cation. Last, but *not* the least, the Langmuir adsorption isotherm refers to gaseous adsorption on non-ionic solid, while adsorption in soil is essentially an ionic process (Sanyal *et al.,* 2009).

3.2.1.2. Freundlich Isotherm

The Freundlich equation (1926), originally an empirical equation, but subsequently derived on a rigorous mathematical model, implies that the bonding energy of the adsorbate on a given adsorbent decreases with fractional coverage of surface area of the adsorbent (Sposito, 1984). This is closer to reality than the assumption of a constant bonding energy inherent in the Langmuir equation. However, the Freundlich isotherm does *not* predict any adsorption maximum. This is one of the most widely used adsorption equations to describe the experimental data on adsorption of ionic or molecular species in soil. Mathematically, it is expressed by Eq. 3.13:

$$x = KC^n \ (n < 1) \tag{3.13}$$

Where, x is the amount adsorbed per unit mass of the adsorbent and C is the equilibrium concentration of the adsorbate. K and n are empirical constants, sensitive to the given adsorbent-adsorbate system and temperature. The linear form of this equation is obtained by taking logarithm of its both sides, when we get expression, Eq.3.14:

$$\log x = \log K + n \log C \tag{3.14}$$

A plot of the experimental data of log x *versus* log C should give a straight line with a slope of n and an intercept of log K. However, the Freundlich isotherm does *not* predict any adsorption maximum. Nevertheless, the Freundlich coefficient, K, may be regarded as a hypothetical index of a given nutrient sorbed by a series of adsorbent surfaces from a solution having unit equilibrium concentration, provided that the factor n (of Eqs. 3.13 and 3.14) is nearly identical for the given adsorption isotherms.

3.2.1.3. Tempkin Equation

Tempkin equation (Tempkin and Pyzhev, 1940) is based on the assumption that the bonding energy of adsorption decreases linearly with increasing surface coverage. For the middle range of surface coverage, the equation reads (Sanyal and De Datta, 1991):

$$(x/x_m) = (RT/a) \ln AC = 2.303 \ (RT/a) \log A + 2.303 \ (RT/a) \log C \tag{3.15}$$

Here, x is the amount adsorbed per unit mass of the adsorbent, A and a are coefficients, R is the universal gas constant, and x_m is the Langmuir adsorption maximum. According to this equation, a plot of x against log C should yield a straight line. Such plots for soils in many cases, however, yielded gentle curves, rather than straight lines, but the agreement with the experimental data was, in general, better over a wider range of concentrations than that with the Langmuir plots.

The driving force or the intensity factor for adsorption of an ion, say phosphate (P) in soil is, in fact, the chemical potential of an ion in equilibrium soil solution. The latter is a logarithmic function of the ion activity (or concentration in a dilute solution) in solution, in agreement with Equation (3.15). Thus, the Tempkin equation is based on the plot of the specific ion adsorption against the logarithm of equilibrium concentration. For instance, the chemical potential (μ) of P in soil solution is directly related to the phosphate concentration in soil solution as,

$$\mu = \mu^\circ + 2.303 \ RT \log a_p \approx \mu^\circ + 2.303 \ RT \log C_p \tag{3.16}$$

Where μ° is the standard chemical potential of phosphate in soil solution of P concentration C_p and activity a_p, since in dilute solutions (such as soil solutions except in case of salt-affected soils), the ionic activity approximates the corresponding concentrations. Hence the Tempkin equation is superior thermodynamically. Furthermore, these plots lend themselves to compute the relative P sorption capacities of several soils corresponding to the empirically observed equilibrium concentrations of, say, P that reportedly lead to the optimum growth of crops [*e.g.* 0.2 ppm P (= 6.45 µmol P/L) concentration in soil solution, while for lowland rice soil, 0.12 ppm P). Such P sorption values roughly follow the order of the P buffering capacities (mentioned earlier) of several P fixing soils (Sanyal and De Datta, 1991). Indeed Klages *et al.* (1988) showed that the P fertilizer requirement for dryland upland wheat was better predicted by such methods based on the P sorption than those based on the Olsen test for soil available P. The P sorption per unit of clay content or organic matter in a number of acidic soils of South and Southeast Asia was also found to decrease with increasing clay or organic matter content of the experimental soils. This tends to suggest that the intimate clay-organic matter complexes, formed through the cationic bridges, render some of the active surfaces of both the colloidal components inaccessible for phosphate sorption. This observation therefore advocates the organic manure incorporation for partially mitigating the P sorption properties in highly P-fixing soils. However, this approach considers only the intensity factor for nutrient in solution while disregarding buffer capacity of soils and the transport of the nutrient in solution to the root zone that also affect nutrient uptake by roots (Sanyal and De Datta, 1991). In addition, this method is of less value when the amount of nutrient already present in soils is large relative to the amount adsorbed. As suggested by Barrow (1978), one way of overcoming this problem is to measure the slope of the adsorption isotherm at the required concentration, rather than the amount adsorbed. The slope will be independent of the amount of the nutrient originally present, besides providing important information about the nutrient adsorption buffer power of the soil that governs the availability of a plant nutrient, including its diffusive mobility.

3.2.1.4. Brunauer, Emmett and Teller (BET) Adsorption Isotherm

The Langmuir equation represents the experimental adsorption data reasonably well in cases where adsorption becomes independent of pressure after the formation of monolayer. For many gases and vapour, however, it fails at higher pressures because more than a monolayer of adsorbed molecules is formed. The theory of Brunauer, Emmett and Teller (BET) extends Langmuir derivation to obtain an equation for multilayer adsorption (Poonia, 2002).

The BET equation is based on the following assumptions (Brunauer *et al.*, 1938):

(i) Heat of adsorption in the second, third,.., nth layers are the same as the heat of adsorption in the monolayer, *i.e.*, heat of liquefaction or condensation of the gas, and

(ii) If the adsorption takes place on a free surface, then at p^0 (saturation pressure of the gas), an infinite number of layers can be built up on the adsorbent.

The BET equation used to define multilayer adsorption at any equilibrium pressure, p, is given in its linear form as in Eq. 3.17:

$$\frac{p}{v(p^0 - p)} = \frac{(C'-1)p}{v_m C' p^0} + \frac{1}{v_m C'}$$

(3.17)

Where, v is the volume of gas adsorbed per unit mass of adsorbent, v_m is the volume of gas required to form a monolayer on the entire adsorbent surface of unit mass, C' is a constant at a given temperature. According to this equation, a plot of the experimental data on $p/v(p^0-p)$ *versus* (p/p^0) should yield a straight line with a slope $[(C'-1)/v_m C']$ and an intercept $(1/v_m C')$ on the Y axis. The values of the slope and the intercept thus yield the values of constant, C', and volume of gas adsorbed when the entire adsorbent surface is covered with a monolayer, v_m. The advantages of BET equation are (Sanyal *et al.*, 2009):

(i) It describes the multilayer adsorption.

(ii) It yields the volume of gas required for the formation of a monolayer.

(iii) It can be used to calculate the surface area of soils (*e.g.*, by the use of Eq. 3.12).

The BET equation is most useful between the relative pressures (p/p^0) of 0.05 and 0.45 (Sanyal *et al.*, 2009). The surface area can be calculated by multiplying v_m with the cross-sectional area of the adsorbate molecule. To determine surface area of soils and minerals, adsorbate molecules such as nitrogen, ethane, water, ammonia or other gases may be used. Weakly adsorbed molecules like nitrogen, however, may not penetrate into all the interlayer surfaces, resulting into an underestimate of total surface area. Use of strongly adsorbed molecules like ammonia, which penetrates into the internal surfaces also, leads to the measured surface area closer to the total surface area (Poonia, 2002).

3.2.2. Hysteresis

It is generally observed that the ions which are adsorbed by a colloidal surface by ionic interactions, through non-specific retention mechanism, are desorbed during the desorption cycle by the expenditure of the same amount of energy as that released during the adsorption pathway. However, for ligand exchange processes, such as those involved in the fixation of phosphate, sulphate, borate, etc. (*see* later), by soil colloids, and also the cation, namely K^+ (the latter being specifically sorbed in the interlayer space of illitic clay minerals), the desorption generally requires much more energy in view of the fact that the retentive forces in such cases are primarily covalent in nature for the anions concerned, while ionic diffusion of K^+

requires its transport through a strongly negatively charged electric field in the illitic interlayer space. This leads to the irreversibility of the adsorption-desorption processes involved, with the degree of such irreversibility depending on the nature of the soil colloids concerned (*e.g.* aluminosilicate clay minerals, hydrous oxide of Fe, Al, etc., in clay-size dimensions), and also the ions concerned. This effect is known as *Hysteresis*. The hysteresis effect in K^+ ionic fixation-release processes depends not only on the amount of illitic minerals present in a given soil, but also on the state of weathering of such minerals. The more weathered the illitic minerals are (*e.g.* in Alfisols/Ultisols), the less will be the extent of hysteresis, as compared to the soils of more recent origin (*e.g.* Inceptisols, Entisols, etc.). Indeed, studies showed that desorption of once-sorbed phosphate is always less than the amount of phosphate sorbed at a given equilibrium P concentration because of such *hysteresis* effect in P sorption–desorption behaviour (Sanyal *et al.*, 1993, 2009). This implies that the sorbed P undergoes further interactions, which impart to it a greater degree of affinity for the surface, and causes difficulty in plant access to residual P build-up in a cropping cycle. In fact, desorption generally requires much more energy to disrupt the retentive forces which are primarily covalent in nature for inner-sphere complex formed between the soil colloid and the added phosphate (*see* later). Further, the desorption equilibrium being a slow process, an apparent re-adsorption during desorption step is also possible. Indeed, such hysteresis effect would lead to an overestimation of the replenishing ability of soils to supply P to soil solution, when P sorption isotherms are used for the purpose. This is true for all other anions undergoing ligand exchange at the soil sorption sites (Sanyal and De Datta, 1991). The desorption isotherm is thus displaced to the left of the corresponding adsorption isotherm. Representative adsorption-desorption Tempkin isotherms of P in two acidic soils are shown in Figure 3.2.

3.2.3. Electrical Diffuse Double Layer Theory

As stated earlier (Section 2.6), the layer-lattice silicate minerals have both the permanent and the pH-dependent charges. The oxides and hydroxides (in clay-sized dimensions) as well as humus components of soil, on the other hand, exhibit only the pH-dependent charge. In 2:1 layer-lattice silicate minerals, the pH-dependent charge is invariably less than the permanent negative charge. In majority of the cases, the soil colloidal complex at pH above 6 possesses a net negative charge. As a result, an electrical double layer is formed at the colloidal-solution interface which has already been explained earlier (Section 2.6).

The main assumptions of the theory of diffuse double layer are the following (Poonia, 2002):

(i) The charged particle has flat surface without edge effect, *i.e.*, the planar surface is effectively infinite in extent.

(ii) The exchange sites are uniformly distributed over the entire planar surface, *i.e.*, the surface charge density is uniform.

(ii) The dielectric constant of water in the aqueous solution phase is constant.

(iv) The ions act as point charges.

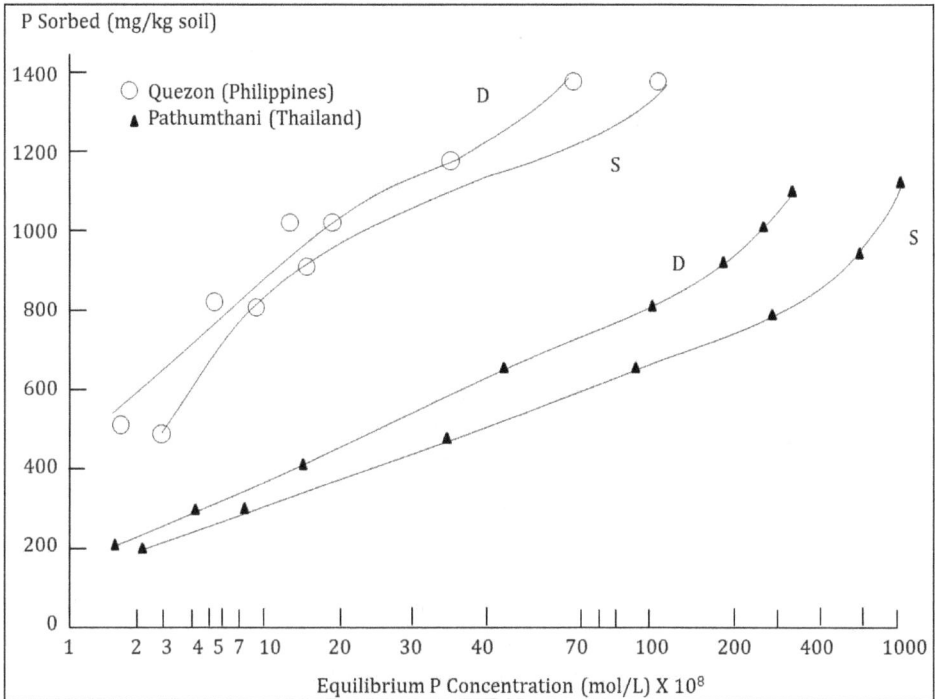

Figure 3.2: Tempkin Isotherms for P Sorption by Acidic Soils Obtained during Adsorption (S) and Desorption (D) Runs in 0.01 Aqueous M CaCl$_2$ at 25°C.
Source: **Sanyal *et al.* (1993).**

(v) The interaction between ions is negligible and the charged surface and the ions interact only electrostatically.

(vi) The charge on the surface is neutralized by the excess of oppositely charged ions in the diffuse double layer, considering the Helmholtz's layer to be too small in extent as compared to the Goüy-Chapman layer (*see* Section 2.6.1).

Let us consider a charged soil colloid surface in contact with the soil olution. The charge density (e_x) at an imaginary lamina, parallel to the colloid surface at a distance x from it (Figure 3.3) can be expressed in two ways, firstly, in terms of the Poisson equation, which, for the x dimension in rectangular coordinates, reads

$$e_x = -\frac{D}{4\pi}\frac{d^2\varphi_x}{dx^2}$$

(3.18)

where φ_x is the outer potential difference between the lamina and the bulk of the solution (taken as $\varphi_{\to\infty} = 0$) and, secondly, in terms of the Boltzmann distribution

$$e_x = \sum_i n_i z_i e_o = \sum_i n_i^0 z_i e_o e^{-z_i e_0\frac{\varphi_x}{kT}}$$

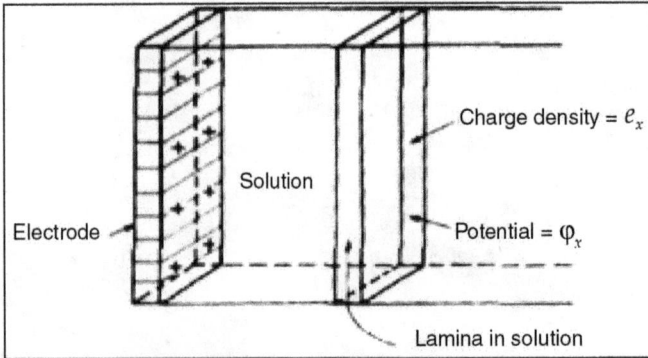

Figure 3.3: Lamina in the Diffuse Layer on the Solution Side of the Electrical Double Layer Parallel to the Charged Soil Colloid Surface (Resembling a charged electrode placed in an ionic solution). *Source*: Bockris and Reddy (1970).

Where n_i and n_i^0 are the concentrations of the ith species in the lamina (at a distance x from the charged colloid surface) and in the bulk of the solution, respectively, z_i is the valence of the species i, and e_0 is the electronic charge. Here, D is the dielectric constant of the medium (*e.g.* water). The factor $(z_i e_0 \varphi_x /kT)$ represents the ratio of the electrical to the thermal energies of an ion at a distance x from the colloid surface. From these two expressions for the charge density e_x, one obtains the Poisson-Boltzmann equation, namely Eq. 3.19 (Shaw, 1970):

$$\frac{d^2\varphi}{dx^2} = \frac{-4\pi}{D} \sum_i n_i^0 z_i e_0 e^{-z_i e_0 \varphi_x / kT}$$

(3.19)

At this stage in the Debye-Hückel theory of ion-ion interactions in homogeneous solution, it was considered that the potential was small enough to *linearize* the equation. This linearization would correspond in the present case to writing

$e^{-z_i e_0 \varphi_x /kT} \approx 1 - z_i e_0 \varphi_x /kT$ (That is, $e^{-x} \approx$ 1-x)

Attempts to achieve so-called "rigorous" solutions by not linearizing the Poisson-Boltzmann equation led to certain inconsistencies. Nevertheless, it has been customary, in diffuse-double-layer treatments, based on the Poisson-Boltzmann equation, to proceed with the solution of the unlinearized differential equation, namely Eq. 3.19.

A simple transformation can now be used. Let us consider the steps (Bockris and Reddy, 1970):

$$\frac{1}{2}\frac{d}{d\varphi}\left(\frac{d\varphi}{dx}\right)^2 = \frac{1}{2}.2\left(\frac{d}{d\varphi}\frac{d\varphi}{dx}\right)\frac{d\varphi}{dx}$$

$$= \left(\frac{dx}{d\varphi}\frac{d}{dx}\frac{d\varphi}{dx}\right)\frac{d\varphi}{dx}$$

$$= \frac{d^2\varphi}{dx^2}$$

The point to be noted is that in the Goüy-Chapman theory – as in the Debye-Hückel theory – one is analyzing potential in one phase only, namely the solution phase. However, what is customary need not necessarily *be* right. A more consistent theory of the diffuse double layer is clearly required.

Thus, the identity, namely

$$\frac{1}{2}\frac{d}{d\varphi}\left(\frac{d\varphi}{dx}\right)^2 = \frac{d^2\varphi}{dx^2}$$

can be used in the differential equation, Eq. 3.19 to give Eq. 3.20, *i.e.*,

$$\frac{d}{d\varphi}\left(\frac{d\varphi}{dx}\right)^2 = -\frac{8\pi}{D}\sum_i n_i^0 z_i e_o e^{-z_i e_0 \frac{\varphi_x}{kT}}$$

(3.20)

Which, by the following rearrangement

$$d\left(\frac{d\varphi}{dx}\right)^2 = -\frac{8\pi}{D}\sum_i n_i^0 z_i e_o e^{-z_i e_0 \frac{\varphi_x}{kT}} d\varphi$$

can be integrated to give

$$\left(\frac{d\varphi}{dx}\right)^2 = -\frac{8\pi}{D}\int\sum_i n_i^0 z_i e_o e^{-z_i e_0 \frac{\varphi_x}{kT}} d\varphi$$

$$= -\frac{8\pi}{D}\sum_i \frac{n_i^0 z_i e_o e^{-z_i e_0 \frac{\varphi_x}{kT}}}{-z_i e_0 / kT} + \text{Constant}$$

$$= \frac{8\pi kT}{D}\sum_i n_i^0 e^{-z_i e_0 \frac{\varphi_x}{kT}} + \text{Constant}$$

(3.21)

The integration constant can be evaluated by considering that, deep in the bulk of the soil solution, *i.e.*, at x→∞, not only is the Volta potential is zero, $\varphi_{x\to\infty} = 0$, but the field ($d\varphi_x/dx$) is also zero. Under these conditions,

$$\text{Constant} = -\frac{8\pi kT}{D}\sum_i n_i^0$$

By introducing this value of the integration constant into Eq. 3.21, the result is

$$\left(\frac{d\varphi}{dx}\right)^2 = \frac{8\pi kT}{D}\sum_i n_i^0\left(e^{-z_i e_0 \frac{\varphi_x}{kT}} - 1\right)$$

For further simplification, let us consider only a z: z-valent electrolyte to be present in the soil solution, with $|z_+| = |z_-| = z$ and $n_+^0 = n_-^0 = n^0$. Thus,

$$\left(\frac{d\varphi}{dx}\right)^2 = \frac{8\pi kT}{D}\sum_i n^0\left(e^{-ze_0\frac{\varphi_x}{kT}} - 1\right)$$

$$= \frac{8\pi kT}{D}n^0\left(e^{ze_0\frac{\varphi_x}{kT}} - 1 + e^{-ze_0\frac{\varphi_x}{kT}} - 1\right)$$

or, $$\left(\frac{d\varphi}{dx}\right)^2 = \frac{8\pi kT}{D}n^0\left[e^{ze_0\frac{\varphi_x}{kT}} - 2\left(e^{ze_0\frac{\varphi_x}{2kT}}\right)\left(e^{-ze_0\frac{\varphi_x}{2kT}}\right) + e^{-ze_0\frac{\varphi_x}{kT}}\right]$$

$$= \frac{8\pi kT}{D}n^0\left[e^{ze_0\frac{\varphi_x}{kT}} - e^{-ze_0\frac{\varphi_x}{2kT}}\right]^2$$

(3.22)

Because $e^{+x} - e^{-x} = 2\sinh x$

Hence, Eq. 3.22 becomes

$$\left(\frac{d\varphi}{dx}\right)^2 = \frac{32\pi kTn^0}{D}\sinh^2\frac{ze_0\varphi_x}{2kT}$$

(3.23)

From this equation, one can get the field $(d\varphi/dx)$ in the solution by taking square roots on both sides. There is, however, a positive and a negative square root. To decide which root is to be taken, one notes that at the negatively charged soil colloid surface, $\varphi < 0$ and $(d\varphi/dx) > 0$, while at the positively charged surface, $\varphi > 0$ but $(d\varphi/dx)\varphi < 0$. Hence it is clear that only the negative root of Eq. 3.23 corresponds to the physical situation (Shaw, 1970; Bockris and Reddy, 1970), *i.e.*,

$$\frac{d\varphi}{dx} = -\left(\frac{32\pi kTn^0}{D}\right)^{\frac{1}{2}}\sinh\left(\frac{ze_0\varphi_x}{2kT}\right)$$

(3.24)

Eq. 3.24 gives the relationship between the electric field or the gradient of potential $[(d\varphi/dx)]$ and the potential (φ_x) at any distance x from the charged surface, according to the diffuse-charge model of Goüy-Chapman.

Instead of the field, it is preferable to have an expression for the total diffuse charge (q_d) in the solution side of the double layer in terms of the potential. Such diffuse charge in the solution side of the double layer is obtained as follows.

Remembering the origin of the field $(d\varphi/dx)$, one recalls that according to the Gauss's law from electrostatics, the charge obtained in a closed surface, or *Gaussian box*, is equal to $D/4\pi$ times the area of the closed surface (taken here as unity) times the component of the field normal to the surface.

That is,

$$q_d = \frac{D}{4\pi}\cdot\frac{d\varphi}{dx}$$

(3.25)

Since $(d\varphi/dx)$ is known from Eq. 3.24, the corresponding q_d can be obtained. Such q_d is the charge enclosed within the closed volume at the surface of which the field is $(d\varphi/dx)$.

To compute the total diffuse charge density (q_d) the *Gaussian box* chosen is a rectangular box with one unit-area side deep in the solution at $x \rightarrow \infty$, where φ_x and ($d\varphi/dx) = 0$, while the other unit-area side "very close to the charged colloid surface, as close as possible", so as not to miss any diffuse charge (Figure 3.4). If the model considered is that of point-charge ions (and this is the Goüy-Chapman model), then the box must be brought up to $x = 0$ (Shaw, 1970; Bockris and Reddy, 1970).

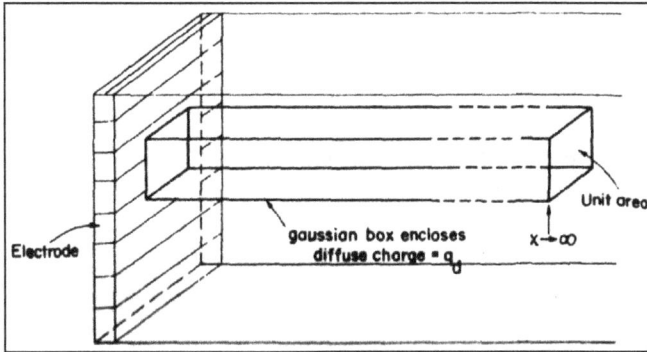

Figure 3.4: A Gaussian Box Containing the Entire Diffuse Charge q_d.
Source: **Bockris and Reddy (1970)**

Hence, the total diffuse-charge density scattered in the solution side of the electrical double layer under the interplay of thermal and electrical forces, q_d , is given from Eqs. (3.24) and (3.25) with $x = 0$, as

$$q_d = -\frac{D}{4\pi}\left(\frac{32\pi kTn^0}{D}\right)^{\frac{1}{2}} \sinh\left(\frac{ze_0\varphi_x}{2kT}\right)$$

$$\text{Or, } q_d = -2\left(\frac{Dn^0kT}{2\pi}\right)^{\frac{1}{2}} \sinh\left(\frac{ze_0\varphi_0}{2kT}\right)$$

$$(3.26)$$

Where φ_0 is the potential at $x = 0$ relative to the bulk of the solution where the potential is taken as zero (*i.e.*, $\varphi_\infty = 0$). Eq.3.26 shows that the total diffuse charge varies (according to the Goüy-Chapman model) with the total potential drop in the solution according to a hyperbolic sine relation.

3.2.3.1. Potential Drop in the Diffuse Layer

The field ($d\varphi/dx$) is the gradient of the potential. By integrating the field, one obtains the variation of the potential with distance. Let this be done with the assumption

$$\sinh\left(\frac{ze_0\varphi_x}{2kT}\right) \approx \frac{ze_0\varphi_x}{2kT}$$

This will be closer to reality for small values of φ_x, that is, for rather small charge on the soil colloidal surface, and a relatively small value of the corresponding potential within the diffuse layer of the solution side of the double layer (Bockris and Reddy, 1970).

Thus, from Eq. 3.24, one has Eq. 3.27:

$$\frac{d\varphi_x}{dx} \approx -\left(\frac{32n^0\pi kT}{D}\right)^{\frac{1}{2}} \frac{ze_0\varphi_x}{2kT}$$

$$\frac{d\varphi_x}{dx} \approx -\left(\frac{8\pi n^0(ze_0)^2}{DkT}\right)^{\frac{1}{2}} \varphi_x$$

$$(3.27)$$

The term within the parenthesis is the familiar $\aleph^2$ of the Debye-Hückel theory, where it was shown that $\aleph^{-1}$ can be considered the effective thickness of the ionic cloud. Assuming the constant within the parenthesis to be represented by the term $\aleph$, Eq. 3.27 becomes

$$\frac{d\varphi_x}{dx} = -\aleph\varphi_x$$

$$(3.28)$$

By integration,

$$ln\ \varphi_x = -\aleph x + \text{Constant} \tag{3.29}$$

To evaluate the constant, the following boundary condition is used:

At $x \rightarrow 0$, $\varphi_x \rightarrow \varphi_0$. It follows, therefore, that

Constant $= ln\ \varphi_0$

Or, $ln\ (\varphi_x/\varphi_0) = -\aleph x$

Or, $\varphi_x = \varphi_0 \exp(-\aleph x)$ (3.30)

$$\text{Where,}\quad \aleph = \left(\frac{8\pi n^0(ze_0)^2}{DkT}\right)^{\frac{1}{2}}$$

$$(3.30a)$$

i.e., the potential decays exponentially into the solution; deep enough inside the outer extent of the double layer close to the bulk solution ($x \rightarrow \infty$), the potential becomes zero. Further, as the solution concentration $n°$ and/or the valency (z) of the counter ions in the diffuse layer increases, $\aleph$ increases (*vide*. Eqs.3.27 and 3.30a), and φ_x falls more and more sharply. In other words, the diffuse double layer is compressed, leading to reduction of the mutual repulsion between the approaching colloidal particles, swarmed by the respective double layers of like charges, and thereby the flocculation of the soil colloids. The latter is facilitated by the addition of electrolytes containing higher-valent ion. Such potential-distance relation (Figure 3.5) is an important and simple result from the Goüy-Chapman model (Shaw, 1970; Bockris and Reddy, 1970).

In the Debye-Hückel ionic-cloud model, it was found that the electrical effect of the cloud on the central ion could be simulated by placing the entire charge of the cloud, $-z_i e_0$, at the distance $\aleph^{-1}$ from the central ion. One wonders, therefore, if

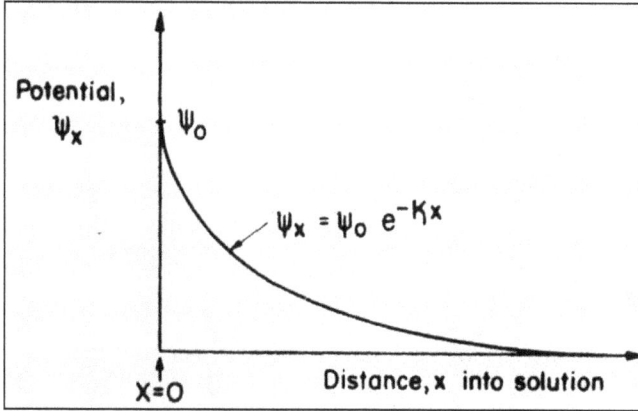

Figure 3.5: The Distance Variation of the Potential in the Diffuse-Charge Region.

the electrical effect of the diffuse-charge region could be simulated by placing the entire Goüy-Chapman charge q_d on a plane parallel to the charged surface, and a distance $\aleph^{-1}$ from it. If this is done, one has in effect a parallel-plate condenser situation, *i.e.*, a charge of $-q_d = q_M$ at the $x = 0$ (charged) surface and the diffuse charge q_d at the $x = \aleph^{-1}$ plate.

The potential difference across a parallel-plate condenser (V) is given by

$$\Delta V = \frac{4\pi}{D} q \delta$$

(3.31)

Where q is the charge on the plates, and δ, their distance apart. By the following substitutions [Eqs. (3.26) and (3.30a)] in Eq. (3.31), namely

$$q_M = -q_d = 2 \left(\frac{Dn^0 kT}{2\pi} \right)^{\frac{1}{2}} \frac{ze_0 \varphi_0}{2kT} \text{ and } \aleph = \left(\frac{8\pi n^0 (ze_0)^2}{DkT} \right)^{\frac{1}{2}}$$

One obtains from Eq. 3.31

$$\Delta V = \varphi_{x=0} - \varphi_{x=\infty} = \varphi_0 - 0 = \varphi_0$$

(3.32)

Which means that the charged surface at $x = 0$ is at a potential φ_0 relative to the $x = \aleph^{-1}$ plate (defining the outer extent or the thickness of the diffuse double layer) at a potential equal to zero. This is indeed the situation (Bockris and Reddy, 1970).

Given such dependence of the diffuse double layer thickness on the concentration and the valency of the counter ions, the same, denoted by as $\aleph^{-1}$ or δ, may also be calculated by Eq. 3.33:

$$\delta = \aleph^{-1} = \left(\frac{\varepsilon \, kT}{2n_0 z^2 e_0^2} \right)^{\frac{1}{2}}$$

(3.33)

where,

$\varepsilon =$ static permittivity of the medium ($\varepsilon = D/4\pi$)

$k =$ Boltzmann constant

$T =$ Absolute temperature

$e_0 =$ Electronic charge

$n^0 =$ Number of ions per m^3 in the equilibrium soil solution, and

$z =$ Numerical value of the valency of the counter ion (cation in case of negatively charged soil colloid surface).

The static permittivity of aqueous medium, ε, is equal to $\varepsilon_0 D$, with ε_0 being the permittivity in vacuum (8.85×10^{-12} C^2. J^{-1}. m^{-1} or F. m^{-1}) and D is the dielectric constant of the medium (78.5 at 298 K for water).

Hence, on substituting the values of ε (= 694.7×10^{-12} C^2. J^{-1}. m^{-1}), k (=1.38×10^{-23} J K^{-1}), T (= 298 K), e_0 (=1.602×10^{-19} C), and n^0 (= $C_0 \times 10^3 \times 6.02 \times 10^{23}$ ions. m^{-3}, where C_0 is the concentration of the equilibrium solution in mol L^{-1}) in Eq.(3.33) at 25 °C (*i.e.*, 298 K), one obtains Equation (3.34):

$$\delta = \frac{3.0 \times 10^{-10}}{z}\left(\frac{1}{C_0}\right)^{\frac{1}{2}} m$$

(3.34)

$$\text{Or, } \delta \approx \frac{0.3}{z\left(C_0\right)^{\frac{1}{2}}} nm$$

(3.35)

For a 1:1 electrolyte (*i.e.*, z =1), and C_0 = 0.1 mol L^{-1}, δ is about 1nm, while at C_0 = 0.001 mol L^{-1}, δ is about 10 nm (Shaw, 1970; Bockris and Reddy, 1970).

3.2.3.2. Factors Affecting the Thickness of Diffuse Double Layer (δ)

It is clear from Eqs. 3.33 and 3.35 that δ depends upon the total electrolyte concentration, the valence of the counter ion, dielectric constant of the medium and temperature (Sanyal *et al.*, 2009).

(i) *Total Electrolyte Concentration* – The higher the total electrolyte concentration, the smaller the thickness of the diffuse double layer (as shown above by substitution of different soil solution concentration values in Eq. 3.35).

(ii) *Valence of the Counter Ion* – The higher the valence of the counter ion, the smaller is the thickness of the diffuse double layer. At a fixed electrolyte concentration, the thickness of the diffuse double layer for divalent ions would be half that for monovalent ions as one would appreciate from Eq. 3.35 (*i.e.*, by substituting z = 1 or 2.).

(iii) *Dielectric Constant of the Medium* – An increase in the dielectric constant of a medium increases the thickness of the diffuse double layer. An increase in temperature decreases the dielectric constant of medium (*e.g.* water of the aqueous soil solution), causing a fall in the thickness of the diffuse double layer due, obviously, to a higher kinetic energy of the counter ions. These inferences were drawn earlier in a qualitative manner (Section 2.6.1) (Poonia, 2002; Sanyal *et al.*, 2009).

3.2.4. Donnan Membrane Equilibrium

If an aqueous solution of an electrolyte consisting of two diffusible ions (a cation and an anion), separated by a membrane (M) from another solution containing a salt with a non-diffusible ion, then at equilibrium (when the flow across the membrane stops at a given temperature), the distribution of the diffusible ions will be unequal on the two sides of the membrane, even though the ionic activity (or concentration in a dilute system) product of equivalent amounts of the oppositely charged ions or ratio of equivalent amounts of the like-charged ions on the two sides of the membrane will be equal. This is known as the Donnan Membrane Equilibrium, while such a membrane of selective permeability is known as a Donnan membrane.

However, if an aqueous solution of NaCl of concentration C_1 is separated from another such solution of concentration C_2 ($C_1 > C_2$) by an ordinary ("open") membrane of equal permeability to Na^+ and Cl^- ions, there will be a flow from the first solution [say, the left-hand side (L) solution with respect to the membrane (M)] to the second one [say, the right-hand side solution (R)] till the NaCl concentrations on both sides of M become equal and an equilibrium is reached. Under such circumstances, at equilibrium

$$(C_{Na}+)_L = (C_{Na}+)_R \qquad \text{3.36 (a)}$$

$$(C_{Cl^-})_L = (C_{Cl^-})_R \qquad \text{3.36 (b)}$$

Obviously then, $\quad (C_{Na}+)_L \, (C_{Cl^-})_L = (C_{Na}+)_R \, (C_{Cl^-})_R \qquad$ 3.36 (c)

However, if the left-hand side solution (L) with respect to the membrane (M) has a solution (of, say, NaCl) with a non-diffusible ion through the given membrane, *i.e.*, a Donnan membrane, *e.g.*, Na^+R^-, where R^- is a large (non-diffusible) resin anion (Figure 3.6), then diffusion of Na^+ and Cl^- ions will take place from L to R across M (Figure 3.6) until an equilibrium (no flow) is attained.

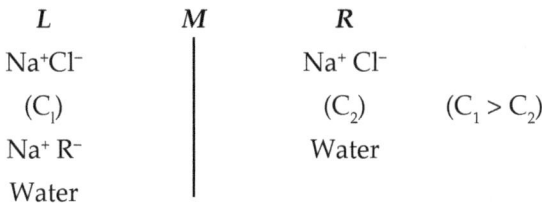

L	M	R	
Na^+Cl^-		$Na^+ Cl^-$	
(C_1)		(C_2)	$(C_1 > C_2)$
$Na^+ R^-$		Water	
Water			

Figure 3.6: The Donnan Membrane System.

At equilibrium, for electrical neutrality on each side of the membrane (M),

$$(C_{Na}+)_R = (C_{Cl^-})_R \qquad (3.37)$$

And, $\qquad (C_{Na^+})_L = (C_{Cl^-})_L + (C_{R^-})_L$ (3.38)

Where the concentration terms refer to those at equilibrium.

According to the Donnan membrane Equilibrium Theory, at equilibrium one obtains

$$(C_{Na^+})_L (C_{Cl^-})_L = (C_{Na^+})_R (C_{Cl^-})_R$$ (3.39)

Substitution of $(C_{Na^+})_R$ in Eq. 3.39 from Eq. 3.37 and $(C_{Na^+})_L$ from Eq. 3.38, one obtains from Eq. 3.39:

$$[(C_{Cl^-})_L + (C_{R^-})_L] (C_{Cl^-})_L = (C_{Cl^-})^2_R$$

Or, $\qquad (C_{Cl^-})^2_L + (C_{Cl^-})_L (C_{R^-})_L = (C_{Cl^-})^2_R$ (3.40)

That is, $\qquad (C_{Cl^-})_L \neq (C_{Cl^-})_R$ (3.41)

And, in the same manner, $\qquad (C_{Na^+})_L \neq (C_{Na^+})_R$ (3.42)

Considering a soil colloidal system, the solid (charged) surface is swarmed by the oppositely charged counter ions on the solution side of the electrical double layer such that the counter ion concentration falls off as the distance from the solid surface increases in the diffuse double layer, while the concentration of the ions of the like charges (as on the soil solid or exchange surface) increases. Thus, there will be a region towards the terminal end of the solution side of the diffuse double layer (close to the bulk soil solution of even distribution of cations and anions) which will be dominated by the anions (for a soil colloid of *net* negative change). This region, intervening between the soil colloidal solid (*i.e.*, exchange) phase and the soil solution phase, will act as if it were a Donnan membrane of selective permeability. Thus, considering the distribution of, say, K^+, Ca^{2+} and Cl^- ions between the soil colloidal (exchange) phase and the solution phase in such a Donnan system, one would have at equilibrium (from Eq. 3.39).

$$(C_{K^+})_{exch} (C_{Cl^-})_{exch} = (C_{K^+})_{soln} \cdot (C_{Cl^-})_{soln}$$ (3.43)

and $(C_{Ca^{2+}})_{exch} (C_{Cl^-})^2_{exch} = (C_{Ca^{2+}})_{soln} \cdot (C_{Cl^-})^2_{soln}$ (3.44)

By combining Eqs. 3.43 and 3.44, one obtains,

$$\frac{(C_{K^+})_{exch}}{(C_{K^+})_{soln}} = \frac{(C_{Cl^-})_{exch}}{(C_{Cl^-})_{soln}} = \sqrt{\frac{(C_{Ca^{2+}})_{exch}}{(C_{Ca^{2+}})_{soln}}}$$

That is, one obtains the basic condition of Donnan Membrane equilibrium as applied to a soil system, dominated by the K^+ and Ca^{2+} ions in the exchange phase and equilibrium soil solution as follows:

$$\frac{C_{K^+ exch}}{\sqrt{C_{Ca^{2+} exch}}} = \frac{C_{K^+ soln}}{\sqrt{C_{Ca^{2+} soln}}} \quad or, \quad \left(\frac{C_{K^+}}{\sqrt{C_{Ca^{2+}}}}\right)_{exch} = \left(\frac{C_{K^+}}{\sqrt{C_{Ca^{2+}}}}\right)_{soln}$$

(3.44a)

This is the essential conclusion, in quantitative terms, in regard to ionic distributions between the exchange and the solution phases at equilibrium in soil, behaving like a Donnan system for reasons explained earlier (Sanyal *et al.*, 2009).

In order that the above noted equilibrium between the exchange and solution phase is maintained, any dilution of the solution phase, for instance, would, therefore, favour release of K^+ ion (monovalent) from the exchange phase to the diluted soil solution phase, and *vice versa* for the divalent Ca^{2+} ion. An increase in concentration of the soil solution will have a reverse effect on the equilibrium distribution of monovalent and divalent cations between the exchange (or adsorbed) and the soil solution phases.

3.2.5. Cation Exchange Equations Based on Law of Mass Action

The formulations expressing the relationships between the cations in the solution and adsorbed phases of soil at equilibrium are termed as cation exchange equations. While describing some empirical relationships, it is assumed that (a) all the exchange sites are freely accessible to solution cations such that cation exchange takes place instantaneously, (b) the surface charge (*i.e.*, cation exchange capacity) remains constant over the entire cationic composition of the exchange complex, (c) the adsorbed phase is homogeneous (*i.e.*, the surface charge is distributed uniformly), (d) the anions are effectively excluded from the monolayer of cations, and finally (e) the surface does not form any chelate with cations. The surfaces of different layer-lattice silicate minerals with *net* negative charge satisfy the above stated conditions only to various degrees (Sanyal *et al.*, 2009).

Such cation exchange formulations may involve homovalent ion exchange or else, heterovalent ion exchange systems.

For homovalent cation exchange equilibria, one has

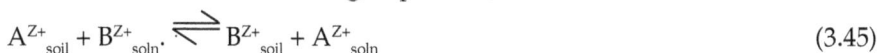

$$A^{Z+}_{soil} + B^{Z+}_{soln}. \rightleftharpoons B^{Z+}_{soil} + A^{Z+}_{soln} \tag{3.45}$$

For instance, one may consider the Na^+- K^+ ion exchange system in soil. Thus,

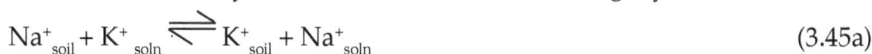

$$Na^+_{soil} + K^+_{soln} \rightleftharpoons K^+_{soil} + Na^+_{soln} \tag{3.45a}$$

However, here K^+_{soil} on the product side does not imply that the entire amount of exchangeable Na^+ ion from the homoionic soil exchanger (rendered initially homoionic with Na^+ ions) will be replaced by K^+ ion at equilibrium so that the exchanger will become K^+ ion-saturated at equilibrium. The exchanger at equilibrium will have both Na^+ ion (the residual Na^+ ion) and K^+ ion (the substituting K^+ ion) as the exchangeable cations, but their proportion will vary, depending primarily on the initial concentration of the bathing electrolyte (*e.g.*, KCl in this case).

The Law of Mass Action leads to the equilibrium constant as:

$$K_m = \frac{[B^{Z+}_{soil}](A^{Z+}_{soln})}{[A^{Z+}_{soil}](B^{Z+}_{soln})} \tag{3.46}$$

Where [ ] denotes the activity of the exchangeable ion concerned on the absorbed phase (*e.g.* soil), while ( ) denotes that in the solution phase at equilibrium.

These equations were derived by Kerr (1928) and Vanselow (1932) who made an assumption that the solid (exchange) phase is an ideal solid solution so that the activities of the adsorbed ions may be approximated by the corresponding concentrations (*i.e.*, the activity coefficients of the adsorbed species in Eq. 3.46 are unity), *e.g.*, the respective mole fractions on the exchange phase at equilibrium. The activity coefficients of the ions in the soil solution phase may be evaluated from the ionic strength of the solution phase, following the Debye-Hückel formulations (Eqs. 3.4 and 3.5) as explained in Section 3.1.2. For this reason, the Kerr-Vanselow formulations (Eq. 3.46) do *not* lend themselves to the evaluation of the corresponding thermodynamic equilibrium constants, and K_m in Eq. 3.46 is termed as the "Selectivity Constant or Coefficient or Quotient". These formulations are mostly applicable to the homovalent ion exchange systems (Sanyal *et al.*, 2009).

Gapon (1933) also considered the hypothesis of ideal solid solution as above, but dealt with the heterovalent systems which are of more practical value than the homovalent ion exchange systems. Gapon preferred to use the ionic concentrations on the adsorbed (exchange) phase on equivalent basis for taking care of the corresponding ionic activities (in practice, the ionic concentrations for reasons stated above).

For a heterovalent ion exchange equilibrium, namely

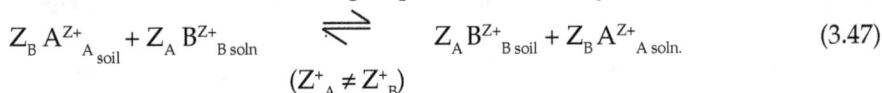

$$Z_B A^{Z+}_{A\ soil} + Z_A B^{Z+}_{B\ soln} \rightleftharpoons Z_A B^{Z+}_{B\ soil} + Z_B A^{Z+}_{A\ soln.} \tag{3.47}$$

$$(Z^+_A \neq Z^+_B)$$

Accordingly, the Kerr-Vanselow formulation (Eq. 3.46) should be modified to

$$K_m = \frac{[B^{Z+}_{B\ soil}]^{Z_A} (A^{Z+}_{A\ soln})^{Z_B}}{[A^{Z+}_{A\ soil}]^{Z_B} (B^{Z+}_{B\ soln})^{Z_A}} \tag{3.48}$$

Where [ ] and ( ) have the same significance as before.

According to Gapon (1933), Eq. 3.48 reads as

$$K_m \ (or\ K_G) = \frac{[CEC - A^{Z+}_{A\ soil}] (A^{Z+}_{A\ soln})^{1/Z_A}}{[A^{Z+}_{A\ soil}] (B^{Z+}_{B\ soln})^{1/Z_B}} \tag{3.49}$$

Where CEC is the cation exchange capacity of the soil.

An example will help to understand the transformation of Eq. 3.48 to Eq. 3.49. Thus,

$$2 Na^+_{soil} + Ca^{2+}_{soln} \rightleftharpoons Ca^{2+}_{soil} + 2 Na^+_{soln.} \tag{3.50}$$

By dividing both sides of Eq. 3.50 by 2, one obtains

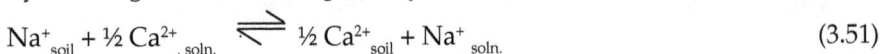

$$Na^+_{soil} + \tfrac{1}{2} Ca^{2+}_{soln.} \rightleftharpoons \tfrac{1}{2} Ca^{2+}_{soil} + Na^+_{soln.} \tag{3.51}$$

In case one starts with a heteroionic soil in respect of Na^+ ion as in Eq. 3.51, then at equilibrium, the soil exchange phase will have only Na^+ and Ca^{2+} ion as exchangeable ions. If the CEC is the cation exchange capacity of the given soil, while the degree of saturation of soil (exchange) phase at equilibrium with respect to Na^+ ion is given as Na^+_{soil}, then obviously, the Ca^{2+} ion saturation of the CEC at equilibrium will be given as

$$Ca^{2+}_{soil} = CEC - Na^+_{soil}$$

provided that the CEC and the degrees of saturation with respect to the exchangeable cations at equilibrium are expressed on equivalent basis. Substitution of these facts in Eq. 3.51 in respect of the exchange equilibrium, as given in Eq. 3.49, would obviously lead to the following form of K_m:

$$K_m \text{ (or } K_G) = \frac{[Ca^{2+}_{soil}]^{1/2}(Na^+_{soln})}{[Na^+_{soil}](Ca^{2+}_{soln})^{1/2}}$$

Or,

$$K_m \text{ (or } K_G) = \frac{[CEC - Na^+_{soil}](Na^+_{soln})}{[Na^+_{soil}](Ca^{2+}_{soln})^{1/2}}$$

(3.52)

A large number of experimental data for homovalent and heterovalent cation exchange equilibria studies have been fitted to the Kerr-Vanselow and the Gapon equations to obtain the corresponding selectivity coefficients (K_m). These values of K_m for, say, a given homovalent ion exchange system (*e.g.* Na^+-K^+, Na^+-Rb^+, Na^+-Cs^+, Na^+-NH_4^+ exchange equilibria), obtained by equilibrating a given amount of a homovalent soil clay (*e.g.* Na^+-clay) with varying initial concentrations of a replacing cation (*e.g.* K^+/Rb^+/Cs^+/NH_4^+), when plotted against the corresponding position of equilibrium (*e.g.* the ionic composition of the solid phase at equilibrium), demonstrate wide variations as shown in Figure 3.7.

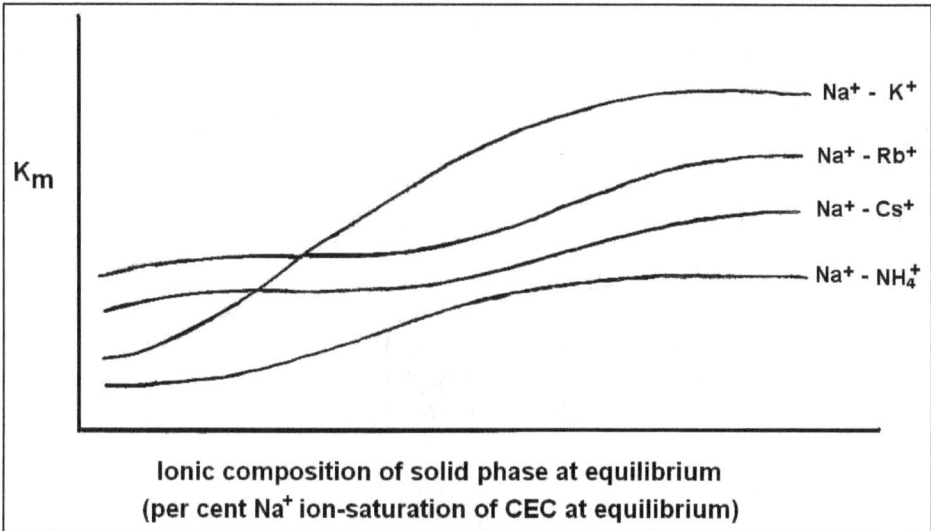

Figure 3.7: Plot of Selectivity Coefficient against Position of Equilibrium in a Homovalent Cation Exchange Equilibria.
Source: Talibudeen (1972).

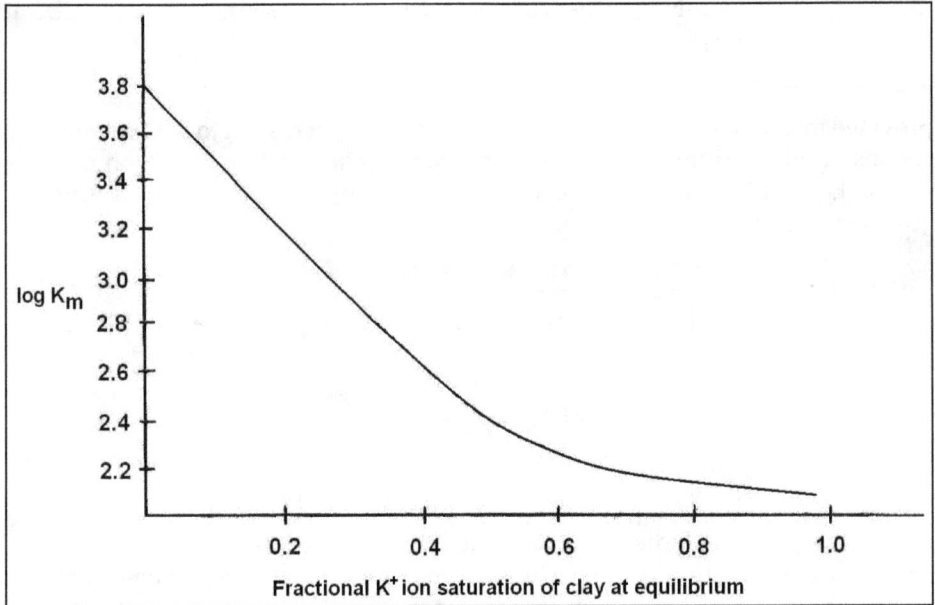

Figure 3.8: Plot of Selectivity Coefficient (log scale) against the Position of Equilibrium in Heterovalent Cation Exchange Equilibria.
Source: Talibudeen (1972)

If K_m were a thermodynamic equilibrium constant, its values should have remained constant with the position of equilibrium, unlike what is apparent from Figure 3.7. When one examines a heterovalent ion exchange equilibrium, such variations of K_m are even wider, as evident from an examination of the experimental data on K_m for K^+ - Ca^{2+} ion exchange (Figure 3.8).

Such variations arise from the assumption of ideal solid solution behaviour of the soil exchanger, thereby neglecting the activity coefficient of the adsorbed ions on the solid (exchange) phase, in the Kerr-Vanselow and the Gapon formulations. This is addressed by the thermodynamic theory of ion exchange which will follow.

3.2.5.1. Thermodynamic Theory (due to Gains and Thomas, 1953)

Let the concept of fugacity of a component (*e.g.* the solute) in a solution be introduced, in brief, before the above mentioned thermodynamic theory is discussed. The fugacity refers to the "escaping tendency" for instance, of a gas in a mixture. The fugacity (ï) of a solute in a binary solution is related to its concentration (c) by the equation,

ï/c = 1 for all values of 'c' in an ideal solution,

while $\lim_{c \to 0}$ ï/c = 1 for a non-ideal solution.

The activity (a) of a solution in a binary solution is related to its fugacity (ï) as,

a = ï/ï⁰

Where $\ddot{\imath}^0$ is the fugacity in the standard state (of unit activity) of the solute. Obviously in the standard state $(a = 1)$, $\ddot{\imath} = \ddot{\imath}^0$.

The standard state of a solute (in which the solute has unit activity) in a binary solution is defined as that hypothetical solution which behaves ideally at a concentration (c) of 1 molar, 1 molal or 1 mole fraction, depending on the scale of concentration used.

However, such concentration (1 molar, 1 molal, etc.) denotes too concentrated a solution to behave ideally as required in the above stated definition of the standard state. Hence, the latter has got *no physical existence*, and it is *hypothetical*. In practice, ideal behaviour may be expected only in dilute solutions of concentration 10^{-4}–10^{-5} M, and at further dilution.

For the above noted standard state of the solute, $a = \ddot{\imath}/\ddot{\imath}_0 = 1$

And the solution being ideal, $\ddot{\imath} = c = 1$ M (say). Hence, $\ddot{\imath} = \ddot{\imath}^0 = 1M = c^0$ (say)

Where c^0 is the standard state concentration of the solute, namely 1 M (say) in this case (Sanyal *et al.*, 2009).

Gains and Thomas' Theory of Cation Exchange in Soil

This theory refers to the general ion exchange equilibrium as given below (Eq. 3.47):

$$Z_B A^{Z+}{}_{A \text{ soil}} + Z_A B^{Z+}{}_{B \text{ soln}} \underset{(Z^+_A \neq Z^+_B)}{\rightleftharpoons} Z_A B^{Z+}{}_{B \text{ soil}} + Z_B A^{Z+}{}_{A \text{ soln.}} \tag{3.47}$$

The selectivity coefficient is given, following earlier arguments, by (Sanyal *et al.*, 2009)

$$K_m \text{ (or } K_C) = \frac{(q_B/q_O)^{Z_A} (C_A/C_O)^{Z_B}}{(q_A/q_O)^{Z_B} (C_B/C_O)^{Z_A}} \cdot \frac{\gamma_A^{Z_B}}{\gamma_B^{Z_A}} \cdot C_O^{(Z_B - Z_A)}$$

where,

q_O = Total number of equivalent negative sites on the exchange phase.

q_A = Equivalent negative sites occupied by A^+ ions on exchange phase at equilibrium.

q_A/q_O = Mole fraction on exchange phase at equilibrium as occupied by A^+ ions.

q_B = Equivalent negative sites occupied by B^+ ions on exchange phase at equilibrium.

q_B/q_O = Mole fraction on exchange phase at equilibrium as occupied by B^+ ions.

γ_A = Activity coefficient of A^+ ions in solution phase at equilibrium.

γ_B = Activity coefficient of B^+ ions in solution phase at equilibrium.

C_O = Total equivalent amount of positive sites in the exchange site (in the diffuse layer). It is equal to q_O, which is in the solid phase ($q_O = C_O$).

C_A = Equivalent number of cationic sites in the solution at equilibrium occupied by A^+ ion.

C_A/C_O = Mole fraction of A^+ ions in the solution phase at equilibrium.

C_B/C_O = Mole fraction of B^+ ions in the solution phase at equilibrium.

The selectivity quotient (K_m or K_c) is **not** a thermodynamic equilibrium constant in that the activity coefficient of the adsorbed ions in the exchange phase at equilibrium is ignored (is set equal to unity). The Gains and Thomas' theory then proceeds to address this issue based on the following two assumptions (Gaines and Thomas, 1953):

(1) The adsorbed ionic layer is treated as a solid solution (**not** ideal) so that the cationic concentrations are obtained from the fractions of the total ion 'sites' on the exchange phase, occupied by cations, A^+ and B^+ ions (q_A/q_0 and q_B/q_0), which are identical to the parameter, namely fractional saturation of the CEC.

(2) The standard state of unit activity of an adsorbed cation is taken to be that state when all the ionic sites on the exchange phase are fully saturated by the cation concerned and the system behaves ideally so that $a_+ = C_+ = 1$. That is, the activity coefficient, $f_+ = 1$. Hence, the bonding energy parameter in the standard state, namely, $- \log f_+$, is given as (Section 3.1.2, Eqs. 3.3 and 3.4)

$$(B.E.)_+ \, \alpha - \log f_+ = 0 \tag{3.54}$$

Thus, all the cations in their respective standard state will exchange with equal ease. Referring to Eq. 3.47, in the standard state for A^+ ion, one has,

$$a_{A+} = f_{A^+} \cdot \frac{q_A}{q_0} = 1$$

But in this state, according to assumption (2), $q_A = q_0$. Hence, $f_A+ = 1$

Thus, the value of the activity coefficient of the adsorbed A^+ ion, in its standard state of unit activity (*i.e.*, f_A+), turns out to be unity, as shown above. Thus, the above relationships, based on the assumption (2) of the Gains and Thomas' theory, are self-consistent, and are also consistent with the definitions of fugacity, activity, activity coefficient and the standard state (Sanyal *et al.*, 2009).

In general, the thermodynamic equilibrium constant (K) for the given ion exchange equilibrium (Eq. 3.47) is related to K_m (or K_c) by the relationship:

$$K = K_C \, (f_B^{Z_A} / f_A^{Z_B}) \tag{3.55}$$

Where f_A and f_B are the activity coefficient of the adsorbed cations, A^+ and B^+, on the exchange phase at equilibrium.

Because the activity coefficient of an ion varies with the ionic strength (I) of the medium [Eqs. 3.4 and 3.5 of Section 3.1.2) for a heterovalent exchange of, say, K^+-Ca^{2+} ion exchange discussed earlier, the ionic strength of the soil solution, I $\left(= \frac{1}{2} \sum_i m_i Z_i^2 ; Eq. 3.1 \right)$, will obviously change with the position of equilibrium, *i.e.*, with more of K$^+$ ion, initially present on the exchange phase, being exchanged at equilibria for Ca$^+$ ions (with the initial concentration of CaCl$_2$ solution used for leaching being higher), the value of I on the exchange phase at equilibrium will increase, and hence the activity coefficient of both K$^+$ and Ca^{2+} ions on the exchange phase will be different compared to those obtained at a higher fractional saturation (at equilibrium) of the soil clay with K$^+$ ion.

This will render K$_C$ (or K$_m$) change for f$_A$ and f$_B$ will change in order that the thermodynamic equilibrium constant (K) remains unaltered (Eq. 3.55). This explains the variations of K$_C$ (or K$_m$) with the position of equilibrium in a heterovalent ion exchange reaction (Figure 3.8).

Even for the homovalent cation exchange (Figure 3.7), the activity coefficients of the monovalent cations involved need not be the same if one examines the Debye-Hückel extended Law (Eq. 3.3), giving the activity coefficient of an ion. Further, the relative accessibility on the exchanger (*e.g.* the clay) of the original and the replacing cations (in a homovalent exchange) need not be the same for a series of replacing cations, depending on the preferential and specific adsorption. Hence the ionic strength on the exchange (solid) phase will vary with the position of equilibrium for each pair of exchanging cations, so also the corresponding ionic activity coefficients, and hence the value of K$_m$ (Figure 3.7). However, the variations in K$_m$ were much more for the heterovalent ion exchange compared to the homovalent cases (Figures 3.7 and 3.8) (Sanyal *et al.*, 2009). Such variations may also be traced to the heterogeneity of the exchange sites in the adsorbed phase in soil. These exchange sites have been considered as homogeneous with respect to their specificity for a given cation in the derivation of the exchange equations. These exchange coefficients vary, though in different proportions, with the cationic saturation of the exchange complex (Sanyal *et al.*, 2009).

3.2.6. Clay-Membrane Electrode and Ionic Activity

A direct method for ionic activity measurement in suspension was developed by Marshall and his co-workers by way of using clay-membrane electrodes for determining the activities of Na$^+$, K$^+$, M$^+$, Mg^{2+}, Ca^{2+} and Ba^{2+} ions with a saturated calomel electrode as a reference electrode. The clay membranes are reversible either to cations, in general, or to monovalent cations only. Because of lack of specificity, the measurement has been limited to homoionic systems and to systems containing monovalent and divalent cations. Marshall (1949, 1964) and his coworkers determined the ionic activity as a function of the degree of saturation (by using the titration curves of H-clay, H-Ca and H-Mg clays), the degree of ionization of exchangeable ions, and the mean free bonding energy (ΔG_M). As per this approach, the bonding energy for a monovalent ion, in particular, a monovalent cation (*e.g.* M$^+$) is obtained from the formula,

$$\Delta G_M = RT \ln \frac{C_M}{a_M}$$

$$(3.56)$$

where,

C$_M$ = Concentration of total monovalent cation (sun of ionized and non-ionized portions)

a_M = Measured activity of the given monovalent cation (MV)

Using the relationship introduced earlier (Eq. 3.2),

$$a_M = f_M C_M \tag{3.57}$$

Where f_M is the activity coefficient of the given MV (M^+), one obtains from Eq. 3.56,

$$\Delta G_M = RT \ln \frac{C_M}{f_M C_M} = -RT \ln f_M = RT \ln f_M^{-1} \tag{3.58}$$

Marshall termed f_M as "fraction active" in that f_M ($= a_M/C_M$) is the fraction, accessible to the ionic activity measurement by the given clay membrane electrode, is capable of participating in the ion exchange equilibrium.

One may wish to examine Eqs. 3.56 and 3.58 little more closely.

Thus, taking a clue from Marshall's idea that the exchangeable ions are only partly ionized, one would envisage the bonding process (and the reverse) of the MV (M^+) on an exchanger, S^- (_e.g._ soil colloid) as follows (Sanyal _et al._, 2009):

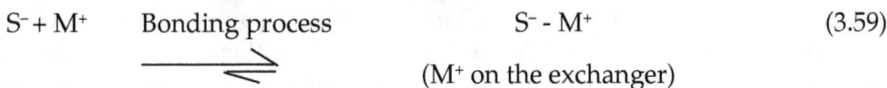

$$S^- + M^+ \quad \text{Bonding process} \qquad\qquad S^- - M^+ \tag{3.59}$$
$$\rightleftharpoons \qquad\qquad (M^+ \text{ on the exchanger})$$

The reverse of the bonding process is given by,

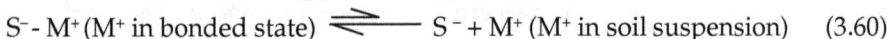

$$S^- - M^+ (M^+ \text{ in bonded state}) \rightleftharpoons S^- + M^+ (M^+ \text{ in soil suspension}) \tag{3.60}$$

Therefore,

$$\Delta G_M(\text{bonding}) = RT \ln \frac{a_{M^+} (\text{bonded})}{a_{M^+} (\text{free in soil suspension})} \tag{3.61}$$

Assuming that the bonded MV (M^+ ion) is only partially ionized in the soil suspension, which is measured by the given clay membrane electrode (reversible to M^+ ions), one has

a_M^+ (free in soil suspension) = a_M (introduced earlier)

and a_M^+ (bonded) $\approx$ Total concentration of MV (M^+) with the assumption that a_M is very small compared to total M^+ concentration = C_M

Hence, one has from Eq. 3.61,

$$\Delta G_M(\text{bonding}) = RT \ln \frac{C_M}{a_M} = RT \ln f_M^{-1} = -RT \ln f_M \tag{3.62}$$

Eqs. 3.58 and 3.62 are identical.

Indeed, it was shown qualitatively earlier (Section 3.1.2), that $- \log \gamma_i$ (where γ_i was the activity coefficient of a counter ion, 'i') provides a measure of the bonding energy of the given exchangeable ion ('i').

The same conclusion follows here (_i.e._, Eq. 3.62) from an altogether different and a rigorous thermodynamic approach (Sanyal _et al._, 2009).

3.2.7. Krishnamoorthy-Davis-Overstreet Equation

Krishnamoorthy *et al.* (1948) introduced another approach, which is based on statistical thermodynamics. For mono-divalent exchange with $Z_A = 1$ and $Z_B = 2$, the general form of the Krishnamoorthy-Davis-Overstreet equation becomes as given by Eq.3.63:

$$K_{KDO} = \frac{a_B}{a^2{}_A} \cdot \frac{m^2{}_A}{m_B[\{m_A + 2 - (2/Y)\}m_B]}$$

(3.63)

Where, m_A and m_B are the number of their moles in the adsorbed phase, respectively. Eq. 3.63 can also be written as Eq. 3.64:

$$K_{KDO} = \frac{a_B}{a^2{}_A} \cdot \frac{M^2{}_A}{M_B\{M_A + 2 - (2/Y)M_B\}}$$

(3.64)

Where M_A and M_B denote the mole fractions of cations A and B in the adsorbed phase, respectively (Poonia, 2002).

3.3. Schofield's Ratio Law

This law states that if a negatively charged surface is in equilibrium with a solution containing certain activities (or concentrations) of monovalent, divalent and trivalent cations, the equilibrium will not be disturbed if the activities of the monovalent cations in the solution phase are altered by a certain factor, of the divalent cations by the square of that factor and of the trivalent cations by the cube of that factor. In other words, in a binary (two cations) system, the ratio of the cations held by a soil depends upon the ratio of the activities of the two cations in the equilibrium solution, in the case of homovalent cations (Na^+ and K^+ or Ca^{2+} and Mg^{2+}); the ratio of the activity of monovalent and square root of the activity of divalent cations, in the case of mono-divalent cations (Na^+ and Ca^{2+} or Na^+ and Mg^{2+}); and the ratio of the activity of monovalent and cube root of the activity of trivalent cations, in the case of mono-trivalent cations (Na^+ and Al^{3+}), and so on. The law thus requires that there is no exchange of cations between the adsorbed and solution phases of the soil so long as a_{Na}/a_K, $a_{Na}/(a_{Ca})^{\frac{1}{2}}$ and $a_{Na}/(a_{Al})^{\frac{1}{3}}$ ratios in soil solution do not alter with changes in the composition, dilution or concentration of the soil solution for $Na^+ - K^+$, $Na^+ - Ca^{2+}$ and $Na^+ - Al^{3+}$ binary systems, respectively.

Indeed, for equilibrium in a binary exchange of cations between an adsorbed (*e.g.*, soil colloidal or exchange) phase and a solution (*e.g.*, soil solution) phase, such as one encounters in Eq. 3.51 (Sanyal *et al.*, 2009), namely,

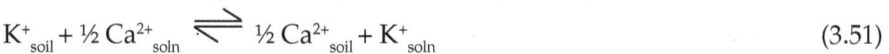

$$K^+{}_{soil} + \tfrac{1}{2}Ca^{2+}{}_{soln} \rightleftharpoons \tfrac{1}{2}Ca^{2+}{}_{soil} + K^+{}_{soln}$$

(3.51)

One would have the corresponding selectivity coefficient (Eq. 3.52)

$$\frac{[Ca^{2+}{}_{soil}]^{\frac{1}{2}} (K^+{}_{soln})}{[K^+{}_{soil}](Ca^{2+}{}_{soln})^{1/2}} = \text{Contant (say, } K_1)$$

(3.65)

Where [ ] denotes the activity (or concentration) in the adsorbed phase, while ( ) that in the soil solution phase. A rearrangement of Eq. 3.65 leads to

$$\frac{[K^+{}_{soil}]}{[Ca^{2+}{}_{soil}]^{1/2}} = \frac{(K^+{}_{soln})}{(Ca^{2+}{}_{soln})^{1/2}} = K = \tfrac{1}{2} \text{ (say)}$$
(3.66)

On irrigation, let us assume that the concentrations in the solution phase decrease 4-fold so that one obtains from Eq. 3.66,

$$\frac{[K^+{}_{soil}]}{[Ca^{2+}{}_{soil}]^{1/2}} = \frac{(K^+{}_{soln}/4)}{(Ca^{2+}{}_{soln}/4)^{1/2}} = \tfrac{1}{2} \times \tfrac{1}{2} = \tfrac{1}{4}$$
(3.67)

Because equilibrium demands that the value of K (= ½) remains constant, as long as the temperature does *not* change, it implies that to keep K = ½, there will be re-distribution of ions and a change of position of equilibrium between the adsorbed and the solution phases. Indeed, there will be release of K^+ ions from the adsorbed phase to the solution phase and fresh adsorption of Ca^{2+} ions present in the solution phase so that the value of K (=$^1/_2$), as in Eq. 3.66, is restored. In other words, when such equilibrium between the above noted two phases is disturbed (*e.g.*, on drying/fertilizer application/plant uptake/irrigation/rainfall, etc.), there will be always a shift of the monovalent cations (MV) from the concentrated to the diluted phase (*e.g.* K^+ ion moving in the above mentioned example from the adsorbed to the diluted solution phase), while the reverse would occur for the divalent cations (DV).

If, however, with the above stated dilution of the solution phase on applying irrigation, appropriate steps had been taken to increase the K^+ concentration in the solution phase (*e.g.* by addition of MOP), in such a way that while Ca^{2+} ion concentration in the solution phase is diluted four times, that of K^+ ions is diluted only two times (*i.e.*, $\sqrt{4}$ times), then Eq. 3.66 would take the form:

$$\frac{[K^+{}_{soil}]}{[Ca^{2+}{}_{soil}]^{1/2}} = \frac{(K^+{}_{soln}/2)}{(Ca^{2+}{}_{soln}/4)^{1/2}} = \frac{(K^+{}_{soln})}{(Ca^{2+}{}_{soln})^{1/2}} = \tfrac{1}{2} \text{ (Constant)}$$
(3.68)

That is the constant of Eq. 3.66 would have remained unaltered and the position of equilibrium between the exchange and the solution phases will not be disturbed, as required by the Ratio Law (Sanyal *et al.*, 2009).

Similar conclusions, as above, may be derived from the Donnan Membrane Equilibrium Theory as applied to ion exchange processes occurring in soil. Indeed, Eq. (3.66) is identical with Eqs. (3.44) and (3.44a) describing the Donnan membrane equilibrium between the exchange and the solution phases of the soil.

Thus an important consequence of the Ratio Law is the applicability of the Donnan Membrane Equilibrium to the description of the distribution of ions between the exchange phase and the soil solution phase at equilibrium with each other. Thus, for a K^+-Ca^{2+} system, as stated earlier (Eq. 3.44a), one has the relationship (Eq. 3.69):

$$\left[\frac{a_{K^+}}{\sqrt{a_{Ca^{2+}}}}\right]_{\text{Exchange or adsorbed phase}} = \left[\frac{a_{K^+}}{\sqrt{a_{Ca^{2+}}}}\right]_{\text{Soil solution phase}}$$
(3.69)

While irrigating a soil with sodic/saline water (brackish water, for instance) having relatively high loading of Na$^+$ ion concentration, compared to the divalent Ca^{2+} and Mg^{2+} ions, the soil solution phase gets concentrated in respect of Na$^+$ ions, which, following the above stated argument, tends to move to the relatively dilute phase (*i.e.*, the exchange phase in this case), while the divalent cations move in the opposite direction. This tends to raise the ESP of the soil, and there is the possibility of dispersion of the exchange phase on continued use of such brackish water. This would cause a poor soil physical tilth.

To counter the above noted alarming trend, pre-treatment of the irrigation source with gypsum (a standard package of practice) is often recommended. This would obviously ensure an increase of Ca^{2+} ion concentration in the irrigation water so as to prevent such rise of ESP of the exchange phase. That is, if the rise in Ca^{2+} ion concentration in the soil solution phase, upon such irrigation with pre-treated brackish water, is such that it is square of the above stated increase in Na$^+$ ion concentration, then according to the Ratio Law, there will be no shift in position of equilibrium between the exchange and the solution phases, and hence the above noted risk of sodification of the exchange phase is avoided (Sanyal *et al.*, 2009).

3.4. Inverse Ratio Law

The Ratio Law considers the disturbance of the position of equilibrium between the exchange and the solution phases due to change of concentrations of the monovalent, divalent ions, etc., in the solution phase due to extraneous factors. However, such position may also be disturbed by the changes in concentrations (or activities) of the ions in the exchange phase. The consequences of the latter are dealt with by the Inverse Ratio Law (due to Matson) (Sanyal *et al.*, 2009).

Thus, given the possibility of occupying the exchangeable positions on the colloid (exchange) phase of a low or high charge-density (*i.e.*, with low or high C.E.C.) clays in a mixture of colloidal dispersions, the monovalent cations are likely to be found in more number on the low charge-density (*i.e.* "dilute" phase in respect of charge density) exchange phase of a colloid (*e.g.* clay), while the reverse will be true for the divalent cations. This ensues from the fact that the ionic fields of the monovalent cations are more compatible with the similar (but oppositely charged) ionic fields of the low-CEC clays, while for the divalent cations, such compatibility results in respect of the ionic fields of high-CEC clays. This bears similarity to what has been stated above in regard to consequences of the applicability of the Ratio Law or the Donnan membrane distribution in case of soil exchange phase - soil solution phase equilibrium (Sanyal *et al.*, 2009).

Thus, when aqueous dispersions of kaolinite and montmorillonite clays are mixed and then allowed to stand to reach a new equilibrium distribution of the corresponding exchangeable ions, the monovalent cations will be found in more numbers on the exchange phase of the kaolinite clay, while the divalent cations will populate mostly the exchange phase of the montmorillonite clay (Sanyal *et al.*, 2009).

3.5. Plant Root Exchange in Soil

Plant roots behave as feeble ion exchangers in soil by virtue of possessing feeble *net* negative charge. Plant roots uptake cations from soil solution in exchange for secretion of equivalent quantity of H^+ ions (and as a result, the rhizosphere pH is generally lower than the bulk soil pH), while the anion uptake occurs in exchange for equivalent amounts of OH^- or HCO_3^- ions. However, because of imbalance in the cation/anion uptake by the plant, the corresponding root system develops a *net* negative charge, and hence an electrical double layer exists at the root-soil solution interfaces (Sanyal *et al.*, 2009).

Among the plants, the monocotyledons (*e.g.* cereals) have a lower root CEC than do the dicotyledon (*e.g.* legumes). Thus, the former root ion exchangers behave more like kaolinite clay, while the dicotyledon behaves like montmorillonite clay. The monocotyledons, therefore, generally exhibit a preference for the monovalent cations over the divalent cation for uptake, while the reverse is true for the dicotyledon (Sanyal *et al.*, 2009).

For instance, if rice and maize (cereals) are grown in the absence of extraneous potash application in a plot of land, marginal in potash supplying capacity, while berseem and rice are grown in an adjacent plot with suitable cultural and management practices (with no potash application), then it is more likely that both the crops of rice and maize would almost simultaneously show potash deficiency symptoms in the first plot, while there will be much delayed appearance of the deficiency symptoms in the crops of the second plot. This obviously arises from the preference of both rice and maize for potash uptake as compared to that of berseem, as explained above (Sanyal *et al.*, 2009).

3.6. Factor Affecting Cation Exchange Capacity

The cation exchange equilibria in soils are affected by two types of factors: (i) soil factors, and (ii) solution factors (Poonia, 2002).

3.6.1. Soil Factors

3.6.1.1. Effect of Clay Minerals

Some of the relevant properties of important clay minerals (Table 3.1) show that CEC and specific surface area (SSA) are in the order: smectite > fine mica > kaolinite. Nevertheless, the trend of the ratio of CEC to specific surface area (*i.e.*, SCD) of these clay minerals is just the reverse. It is important to mention that as per the Diffuse Double Layer (DDL) equation, the main emphasis should be given to SCD rather than to CEC of clay minerals while describing the mono-divalent cation exchange equilibria data (Poonia, 2002).

3.6.1.2. Effect of Organic Matter

Organic matter in a soil affects the cation selectivity in two opposing directions (Poonia, 2002). Closely spaced carboxyl and phenolic groups present in organic compounds cause an increase in SCD of soil and thus selectivity towards multivalent cations. Contrary to this, an increase in organic matter in the soil results in the

Table 3.1: Cation Exchange Capacity (CEC), Surface Charge Density (SCD) and Specific Surface Area (SSA) of some Clay Minerals

Property	Clay Minerals		
	Kaolinite	Fine Mica	Smectite
Cation exchange capacity (CEC) [cmol (p+)kg⁻¹]	3–15	15–40	80–120
Specific surface area (SSA) (m² g⁻¹)	5–20	100–120	700–800
Surface charge density (SCD) (mol. m⁻²)*	6.7×10^{-6}	1.65×10^{-6}	1.25×10^{-6}

* Surface Charge density (SCD) = CEC/SSA

Source: Poonia (2002).

increase of internal: external surface area/exchange sites (Poonia, 2002). This increases the preference of the exchanger for those monovalent cations like K^+ and NH_4^+, which are mainly influenced by the geometry of adsorption sites (Poonia and Niederbudde, 1990).

3.6.2. Solution Factors

3.6.2.1. Effect of Total Electrolyte Concentration (TEC)

The relative preference of soil for a divalent cation over monovalent cation increases with the increase in total electrolyte concentration (TEC) of the equilibrium solution (Poonia *et al.*, 1984).

3.6.2.2. Effect of Type of Cations

Cations of the same valency but of different hydrated size may have considerable difference in their specificity for exchange sites. This is demonstrated by the data presented in Table 3.2 pertaining to Na–(Ca+Mg) exchange equilibria for soils having different physico-chemical properties (Poonia and Raj Pal, 1979; Poonia *et al.*, 1980). The relatively larger values of ΔG_r and smaller values of K_G in Ca- as compared to Mg-dominated system suggest higher preference of the soils for Ca over Mg (Poonia, 2002).

Table 3.2: Effect of Ca : Mg Ratio on ΔG^o_r and K_G for Na–(Ca+Mg) System

Soil	CEC [cmol (p+) kg⁻¹]	O.C. (per cent)	Ca: Mg	ΔG^o_r (kJ. eq⁻¹)	K_G (mol⁻¹/² L¹/²)
1.	13.32	1.37	3:1	0.69	0.512
			1:3	0.60	0.528
2.	20.27	0.60	3:1	1.07	0.477
			1:3	0.87	0.498
3.	11.19	0.27	3:1	1.97	0.387
			1:3	0.96	0.447

Source: Poonia *et al.* (1980).

3.6.2.3. Effect of Type of Anions

The cation exchange equations do not take into account the nature of anions. The formation of more extensive and stronger undissociated ion pairs by divalent anions with divalent cations (*e.g.*, $CaSO_4^0$ and $MgSO_4^0$) as compared to monovalent cations (*e.g.* $NaSO_4^-$) influence mono-divalent cation exchange equilibria in soils (Mehta *et al.*, 1983; Poonia, 2002).

3.7. Negative Adsorption

The interaction between anions and clay minerals, in principle, comprises two different phenomena, *viz.* the bonding of the anions by positive charges, presumably located at the broken edges of the clay, and the repulsion between the anions and the negative charges, mainly on the planar surfaces of the clay particles. Both the electrostatic attraction (between the anions and positively charged sites) and by the chemical (covalent) bonding may contribute to the adsorption of anions in soil. The repulsion of anions by the negatively charged sites leads to a local deficit of anions near a clay particle in comparison to the equilibrium solution. This deficit has been termed as 'negative adsorption'. As almost all the soils found in nature possess *net* negative charge, negative adsorption or 'anion exclusion' is exhibited to a certain extent by all soils (Poonia, 2002).

3.8. Anion Exchange

As has been discussed above, positively charged sites exist on the edges of layer-lattice silicate minerals (1:1 type in particular) and surfaces of oxides and hydroxides (with low specific surface area), mainly under acidic conditions (*i.e.*, at pH below the corresponding ZPC of these variable-charge components in soil (*see* Section 2. 5.2). The total amount of anions held exchangeably by a unit mass of soil, termed as its anion exchange capacity (AEC), is therefore much less than the cation exchange capacity of the soil. Further, contrary to cation exchange, the capacity of soil to adsorb anions increases as pH decreases or acidity increases which increases the pH - dependent positive charge of soil colloids. In addition to pH, the adsorption of anions depends upon the concentration of anions. Higher the concentration of an anion, the greater is its adsorption. As the ratio of cation exchange capacity to anion exchange capacity of layer-lattice silicate mineral increases, the adsorption of anions decreases. Soils containing montmorillonite (where the CEC is mainly due to isomorphous substitution) as the dominant clay mineral generally exhibit low AEC than those dominant in kaolinite and hydrous oxides of Al and Fe. Similarly, acid soils in the tropical and subtropical regions containing hydrous oxides of Al and Fe exhibit much higher AEC than alkaline and calcareous young soils of arid and semiarid regions (Poonia, 2002). Because humic colloids always carry negative charge, they can compete with other anions for the adsorption sites. Also, the selectivity of the positive sites with respect to different anions, as a rule, is much larger than that of the negative sites for cations. On the basis of experimental data, the preference of different anions for the positively charged sites has been found to follow the order (Poonia, 2002):

$$SiO_4^{4-} > PO_4^{3-} \gg SO_4^{2-} > NO_3^- = Cl^-$$

Consequently, SO_4^{2-}, NO_3^- and Cl^- ions are *not* adsorbed even if PO_4^{3-} ions are present at very low concentration in comparison to concentrations of other anions. At low pH values (< 6) and in the absence of PO_4^{3-} from a system, adsorption of SO_4^{2-} and Cl^- may also occur (Poonia, 2002).

So far the adsorption of anions on layer-lattice silicates and oxides has been considered as being due to electrostatic force only. But some anions are bound in natural soils by mechanisms other than electrostatic. These mechanisms pertain to isomorphous displacement or substitution of hydroxyl ions from the lattice of clay minerals or hydrous oxides and the formation of chemical bonds with edge Al groups of layer silicates. Fluoride ion (F^-), for example, has the same size as OH^- ion and may therefore substitute OH^- ion in an isomorphous manner as shown below:

$$ROH + F^- = RF + OH^-$$

There is also evidence that the phosphate ion, despite having a different size from that of OH^- ion, is also bound as a structural non-diffusible unit by displacement of lattice OH^-, as the so-called 'ligand-exchanged' anion (*see* below).

During decomposition, soil organic matter liberates organic compounds, which acquire negative charges due to the dissociation of protons from the carboxylic acids and phenolic hydroxyl groups. These negatively charged compounds displace phosphate ions from the adsorption sites and thus increase phosphate ion concentration in soil solution (Sanyal and De Datta, 1991). This phenomenon generally enhances phosphate availability to plants, even though organic matter in acid soils, rich in sesquioxides and amorphous Fe and Al oxides, sometimes reduces the plant availability of phosphate in soil (Poonia, 2002). The humic colloids also chelate Al, Fe, Mn ions into soluble chelates which are thereby deactivated in immobilizing the phosphate ions in the soil solution (Sanyal and De Datta, 1991).

3.9. Ligand Exchange

A ligand, which may be an anion or a neutral molecule, satisfies the secondary valency or coordination number (Table 2.2) of a central cation in a crystal lattice. If it is an anion, then it simultaneously satisfies the primary valency as well. Thus, Fe^{3+} ion has got a primary valency of +3 and a secondary valency of 6 (Table 2.2). In $FeCl_3$, three Cl^- ions satisfy the primary valency (+3), while in $[Fe^{III}(CN)_6]^{3-}$, six CN^- ions satisfy the primary valency (+3) and the secondary valency (6) of Fe^{3+} ion. However, in $[Fe(H_2O)_6]^{3+}$ ion, six water molecules (neutral species) satisfy the secondary valency (6) of Fe^{3+} cation, while the primary valency (+3) remains *yet* to be satisfied. In the complex ion, $[Fe^{III}(H_2O)_5(OH)]^{2+}$, one OH^- ion simultaneously (and partly) satisfies both the primary valency and the secondary valency of Fe^{3+} ion, while the five neutral water molecules satisfy the remaining (5 of 6) secondary valency of Fe^{3+} cation. In this case, OH^- ions and water molecules are both ligand to Fe^{3+} cation, and are liable to undergo exchange with the extraneous anions which would form covalent bonds with Fe^{III} ion (Mott, 1981; Sanyal *et al.*, 2009).

The conjugate base of strong mineral acids (*e.g.*, HCl, HNO_3, HBr, $HClO_4$, etc.), namely the monovalent anions derived from these acids, *e.g.* Cl^-, NO_3^-, Br^-, ClO_4^-, etc., undergo anion exchange and are held at the positively charged soil colloidal sites by virtue of electrostatic forces. Such retention of anions is also referred to as the non-specific anion retention. This may possibly be described as the physical adsorption in that *no* chemical bond is formed between the soil solid surface and the anions in the same sense that the exchangeable cations (*e.g.* in the diffuse double layer) swarm around a mica-type surface which has been discussed earlier in Section 2.6.1 (Sanyal *et al.*, 2009).

Several other anions, including the organic anions, are generally specifically adsorbed to the surface of clays or clay-sized primary minerals. These anions may be derived from both strong and weak acids, and are of mono- and polyvalent type. In this case, the anions, held at the specific sites, form a chemical (covalent) bond with the surface group. This phenomenon is also known as ligand exchange in that the extraneous anion, acting as a ligand to the centrally located cation (in the clay lattice), substitutes an existing surface ligand (*e.g.* an OH^- ion or a H_2O molecule) (Sanyal *et al.*, 2009).

While anion exchange takes place only at the positively charged colloid surface, *e.g.* for a variable-charge surface, namely hydrous oxides of Fe^{III}, such as goethite [α-$Fe^{III}OOH$], at a pH below the corresponding ZPC (pH 8.5, vide. Section 2.5.2), such oxides, having only the pH-dependent charge, these oxides may, however, undergo ligand exchange at pH values below, above and *even* at the ZPC pH of the given variable-charge colloid surface. The same would be true for other types of soil colloid surfaces of variable-charge as well (Sanyal *et al.*, 2009).

The ligand exchange is distinct from the simple anion exchange in that the phosphate, for example, in acid soil readily gets immobilized and removed from the soil solution, following a generous application of soluble phosphatic fertilizer, even in the presence of a strong (say 1M) solution of NaCl. This tends to suggest that phosphate and chloride anions compete for different types of adsorption sites. The former is chemisorbed in soil, while the latter undergoes electrostatic (physical) retention (Sanyal *et al.*, 2009).

3.9.1. Adsorption Envelope

Recalling the Langmuir adsorption isotherm (Eq. 3.9), one obtains

$$\frac{C}{x} = \frac{C}{x_m} + \frac{1}{x_m K} = \frac{KC+1}{x_m K}$$

(3.70)

Or,

$$\frac{x}{x_m} = \frac{KC}{1+KC}$$

(3.71)

where,

K = Langmuir constant related to bonding energy

x_m = Adsorption maximum

C = Equilibrium concentration in soil solution

x = Amount adsorbed per unit mass of the adsorbent.

On studying the adsorption in 0.1M NaCl solution of a series of anions such as phosphate, fluoride and silicate, etc., on goethite at a range of pH from 2.5 to 12.5 and a given temperature, followed by plotting of the corresponding x_m against the pH of the study for each adsorbate, the following adsorption envelopes are obtained (Figure 3.9).

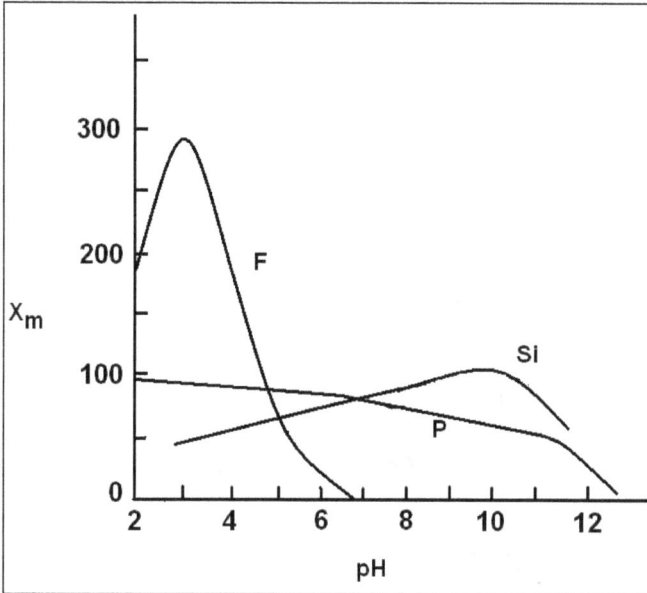

Figure 3.9: Adsorption Envelope of Specifically Adsorbed Anions.
Source: **Mott (1981).**

The x_m value for F$^-$ is the highest at pH ~ 3.0, silicate at pH ~ 9.6 (a broad maximum), while for phosphate, two inflexion points are obtained at pH ~ 6.7 and ~ 11.3 where the phosphate curve changes its slope (Mott, 1981; Sanyal *et al.*, 2009).

Incidentally, the pK_a of HF is 3.2 (at 25°C), that of silicic acid is 9.7, while the pK_2 and pK_3 of phosphoric acid are very close to (and little higher than) 6.7 and 11.3, respectively. Presumably, if the experiment could have been continued below pH 2.0 (where goethite starts dissolving and losing its form), the phosphate curve would show a decreasing trend at a pH close to its pK_1, namely 2.1 (Sanyal *et al.*, 2009).

One would recall from the Henderson equation, giving the pH of a buffer mixture of a weak acid and its salt, namely (Sanyal *et al.*, 2009)

$$pH = pK_a + \log (C_{salt}/C_{acid}) \tag{3.72}$$

Thus, at a pH ≈ pKa, $C_{salt} = C_{acid}$

i.e., $C_{anion} = C_{acid}$ $\tag{3.73}$

Keeping in view the nature of the adsorption envelopes presented in Figure 3.9, this tends to suggest that to cause maximum ligand exchange retention on goethite surface, *not only* the given anion, *but* the corresponding *undissociated* acid molecule has also an equally important role to play.

Let us take a look at the possible pathway of such ligand exchange reaction on goethite surface (which can be extended to cover other soil colloidal surfaces as well) (Mott, 1981; Sanyal *et al.*, 2009).

At a pH below the ZPC of goethite (pH < 8.5),

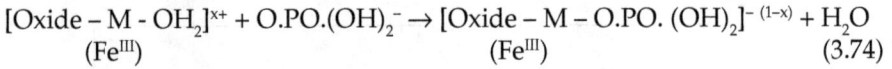

$$[\text{Oxide} - M - OH_2]^{x+} + O.PO.(OH)_2^- \rightarrow [\text{Oxide} - M - O.PO.(OH)_2]^{-\,(1-x)} + H_2O$$
$$(Fe^{III}) \qquad\qquad\qquad\qquad (Fe^{III}) \qquad\qquad\qquad\qquad (3.74)$$

At the ZPC (pH = 8.5),

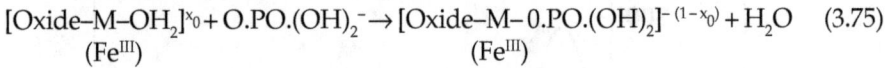

$$[\text{Oxide}-M-OH_2]^{x_0} + O.PO.(OH)_2^- \rightarrow [\text{Oxide}-M-0.PO.(OH)_2]^{-\,(1-x_0)} + H_2O \quad (3.75)$$
$$(Fe^{III}) \qquad\qquad\qquad\qquad (Fe^{III})$$

At a pH above the ZPC (pH > 8.5),

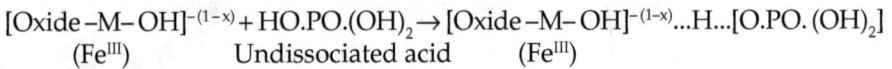

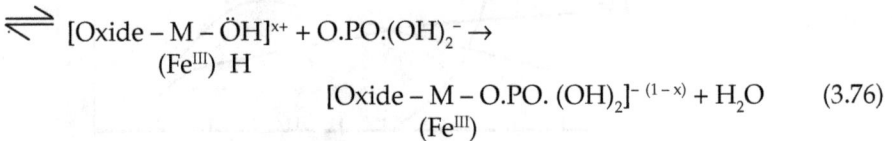

$$[\text{Oxide}-M-OH]^{-\,(1-x)} + HO.PO.(OH)_2 \rightarrow [\text{Oxide}-M-OH]^{-\,(1-x)}...H...[O.PO.(OH)_2]$$
$$(Fe^{III}) \qquad\qquad \text{Undissociated acid} \qquad (Fe^{III})$$

$$\rightleftharpoons [\text{Oxide} - M - \ddot{O}H]^{x+} + O.PO.(OH)_2^- \rightarrow$$
$$(Fe^{III})\ \ H$$
$$[\text{Oxide} - M - O.PO.(OH)_2]^{-\,(1-x)} + H_2O \qquad (3.76)$$
$$(Fe^{III})$$

Thus at a pH > ZPC, the undissociated phosphoric acid molecule plays a key role in supplying the proton from one of its dissociable (OH) groups to protonate the OH^- ligand of Fe^{III} cation in goethite, thereby converting it to a neutral (H_2O) ligand, which is then replaced by the negatively charged $H_2PO_4^-$ ion, causing ligand exchange (Sanyal *et al.*, 2009).

It is also apparent that at pH equal to or less than the ZPC, the *net* negative charge of the soil colloid would increase as a result of ligand exchange, involving, say, phosphate. This will raise the CEC of the soil. Indeed, in the volcanic ash soils of, for instance, the East Indies, allophanic clay minerals, being amorphous with high specific surface charge density, have very high phosphate fixation capacity. To these soils, often a massive dose of phosphatic fertilizer (*e.g.* 1000 kg P_2O_5. ha^{-1}) is added to satisfy the phosphate hunger of the soil, and in the subsequent crop seasons, moderate dose of such fertilizer is added, which is supplemented by a slow release of small amounts of phosphate from the (earlier) immobilized phosphate in soil. Furthermore, such massive application of phosphatic fertilizer causes a rise in the CEC of the soil due to ligand exchange process as explained above. In such instances, phosphate plays the role of *not only* a fertilizer, *but also* that of a soil amendment (Sanyal and De Datta, 1991; Sanyal *et al.*, 2009).

3.10. Quantity-Intensity Relations

Recalling the Gapon equilibrium (Eq. 3.51),

$$[Soil]^- \frac{Ca^{2+}/2}{Mg^{2+}/2} + K^+_{soln.} \rightleftharpoons [Soil]^- K^+ + \frac{1}{2}(Ca^{2+} + Mg^{2+})_{soln.}$$

One has the corresponding selectivity coefficient, or the Gapon exchange constant, K_G (or K_m), given by (*vide.* Eq. 3.52),

$$K_G \text{ (or } K_m) = \frac{[K^+_{soil}](Ca^{2+} + Mg^{2+})^{\frac{1}{2}}_{soln.}}{[CEC - K^+_{soil}](K^+_{soln.})} \tag{3.77}$$

Where [ ] denotes the activity (or concentration) of the exchangeable ions concerned on the adsorbed phase and ( ) denotes the activity in the soil solution phase at equilibrium.

Rewriting Eq. (3.77), one obtains

$$K_G = \frac{[Exch. K^+](Ca^{2+} + Mg^{2+})^{\frac{1}{2}}_{soln.}}{[Exch. Ca^{2+} + Mg^{2+}](K^+_{soln.})} \tag{3.78}$$

Or, $[Exch. K^+] = K_G \cdot [Exch. Ca^{2+} + Mg^{2+}] \cdot AR^K$ (3.79)

Where AR^K is the ionic activity ratio in the equilibrium soil solution, and is given by,

$$AR^K = \frac{a_{K^+}}{\sqrt{a_{Ca^{2+}} + a_{Mg^{2+}}}} \tag{3.80}$$

Assuming that the change of activity of the exchangeable $[Ca^{2+} + Mg^{2+}]$ in the exchange phase is small due to such ion exchange process, the ionic activity (or concentration) of these ions on the exchange phase may be assumed to remain virtually constant during the process so that it follows from Eq. 3.79,

$[Exch. K^+] \approx Constant \times AR^K$ (3.81)

Hence, $\Delta [Exch. K^+] = Constant \, \Delta AR^K$ (3.82)

so that a plot of $\Delta [Exch. K^+]$ or $\Delta Exch. K^+$ against the corresponding AR^K at equilibrium for various initial ratios of K^+ to $(Ca^{2+} + Mg^{2+})$ ionic activities in the equilibrating solutions should yield a straight line, passing through the origin (Sanyal *et al.*, 2009). In actual practice, however, deviation is noted from the expected linearity as demonstrated in the following plot (Figure 3.10).

Such deviation from linearity is generally exhibited by a soil at the lower ranges of the AR^K values. This arises from the release of K^+ ions from the nonexchangeable pool of K^+ ions in the soil when the solution level (and hence the exchangeable K level) is low, causing thereby a small value of AR^K in the soil solution. Such release, in turn, leads to a new position of equilibrium between the exchange and the soil solution phases (Sarkar *et al.*, 2014).

Because the Gapon equation (Eqs. 3.47, 3.49, 3.51 and 3.52) considers the equilibrium distribution of cations only between the exchange and the soil solution phases, the intervention by the non-exchangeable pool of K^+ ions, thereby disturbing

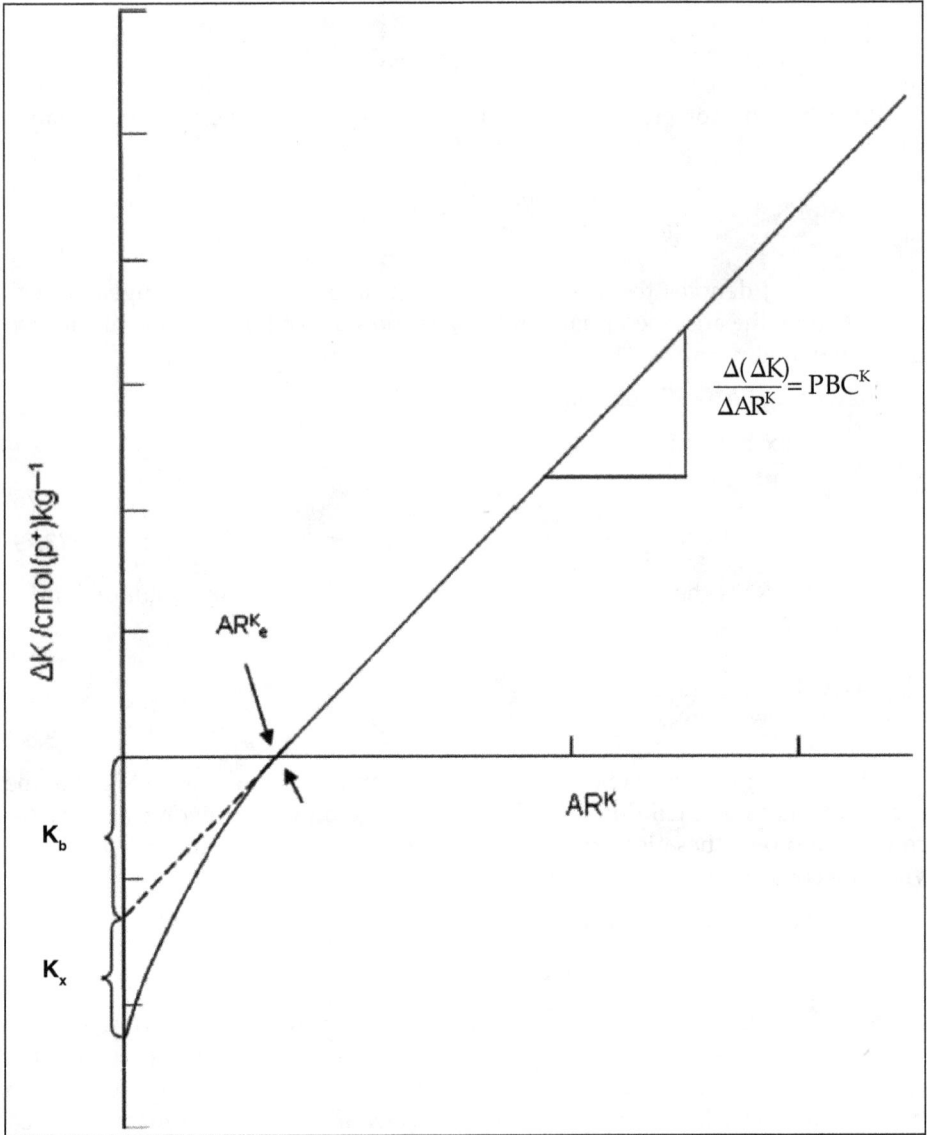

$$\frac{\Delta(\Delta K)}{\Delta AR^K} = PBC^K$$

Figure 3.10: Quantity-Intensity Plot of Soil Potassium.
Source: **Sarkar *et al.* (2014).**

the above noted equilibrium, would cause deviation from the linearity (as observed), which is demanded by the validity of the Gapon equation.

The plot in Figure 3.10 is referred to as the Quantity-Intensity (Q/I) relationship (Beckett, 1964), and brings out the changes in the quantity of, for instance, adsorbed potassium, in relation to the corresponding value of the activity ratio, AR^K [= a_K+/ ($\sqrt{a_{Ca}^{2+}} + a_{Mg}^{2+}$)] in the equilibrium soil solution (Sanyal *et al.*, 2009).

The Q/I curves may be obtained by equilibrating a number of soil sub-samples with a series of solutions containing increasing amounts of KCl in $CaCl_2$ solution of fixed concentration (say 0.01 mol L^{-1}). For each suspension, the difference between the potassium concentrations in the initial and equilibrated solution gives the amount by which the soil gains or loses potassium (ΔK) in reaching the equilibrium. The activity ratios, AR^K, are calculated from the cationic composition and the activity coefficients of the relevant cations in the resultant equilibrium soil solution (Sarkar *et al.*, 2014).

Several (Q/I) parameters are defined from Figure 3.10. These include the equilibrium activity ratio (AR^K_e) which refers to that value of the activity ratio in the equilibrium soil solution where the exchange phase neither gains nor releases exchangeable K$^+$ ions from or to the soil solution phase in contact (Sarkar *et al.*, 2014). This value (AR$_e^K$) provides a measure of the inherent equilibrium soil solution activity ratio of the soil, and hence the inherent K$^+$ availability (to plants) compared to Ca^{2+} and Mg^{2+} availability (Sanyal *et al.*, 2009; Sarkar *et al.*, 2014).

The parameter, K_x, is related to the number of sites (on the adsorbed phase in soil) contributing towards specifically held K$^+$ ions. It was estimated from the difference between the intercept made by the interpolated linear portion of the (Q/I) plot and the intercept actually made by the (Q/I) plot on the Y-axis of the curve. Such K_x value is generally much higher for the illitic soil (with interlayer K$^+$ reserve as the nonexchangeable K pool of soil) as compared to the kaolinitic or the montmorillonitic soils (Sanyal *et al.*, 2009).

The parameter K_b is defined as a measure of the labile K that the soil holds. The linear potential buffering capacity for soil K (PBC^K) is obtained from the slope of the linear portion of the (Q/I) plot (Figure 3.10). This refers to the inherent buffer capacity of a soil to maintain a given value of AR^K, *i.e.*, the greater its value [*i.e.*, the steeper the (Q/I) plot], the more will be the ability/tendency of the given soil to release K$^+$ from the exchange phase to the soil solution when K$^+$ is lost (through plant uptake and/or leaching), and *vice-versa* for K$^+$ gain by the soil solution (*e.g.* on fertilizer application and/or irrigation water rich in potassium). All these (Q/I) parameters are therefore related to the chemistry and the fertility relationships of soil potassium (Sanyal *et al.*, 2009).

3.11. Methodologies to Study Ion Exchange Equilibria

3.11.1. Equilibrium Procedure

The determination of cations on the exchange complex is generally carried out by two methods, *viz.* stripping technique and isotopic dilution technique (Baruah and Patgiri, 1996; Sanyal *et al.*, 2009).

Stripping Technique

In this technique, two approaches are adopted: (a) the soil in equilibrium with the solution is washed free of soluble salts with alcohol and then it is extracted in a salt solution ($\approx 1N$) (free from the cations on the exchanger); or (b) the soil with the equilibrium solution is centrifuged, the supernatant liquid is collected for analysis and the equilibrium solution left in the soil is measured. The total cations extracted in the salt solution minus the cations in the left-over equilibrium solution are designated as the exchangeable cations (Baruah and Patgiri, 1996; Sanyal et al., 2009).

Isotopic Dilution Technique

A given amount of soil (say 1 g) with a known volume of untagged equilibrium solution (say 5 mL) is mixed with an equal volume of tagged solution of the same composition. The contents are fully dispersed and equilibrated for 48 h on a shaker. The equilibrated soil suspension is centrifuged and the radioactivity is measured in the supernatant liquid, using a suitable radiation-measuring device (Baruah and Patgiri, 1996; Sanyal et al., 2009). The amount of adsorbed cation (A_x) is calculated using Eq.3.83:

$$A_X = \left[\frac{CPM_{control}}{CPM_{test}} - 1\right][A]$$

(3.83)

where,

CPM = Counts per minute

[A] = Total amount of species x in solution

3.12. Kinetics of Ion Exchange

Notwithstanding the thermodynamic principles, used to characterize exchange equilibria on clay and on soil surfaces, the use of a chemical kinetic approach to determine thermodynamic parameters has received increasing attention recently (Sanyal and Majumdar, 2001; Sanyal et al., 2009). This is particularly true for potassium exchange in clays and in soils. Although most studies on the kinetics of cation exchange in soils have been concerned with potassium release, the general principles for cation release are similar (Sanyal and Majumdar, 2001).

Techniques of common use for the study of kinetic reaction in clays and soils include Batch, Vigorously shaken batch and Miscible displacement or flow techniques (Baruah and Patgiri, 1996; Sanyal et al., 2009).

For the kinetic model, the reaction written in exact stoichiometric expression is given as follows:

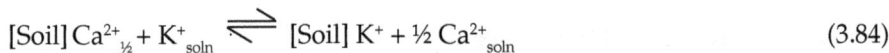

$$[Soil]\, Ca^{2+}{}_{\frac{1}{2}} + K^+{}_{soln} \rightleftharpoons [Soil]\, K^+ + \frac{1}{2}\, Ca^{2+}{}_{soln}$$

(3.84)

The rate and direction of these reactions determine whether applied K will be leached into lower soil horizons or absorbed by plants or converted into unavailable forms, or else released into available forms. Knowledge of rates of reaction between solution and adsorbed phases of soil K is essential in order to predict the fate of the applied K fertilizer in soils and make appropriate K fertilizer recommendations (Sanyal et al., 2009).

Miscible Displacement Technique

Miscible displacement technique refers to the mixing of soil solution initially present both in macro (non-capillary) and micro (capillary) pores in mobile and immobile regions with the incoming solution, and consequently the displacement of the original solution from the system on the basis of equivalent pore volume. The miscible displacement technique simulates solute movement in soils under field conditions (Baruah and Patgiri, 1996; Das *et al.*, 2014).

Prior to kinetic measurements on adsorption and desorption processes, soil samples should be passed through a 60-mesh stainless steel sieve and saturated with Ca using I N CaCl$_2$. Excess electrolyte should be removed with distilled water and with a 1:1 acetone-to-water mixture, until a negative test for Cl$^-$ is obtained using AgNO$_3$ (Baruah and Patgiri, 1996; Sanyal *et al.*, 2009).

For kinetics of K adsorption, 1 g of Ca-saturated soil is used and mixed with 50 ml of deionized water. The suspension is transferred into a 47-mm nucleopore-filter column which is attached to a fraction collector on the terminal side and to a peristaltic pump on the delivery side. The pump is ideally required to deliver the K solution through the soil at a uniform flow rate of 1.0 mL min^{-1}. The aliquots are collected, preferably at 10 minute intervals until the concentration of the leachate equals that of the initial K concentration (Baruah and Patgiri, 1996; Sanyal *et al.*, 2009).

For kinetics of K desorption, the tubes, connected to the column, are rinsed thoroughly with distilled water and then the K desorption is initiated using 0.01M CaCl$_2$. This solution is passed through the soil until no K appears in the leachate. The quantity of K in solution for both the adsorption and desorption studies is measured. The kinetic experiments should also be carried out ideally at 25±1°C (Baruah and Patgiri, 1996; Sanyal *et al.*, 2009). The apparent adsorption and desorption rate constants are determined using the first-order kinetics as elucidated below.

Batch Technique

The miscible displacement technique suffers from the fact that the diffusion-controlled ion exchange is prominent, due to the prevalence of steady-state condition during the course of fluid dynamics and ion interaction. Thus, ion exchange may be overestimated under such situations. Batch technique is an alternative to overcome this problem (Baruah and Patgiri, 1996; Sanyal *et al.*, 2009).

Similar to the miscible displacement technique, the soil samples are saturated with Ca using 1 N CaCl$_2$ solution; the excess salts are washed out with distilled water first and with 1:1 acetone-to-water mixture, as long as chloride is present.

One g samples of Ca-saturated soils are equilibrated with 50 mL of 50 µg K mL^{-1} of solution in centrifuge tubes at 25±1°C on a reciprocating shaker. After sampling at 10 minute intervals, the suspension from each tube is centrifuged for 3 minutes at 2,000 rpm; then the supernatant solution is analyzed for K flame-photometrically. From the solution phase measurements of K as a function of time, the adsorption and desorption rate constants are worked out using the equations to be outlined below (Baruah and Patgiri, 1996; Sanyal *et al.*, 2009).

3.12.1. Theoretical Approach for Kinetic Studies

Referring to the K^+ - Ca^{2+} exchange reaction in soil (Eq. 3.84), let us deal with the kinetic frame-work for the first-order K release and fixation reactions in soil (Sanyal and Majumdar, 2001).

For K release, this equation assumes that the K concentration at the exchange sites of the soil colloid is the determining factor for the release rate of K into soil solution. In the batch technique, the rate is given by

$$dK_t/dt = k_d (K_0 - K_t) \qquad (3.85)$$

Where K_t is the amount of K^+ ion released in time t, K_0, the amount of K^+ ion that could be released at equilibrium, while k_d is the desorption rate coefficient.

For a miscible displacement or flow technique, the rate is given by

$$d (K_t/K_0)/dt = - k_d (K_t/K_0) \qquad (3.86)$$

Where K_t and K_0 denote, respectively, the amount of K^+ ion on the exchange sites of the soil colloid at time t and zero time of desorption.

On suitably integrating Eqs. 3.85 and 3.86 and utilizing the initial and boundary conditions, one arrives at the following integrated forms, namely

For batch technique:

$$\ln (K_0 - K_t) = \ln K_0 - k_d t \qquad (3.87)$$

$$\text{Or } \log (K_0 - K_t) = \log K_0 - (k_d/2.303).t \qquad (3.88)$$

For miscible displacement or flow technique:

$$\ln (K_t/K_0) = - k_d t \qquad (3.89)$$

$$\text{Or, } \log (K_t/K_0) = - (k_d/2.303) t \qquad (3.90)$$

A plot of $\log (K_0 - K_t)$ against t leads to a linear plot from the slope of which the specific release (desorption) reaction rate (k_d) is obtained in a batch technique. For flow technique, a plot of $\log (K_t/K_0)$ against t is used.

The first-order kinetics for K-fixation by soil/clays may also be described, *e.g.* for batch technique,

$$d (K_0 - K_t)/dt = k_a K_t \qquad (3.91)$$

Where K_t and K_0 are the concentrations of K in solution at time t and zero time, while k_a is the adsorption rate coefficient.

For miscible displacement or flow technique, one has

$$d(K_t/K_\infty)/dt = k_a (K_\infty - K_t)/K_\infty = k_a (1 - K_t/K_\infty) \qquad (3.92)$$

Where K_t and K_∞ are the amounts of K at the exchange sites of the colloid at time t and at equilibrium.

The integrated forms are:

$$\ln K_t = \ln K_0 - k_a t \tag{3.93}$$

$$\text{Or,} \log K_t = \log K_0 - (k_a/2.303).t \tag{3.94}$$

For miscible displacement or flow technique:

$$\ln (1 - K_t/K_\infty) = -k_a t \tag{3.95}$$

$$\text{Or,} \log (1 - K_t/K_\infty) = -(k_a/2.303).t \tag{3.96}$$

Thus, a plot of $\log K_t$ against t leads to k_a from the slope of the resulting linear graph in a batch technique. The adsorption rate coefficient (k_a), in miscible displacement technique, is obtained from the slope of the linear plot of $\log (1 - K_t/K_\infty)$ *vs.* time (Sanyal and Majumdar, 2001; Sanyal *et al.*, 2009).

The first-order kinetics have been used by a number of workers to describe the exchange-kinetics and K release from clay minerals, as well as surface and subsurface horizons of a number of soils (Sanyal and Majumdar, 2001).

The equilibrium constant (K) for adsorption-desorption reactions in soil may also be derived from the corresponding adsorption-desorption rate coefficients, namely k_a and k_d introduced above. Thus, it can be shown that for adsorption (fixation) process in soil, the corresponding equilibrium constant (K_a) is given as,

$$K_a = k_a/k_d \tag{3.97}$$

While for the desorption (release) process in soil, the equilibrium constant (K_d) is obtained as ,

$$K_d = k_d/k_a \tag{3.98}$$

Many reactions in soil are multi-step ones, each step (elementary reaction) being characterized by the corresponding equilibrium (K) as well as the kinetic (k) constants. In such case, for instance, in a two-step process of adsorption, the overall equilibrium constant (K_a) is given as (Sanyal and Majumdar, 2001; Sanyal *et al.*, 2009),

$$K_a = K K' = k_a k'_a/k_d k'_d \tag{3.99}$$

Where K and K' are the equilibrium constants for the two elementary reactions, leading to the overall adsorption process, while k_a and k'_a are the corresponding adsorption rate coefficients and k_d and k'_d , the corresponding desorption rate coefficients.

3.12.2. Usefulness of Adsorption-Desorption Kinetic Data

There are two salient reasons for studying the rate of soil chemical processes given below ((Baruah and Patgiri, 1996; Sanyal *et al.*, 2009) :

1) To predict how quickly reactions approach equilibrium or quasi-state equilibrium.

2) To investigate reaction mechanism.

Many reactions in soils are slow, yet they proceed at a measurable rate. Kinetic data on slow reactions may be of importance with regard to plant nutrition. Also information about the reaction mechanism and processes occurring may be obtained from the kinetic data (Sanyal *et al.*, 2009).

Analysis of adsorption isotherm data allows some prediction to be made on the field dosages of a chemical (*e.g.* fertilizer) to obtain a particular biological effect. The adsorption-desorption characteristics of a chemical, evaluated under laboratory conditions, and/or even in pot-culture experiments under greenhouse conditions (Hance *et al.*, 1968), merely provide an indication of how that chemical will perform *in situ*. The shift of the field performance from laboratory results is due to the use of excess water to soil and extensive mixing in laboratory studies on one hand, and the effects of climatic variations, biological activity as well as agricultural practices, followed (in the field) on the continued changes in the properties of soils, on the other ((Baruah and Patgiri, 1996; Sanyal *et al.*, 2009).

In transferring results to field situations, it remains to be understood whether the kinetics of adsorption are comparable and whether the same surfaces are available in natural soils. Generally, the soils in the field are aggregated and the water content seldom exceeds the field capacity during the crop growing season (Baruah and Patgiri, 1996; Sanyal *et al.*, 2009).

References

Barrow, N.J. (1978). The description of phosphate adorption curves. *J. Soil Sci.*, **29**, 447-462.

Baruah, T. C. and Patgiri, D. K. (1996). *Physics and Chemistry of Soils*. New Age International Publishers Private Limited, New Delhi.

Bear, F. E. (Ed.) (1976). *Chemistry of the Soil*, Second Edition, Third Indian Reprint, Oxford and IBH Publishing Co., New Delhi, pp. 30-32.

Beckett, P.H.T. (1964). Studies on soil potassium: 2. The immediate Q/I relations of labile potassium in soils. *J. Soil Sci.*, **15**, 9-12.

Bockris, J. O. M. and Reddy, A. K. N. (1970). *Modern Electrochemistry*, Volume 2, Plenum Press, New York.

Brunauer, S., Emmett, P.H. and Teller, E. (1938). Adsorption of gases in multimolecular layers. *J. Amer. Chem. Soc.*, **60**, 309-319.

Das, I., Ghosh, K., Das, D.K. and Sanyal, S.K. (2014). Transport of arsenic in some affected soils of Indian sub-tropics. *Soil Res.*, CSIRO Publishing, **52**, 822-832.

Freundlich, H. (1926). *Colloid and Capillary Chemistry*. Translated from the third German edition by H. Stafford Hatfield, Methuen and Co. Ltd, London.

Gaines, G.L. and Thomas, H.C. (1953). Adsorption sites on clay minerals. II. A formulation of the thermodynamics of exchange adsorption. *J. Chem. Phys.*, **21**, 714-718.

Gapon, E.N. (1933). On the theory of exchange adsorption in soils. *J. General Chem., U.S.S.R.*, **3**, 144152.

Hance, R. J., Holcombe, S. D. and Holroyd, J. (1968). The phytotoxicity of some herbicides in field and pot experiments in relation to soil properties. *Weed Res.*, **8**, 136-144.

Kerr, H.W. (1928). Nature of base exchange and soil acidity. *J. Am. Soc. Agron.*, **20**, 309-335.

Khasawneh, F. E. and Copeland, J. P. (1973). Cotton root growth and uptake of nutrients: Relation of phosphorus uptake to quantity, intensity, and buffering capacity. *Soil Sci. Soc. Am. Proc.*, **37**, 250-254.

Klages, M. G., Olsen, R. A. and Haby, V. A. (1988). Relationship of phosphate isotherms to NaHCO3-extractable phosphorus as affected by soil properties. *Soil Sci.*, **146**, 85-91.

Krishnamoorthy, Ch., Davis, L.E., and Overstreet, R. (1948). Ion exchange equations derived from statistical thermodynamics. *Science, New York*, **108**, 439-440.

Langmuir, I. (1918). The adsorption of gases on plane surfaces of glass, mica and platinum. *J. Amer. Chem. Soc.*, **40**, 1361-1403.

Marshall, C. E. (1949). *The Colloid Chemistry of Silicate Minerals*. Academic Press, New York.

Marshall, C. E. (1964). *Physical Chemistry and Mineralogy of Soils*, Volume 1, John Wiley and Sons, New York.

Mehta, S.C., Poonia, S.R. and Pal, R. (1983). Sodium-Ca and Na-Mg exchange equilibria in soils from chloride and sulphate dominated systems. *Soil Sci.*, **136**, 339-346.

Mott, C. J. B. (1981). Anion and ligand exchange. **In:** *The Chemistry of Soil Processes* (D. J. Greenland and M. H. B. Hayes, Eds), John Wiley and Sons Limited, New York, U.S.A., p. 188.

Poonia, S. R. (2002). Ion exchange in soils. **In:** *"Fundamentals of Soil Science"* (G. S. Sekhon, P. K. Chhonkar, D. K. Das, N. N. Goswami, G. Narayanasamy, S. R. Poonia, R. K. Rattan and J. L. Sehgal, Eds.), Indian Society of Soil Science, New Delhi, pp. 261-279.

Poonia, S.R. and Niederbudde, E.A. (1990). Exchange equilibria of potassium in soils. VI. Effect of natural organic matter on KCa exchange. *Geoderma*, **47**, 233-242.

Poonia, S.R. and Pal, R. (1979). The effect of organic manuring and water quality on water transmission parameters and sodification of a sandy loam soil. *Agril. Water Managmt*, **2**, 163-175.

Poonia, S.R., Mehta, S.C. and Pal, R. (1980). Calciumsodium and magnesiumsodium exchange equilibria in relation to organic matter in soil. **In:** *Proc. Intern. Symp. Salt affected Soils.* CSSRI, Karnal, pp. 134-141.

Poonia, S.R., Mehta, S.C. and Pal, R. (1984). The effect of electro-lyte concentration on calciumsodium exchange equilibria in two soil samples of India. *Geoderma*, **32**, 63-70.

Sanyal, S. K. and De Datta, S. K. (1991). Chemistry of phosphorus transformations in soil. *Adv. Soil Sci.*, **16**, 1-120.

Sanyal, S. K. and De Datta, S. K. and Chan, P.Y (1993). Phosphate sorption desorption behaviour of some acidic soils of South and Southeast Asia. *Soil Sci. Soc. Am. J.*, **57**, 937-945.

Sanyal, S. K. and Majumdar, K. (2001). Kinetics of potassium release and fixation in soils. In: *Potassium in Indian Agriculture*. International Potash Institute, Berne, Switzerland and Potash Research Institute of India, Gurgaon, India, pp. 9-31.

Sanyal, S. K., Poonia, S. R. and Baruah, T. C. (2009). Soil colloids and ion exchange in soil. In: *Fundamentals of Soil Science* (N. N. Goswami, R. K. Rattan, G. Dev, G. Narayanasamy, D. K. Das, S. K. Sanyal, D. K. Pal and D. L. N. Rao, Eds.) Second Edition, Indian Society of Soil Science, New Delhi, pp. 269-315.

Sarkar, G. K., Debnath, A., Chattopadhyay, A. and Sanyal, S. K. (2014). Depletion of soil potassium under exhaustive cropping in Inceptisol and Alfisols. *Commun. Soil Sci. Plant Anal.*, **45 (1)**, 61-72.

Shaw, D. J. (1970). *Introduction to Colloid and Surface Chemistry*. Butterworths, London.

Sposito, G. (1984). *The Surface Chemistry of Soils*. Clarendon Press, Oxford, U.K.

Talibudeen, O. (1972). Exchange of potassium in soils in relation to other cations. *Proc. 9th Colloq. Intern. Potash Inst.*, Berne, Switzerland, pp. 97-112.

Tempkin, M.I. and Pyzhev, V. (1940). Kinetics of ammonia synthesis on promoted iron catalysts. *Acta Physiochim.*, **12**, 327-356.

Vanselow, A.P. (1932). Equilibria of the base exchange reactions of bentonite, permitites, soil colloids and zeolites. *Soil Sci.*, **33**, 95-113.

Soil Acidity and Salt-Affected Soils

4.1. Ionic Environment of Acid Soils

Acid soils occur in nearly one-third of the cultivated area of India. These soils are formed mainly due to weathering under humid climate and heavy rainfall. Laterization, podsolization, intense leaching and accumulation of undecomposed organic matter under marshy conditions are the processes contributing to acid soil development (Sanyal, 1995). In addition, soil acidification due to prolonged application of fertilizers, particularly ammoniacal nitrogen sources (*e.g.* ammonium sulphate and ammonium nitrate), and leaching with acid precipitation (the result of industrial pollution) are also contributory factors (Haynes, 1984).

Acid soils belonging to laterite and lateritic, red, and yellow soils (predominantly Alfisols and Ultisols) contain mainly kaolinitic type of clay minerals, and in some cases, mixed with illite. In Arunachal Pradesh, Bihar, and parts of West Bengal, smectite is the dominant mineral. In parts of Assam and Tripura, the dominant clay fraction is kaolinitic (Panda and Koshy, 1982). The clay minerals in the acid alluvial soils (predominantly Inceptisols) were reported to be predominantly illite, smectite, or kaolinite.

Acid soils are often characterized by low cation exchange capacity (CEC), intermediate texture, ranging from sandy loam to loam, low organic matter content (except in case of hill and *Terai* soils, and also soils under forest), and low phosphorus content, while the nitrogen content is variable. Acid soils contain relatively high amounts of Fe^{3+} and Al^{3+} ions in soil solution, and also have high exchangeable H^+ and Al^{3+} ions. Conventionally, acid soils have been defined in terms of soil pH and percent base saturation both of which are low. Soil acidity is a complex phenomenon depending upon the interplay of several factors, among which decline in the absolute amount of exchangeable bases with pH decrease, resulting in low Ca and Mg status, is important. Soil acidity was grouped by Jackson (1963) as follows:

1. Strong acids, soil pH 4.2 and less.
2. Weak acids, soil pH 5.0 or 5:2 and less.
3. Very weak acids, soil pH 5.2 to 6.5 or 7.0
4. Very very weak acids, soil pH 6.5 or 7.0 to 9.5.
5. Extremely weak acids, soil pH more than 9.5.

In order to appreciate the intricacies of the ionic environment of acid soils, and also to manipulate the latter to farmer's advantages, it may be appropriate to review the various concepts, put forward from time to time, relating to the properties of acids and bases (Sanyal, 1995).

4.2. Concepts of Acids and Bases

4.2.1. Arrhenius' Concept (1880-1890 A.D.)

This defined an acid as a substance, containing hydrogen, which releases protons (H^+) in an aqueous solution. A base was defined as a substance that produces hydroxyl ion (OH^-) in solution. This concept thus excludes a number of non-hydroxylic bases such as ammonia, amines and several organic substances.

4.2.2. Brønsted-Lowry Concept (1923 A.D.)

J.M. Brønsted and T.M. Lowry (1923) simultaneously and independently suggested that an acid is a proton donor while a base is a proton acceptor.

For instance, the behaviour of HC1 in water may be represented by the acid-base reaction equilibrium:

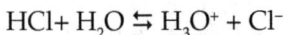

$$HCl + H_2O \rightleftharpoons H_3O^+ + Cl^-$$

HC1, being a proton donor (to H_2O) is an acid, while Cl^- is a base, but a weaker base than H_2O all the same. Cl^- is termed as the conjugate base of HCl. The conjugate base of a strong acid is thus a weak base, and *vice versa*.

4.2.3. Lux-Flood Concept (1939-1947 A.D.)

This theory describes acid-base behaviour in terms of oxide ion in place or proton. The base is an oxide ion donor and an acid is an oxide ion acceptor. Thus,

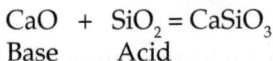

$$CaO + SiO_2 = CaSiO_3$$
Base   Acid

An acid is thus an acid anhydride, *i.e.*,

$$SiO_2 + 2H_2O = H_4SiO_4$$

(Acid anhydride of silicic acid)

4.2.4. Lewis Concept (1923 A.D.)

A Lewis acid is one which accepts a pair of electrons, and a Lewis base is one which donates a pair of electrons. Thus

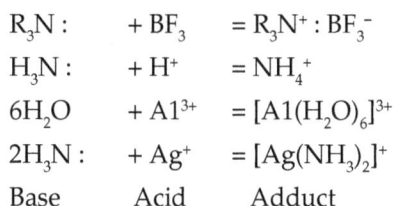

$$R_3N: \quad + BF_3 \quad = R_3N^+ : BF_3^-$$
$$H_3N: \quad + H^+ \quad = NH_4^+$$
$$6H_2O \quad + Al^{3+} \quad = [Al(H_2O)_6]^{3+}$$
$$2H_3N: \quad + Ag^+ \quad = [Ag(NH_3)_2]^+$$
$$\text{Base} \qquad \text{Acid} \qquad \text{Adduct}$$

This is a very general definition of acids and bases, and is useful in relation to soil systems.

4.2.5. Generalized Acid-Base Concept

Acidity may be defined as a positive character of a chemical species which is decreased by reaction with a base; similarly basicity is a negative character of a chemical species which is decreased by reaction with an acid.

The processes of hydration and hydrolysis may also be reviewed in terms of such concepts. It is known that a large charge-to-radius ratio (*i.e.*, ionic potential) for cations results in an increase in hydration energy. Closely related to hydration is the process of hydrolysis. In general, we speak of hydration if no reaction beyond coordination of water molecules to cation occurs:

$$K^+ + nH_2O \rightarrow [K(H_2O)_n]^+$$

In case of hydrolysis, the acidity (*i.e.*, ionic potential) of cation is so great as to cause the rupture of the H-O bond, followed by ionization of the hydrate to yield hydronium ions (protons)

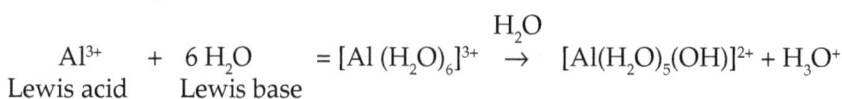

$$\overset{H_2O}{\underset{\text{Lewis acid}}{Al^{3+}} + \underset{\text{Lewis base}}{6\,H_2O} = [Al(H_2O)_6]^{3+} \rightarrow [Al(H_2O)_5(OH)]^{2+} + H_3O^+}$$

Cations that hydrolyse are either small (*e.g.* Be^{2+}), and/or are highly charged (*e.g.* Al^{3+}, Fe^{3+}, Sn^{4+}). From this view point, a bare H^+ ion should have the highest charge-to-radius ratio, and hence the highest acidity. Of course, this is not to imply that the protons exist uncoordinated in chemical systems; they attach themselves to any chemical species containing electrons. It is too strong an acid to coexist with any base without reacting.

4.2.6. Certain Other Terms and Concepts

pH

The acidity status of a solution or colloidal dispersion is expressed as the corresponding pH given by,

$$pH = -\log a_H^+ \approx -\log C_{H^+}$$

Where a_{H+} is H^+ ion activity in the solution or dispersion which may be approximated by the corresponding concentration (C_H^+) for dilute systems.

Buffer Mixtures

These refer to systems that are capable of maintaining the pH of a solution when small amounts of acids or alkalis are added. Buffer mixtures generally contain weak acids or bases and their salts. For such a mixture (of, say, HA and its salt A^-, *e.g.* CH_3COOH and $CH_3COO^-Na^+$), the following dissociation equilibrium (of HA) exists:

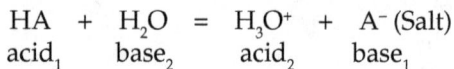

$$HA + H_2O = H_3O^+ + A^- \text{ (Salt)}$$
$$\text{acid}_1 \quad \text{base}_2 \quad \text{acid}_2 \quad \text{base}_1$$

The dissociation constant, K, of HA is given as,

$K = [(a_{H3O+}. a_A^-)/a_{HA}]$, by neglecting the activity of water which is nearly constant in a dilute solution.

Or, $pK = -\log K = pH - \log (a_A^-/a_{HA})$

Or, $pH = pK + \log \dfrac{a_{A^-}}{a_{HA}} = pK + \log \dfrac{C_{salt}}{C_{acid}}$

This equation is known as the Henderson equation. Thus, the buffering action is maximum when the ratio of the activity of the salt (*i.e.*, conjugate base) to that of the acid is unity, leading to a fixed value (= pK) of the pH of the mixture. Furthermore, the determination of pH of the titration mixture at half-neutralization point of the acid, HA (when $C_A - = C_{HA}$), would provide a measure of the pK of the given acid.

4.3. True and Colloidal Acids

As mentioned briefly earlier in Section 2.7.1, the gradual neutralization of an aqueous acid, when titrated against a base, can be followed by plotting pH *versus* the volume of the titrant, which leads to the pH-metric titration curves. The latter show characteristic buffering and inflexion points. An n-basic acid may show n-inflexions provided that the dissociation constants of the $(n-1)^{th}$ and the nth stages are widely separated [typically, the ratio, K_{n-1}/K_n being $\sim 10^4$]. In actual practice, however, the dissociation constants of a polybasic acid are often closer together (such as orthophosphoric acid, H_3PO_4), and neutralization at one stage may overlap partially with the preceding or the succeeding stages, giving rise to 'statistically monobasic' acid behaviour with an average dissociation constant (Mukherjee, 1974). The total acidity of such an acid solution will be independent of the base chosen to measure the former. If, however, the anion of an acid is of colloidal dimension, the protons form altogether a different phase from the former, leading to a poly-phase system. Examples are provided by, among others, acid clays of soil. Towards the H/glass, or any proton-selective electrode, these colloidal acids will doubtless register H^+ ion activity, but only a small fraction of the total quantity does so, behaving like a weakly dissociating acid. The H^+ ions cannot be separated out into the intermicellary liquid, for instance, by ultrafiltration. On conducting titration with a base, the pH-metric titration curves not only show differences in their features, depending on the base used, but also register variations in total acidity. This is because the neutralization reaction here is a two-step process, namely (i) an exchange reaction between the cation of the added base and H $^+$ ion on the colloidal anion, followed by (ii) neutralization of H$^+$ ions by OH$^-$ ions in the solution phase (*see* Section 2.7.1). Furthermore, the acidity of a colloidal acid increases when titrated in the presence of neutral salts unlike that of true acids. The features of the titration curve also change in presence of such salts. This, again, obviously results from an exchange of H$^+$ (also Al^{3+}) for cations of the added salt preceding neutralization.

4.4. Nature of Soil Acidity

As stated in Section 2.5, the aluminosilicate clay minerals (typified by kaolinite and montmorillonite as members of 1 : 1 and 2 : 1 type layer-lattice silicates) contribute to soil acidity by way of holding on to H$^+$ (and Al^{3+}) ions to neutralize the negative charge on them. The latter arises from (i) isomorphous substitution within Al-octahedral and Si-tetrahedral layers of higher-valent cations for lower-valent cations having similar ionic sizes, leading to permanent negative charge, and (ii) the dissociation of proton from the exposed hydroxyl group or bound water of constitution, both of which are structural components in crystal lattice, leading to the development of negative charge which, naturally, depends on pH of the surrounding solution. Besides, the hydrous oxides of Fe and Al (also Mn), *i.e.*, the sesquioxides panicles of clay-size dimensions may also contribute to the pool of pH-dependent negative charge in soil. These are demonstrated below with the following equations (as shown earlier in Section 2.5.2):

(i) On hydrous oxides of Fe and Al (pH-dependent negative charge):

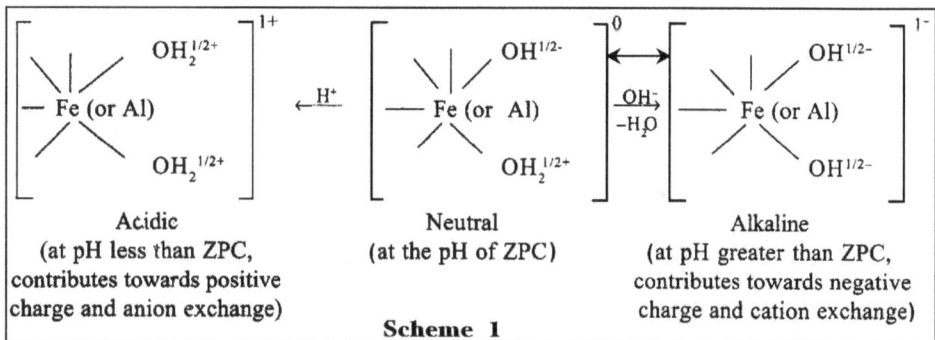

Acidic	Neutral	Alkaline
(at pH less than ZPC, contributes towards positive charge and anion exchange)	(at the pH of ZPC)	(at pH greater than ZPC, contributes towards negative charge and cation exchange)

Scheme 1

(ii) On clay minerals (pH-dependent negative charge): see the Scheme 2.

(iii) On clay minerals (permanent negative charge): Mg^{2+}, Fe^{2+}ions replacing Al^{3+} ions in octahedral layer; Al^{3+} ions replacing Si^{4+} in tetrahedral layer.

Another constituent of the soil which can contribute to its acidity in various ways is humus. The latter contains reactive carboxylic and phenolic (OH) groups that behave as weak acids. The dissociation of the latter, contributing to soil acidity, would depend on the dissociation constant of the acid, and the pH of the surrounding medium. The substituted

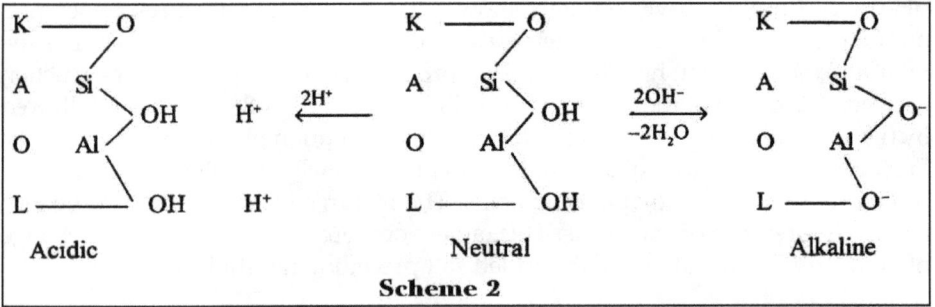

Scheme 2

phenols such as nitrophenols in humus are often stronger acids than simple phenolic components themselves owing to their π-electron withdrawing character of the nitro group. For instance, symmetrical trinitro phenol (picric acid) is as strong an acid as a carboxylic acid, liberating, like the latter, CO_2 from $NaHCO_3$. The acidity contribution from humus is also pH-dependent as shown below:

(iv) On organic matter (pH-dependent negative charge):

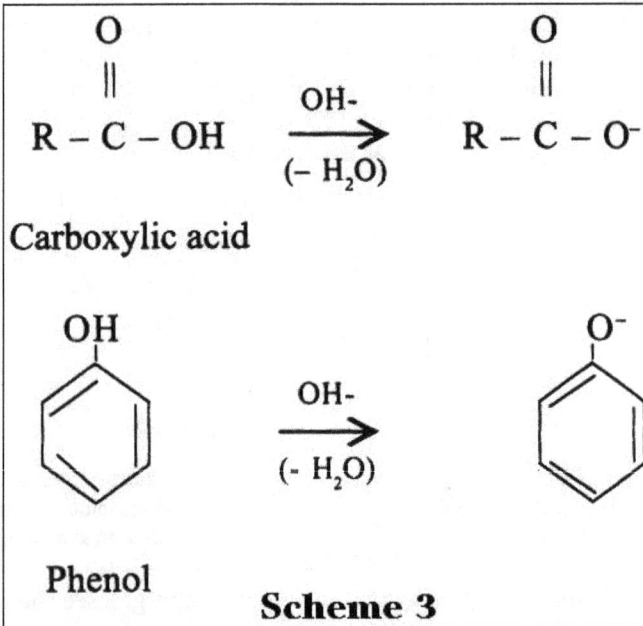

Scheme 3

4.5. Exchangeable Aluminium in Acid Soil

As mentioned briefly earlier in Section 2.7.2, evidence is available to suggest that an acid soil is actually a H-Al system. The presence of Al^{3+} ions could be due either to dissolution from silicate and other minerals, or to the presence of exchangeable Al^{3+} in the silicate exchangers. Although the amount of Al^{3+} ions liberated by neutral salts is greater, the higher the acidity of the soils and clays, the strength of the acid liberated from such soils and clays is not sufficient to cause dissolution of Al from the silicate minerals (Mitra and Kapoor, 1969; Mukherjee, 1974). There is strong evidence, on the other hand, to suggest that Al^{3+} along with H^+ ions are exchanged for the cations of the neutral salt (Mukherjee, 1974).

As to the origin of these exchangeable Al^{3+} ions, the H-soil or clay when immediately titrated after preparation (on passage of the suspension through a column of H-resin), shows strong acid character and a buffering at low pH (Sanyal, 1995). The neutral salt extract contains little or no aluminium. But on heating the suspension to 95°C, the titration curves change markedly to those of weak acids, buffering occurs at higher pH with the total acidity remaining unchanged, the neutral salt extract showing the presence of an increasing amount of Al^{3+} ions with the time of aging (or heating). Such a conversion of H-soil or clay to mixed H-A1 system can be arrested by keeping the suspension in a non-aqueous solvent such as methanol or acetone of relatively low dielectric constant (Mukherjee, 1974; Sanyal, 1995; Sanyal *et al.*, 2009).

It has been suggested that Al^{3+} ions in the octahedral position are mobilized laterally by the exchangeable H^+ ions, and exchange positions for the latter (Mukherjee, 1974). This slow process is accelerated on heating. Acid clays and soils, prepared in the usual way, age considerably during preparation so as to contain appreciable amounts of mobilized Al^{3+} ions. If the aged clays are once again passed through a column of H-resin, the Al^{3+} ions will be removed from the exchangeable positions by the H^+ ions. The resulting clay on aging would again mobilize $A1^{3+}$ ions. Thus, if the processes of aging and recovering the H-clay are repeated, the resulting H-clay would reach a stage when the clay would show signs of degradation due to continued depletion of Al from the clay structure (Mukherjee, 1974). Indeed, such degradation has been confirmed by means of electrometric titration, X-ray and DTA studies (Majumdar and Mukherjee 1979).

4.5.1. Aluminium and Iron Polymers in Acid Soils

The exchangeable $A1^{3+}$ ions, displaced by cations from clay minerals, are hydrolyzed to monomeric and polymeric hydroxy-aluminium complexes. As stated earlier (*see* Section 4.2.5), the high charge-to-radius ratio, *i.e.*, high acidity of Al^{3+} ion enables it to rupture the O-H bond in the $[Al(H_2O)_6]^{3+}$ hydrate, causing ionization of the hydrate, thereby yielding protons. This reaction will be facilitated by the presence of a proton acceptor in soil solutions. At pH less than 4.0 to 4.7, when the C_{H+} in solution is already high, the water molecules, the normal proton-acceptor in aqueous solutions, are not available to accept fresh proton generated on hydrolysis of $[A1(H_2O)_6]^{3+}$ (*viz.* Reactions 1 to 6 below). The latter is, therefore, the predominant Al species at pH below 4.0 to 4.7. The stepwise hydrolysis of monomeric $A1^{3+}$ forms, occurring progressively at higher pH values, is shown below (Reactions 1 to 6):

(1) $[Al(H_2O)_6]^{3+} + H_2O = [A1(H_2O)_5OH]^{2+} + H_3O^+$

(2) $[Al(H_2O)_5OH]^{2+} + H_2O = [Al(H_2O)_4(OH)_2]^+ + H_3O^+$

(3) $[A1(H_2O)_4(OH)_2]^+ + H_2O = [Al(H_2O)_3(OH)_3]^0 + H_3O^+$

(4) $[Al(H_2O)_3(OH)_3]^0 + H_2O = [Al(H_2O)_2(OH)_4]^- + H_3O^+$

(5) $[Al(H_2O)_2(OH)_4]^- H_2O = [Al(H_2O)(OH)_5]^{2-} + H_3O^+$

(6) $[Al(H_2O)(OH)_5]^{2-} + H_2O = [Al(OH)_6]^{3-} + H_3O^+$

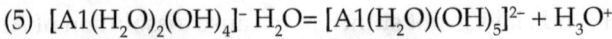

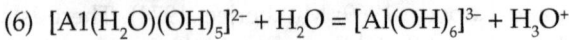

Reactions 1 to 6 occur as the pH progressively increases, *i.e.*, as there are increasing amounts of proton acceptors available for the hydrogen ions generating from the hydrolysis of monomeric hydroxy Al^{3+} species. Between pH 4.7 and 6.5, $[Al(OH)_2(H_2O)_4]^+$ is the principal species. Between pH 6.5 and 8.0, $[(Al(OH)_3(H_2O)_3]°$ is the predominant form, and $[Al(OH)_4(H_2O)_2]^-$ is the major species above pH 8.0. The solubilities of these monomeric species are shown in Figure 4.1 (McLean, 1976). Thus, liming acid soils to higher pH reduces exchangeable and soluble aluminium concentrations as a result of precipitation of Al as insoluble hydroxy Al species. At pH above 6.5, Al becomes increasingly soluble as negatively charged aluminate forms, Al^{3+} being amphoteric in nature [*i.e.*, $Al(OH)_3 + NaOH = Na^+ \{Al(OH)_4\}^-$].

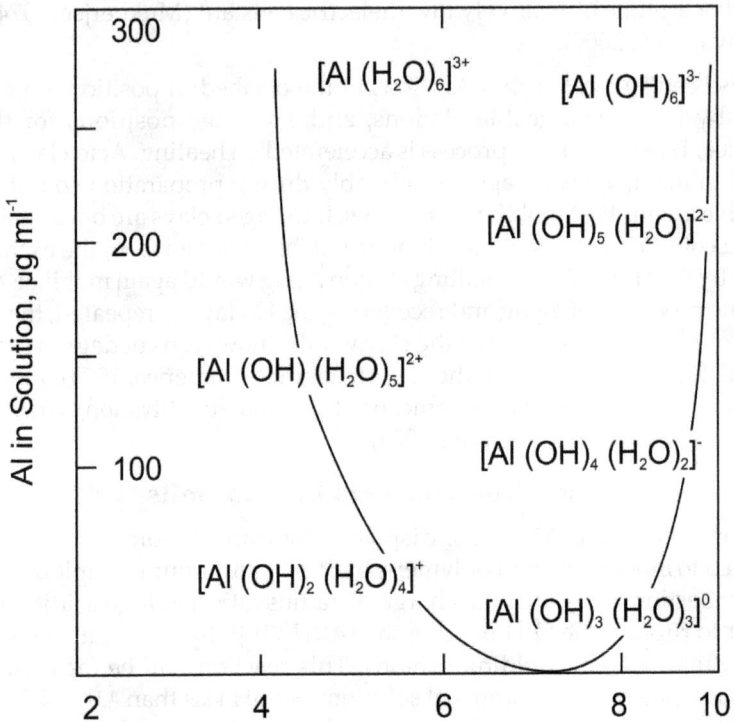

Figure 4.1: Solubilities of Monomeric Al³⁺ Species.
Source: McLean (1976).

The monomeric positively charged, $[Al(OH)(H_2O)_5]^{2+}$ and $[Al(OH)_2(H_2O)_4]^+$ species can polymerize to form both large and small positively charged poly-nuclear complexes (Haynes, 1984). Such polymerisation is surface-catalyzed by soil colloids having large specific surface area. The poly-nuclear compounds, having the general formula, $[Al_m(OH)_{2m-2}(H_2O)_{2m+4}]^{(m+2)+}$, are formed through sharing of hydroxyl groups; further these polymers often age, releasing proton as shown below:

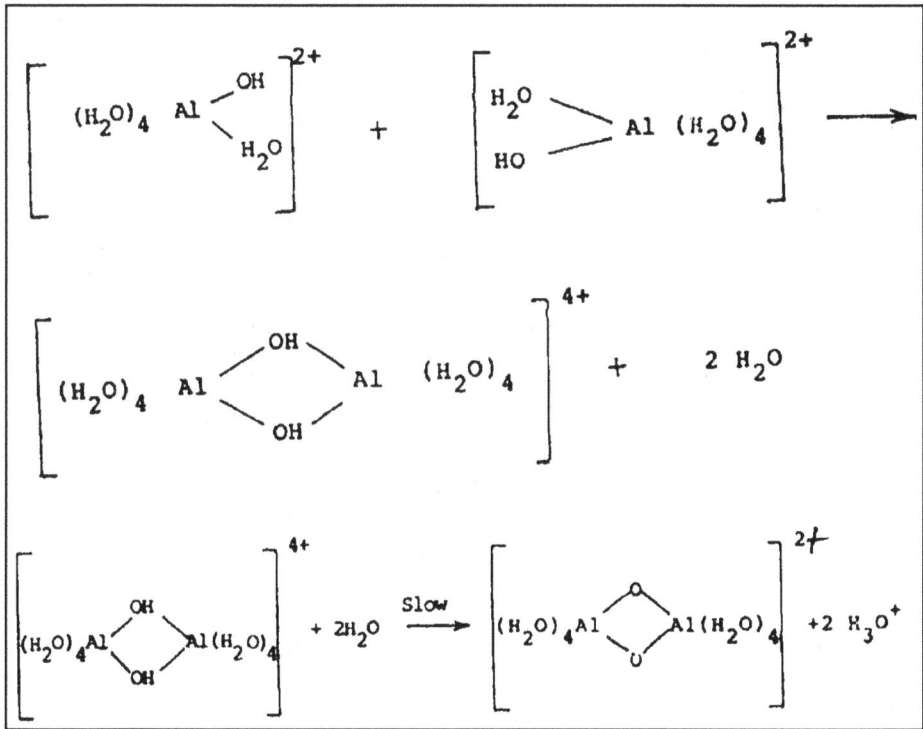

Figure 4.2: Polymerisation of Monomeric Aluminium Species in Acid Soils.
Source: **Adapted from Tisdale *et al.* (1995).**

The given poly-nuclear aluminium compounds may also polymerize further and become firmly bound to the colloid surfaces (Tisdale *et al.*, 1995).

In acid soil (pH less than 5.0), some of the aluminohexahydronium ions, $[Al(H_2O)_6]^{3+}$, remain in solution, but most of them are adsorbed on soil cation exchange sites from which they are easily displaced by the unbuffered salt solutions such as 1M KCl.

On the other hand, hydroxy-Al polycations can be adsorbed on to clay surfaces in forms that are both exchangeable and non-exchangeable with unbuffered salt solutions (Oades, 1979). The sorption of hydroxy-Al polymers by clay minerals is important in soils since small quantities of such polymers distributed over the surfaces of colloidal particles can dramatically change the surface chemistry of the soil. This is because most aluminosilicates have a rather low zero-point of charge (ZPC), and a large negative charge. In contrast, the various hydroxy Al and Fe polymers have a high ZPC and exhibit a large *net* positive charge at the pH of soils unless this is neutralized by sorbed anions such as phosphate. Rengasemy and Oades (1978) showed that sorption of Al and Fe hydroxy cations could cause charge reversal on clay particles.

Hydroxy-Al polycations can also form large, strongly held, interlayer surfaces in expansible lattice clays. A reduction in the CEC in smectites has been correlated with the amounts of interlayer hydroxy-Al species (Sanyal, 1995).

For acidic soils containing appreciable quantities of amorphous hydroxy-Al, the exchangeable Al fraction (*e.g.* 1M KCl-extractable) contains both exchangeable Al^{3+} *per se*, and a variable amount of hydroxyl-Al. The amount of the latter fraction extracted is dependent on soil pH and electrolyte concentration. Thus, exchangeable Al does *not* represent an absolute measurable characteristic of acid soils, but rather it is a mixture of unbuffered salt-extractable forms of Al.

Iron hydrolysis in acid soil is similar to that of aluminium as shown above, *e.g.*

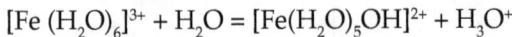

$$[Fe (H_2O)_6]^{3+} + H_2O = [Fe(H_2O)_5OH]^{2+} + H_3O^+$$

4.5.2. Aluminium-Organic Matter Complexes

Aluminium ions (Al^{3+}) form tightly held, stable complexes with organic matter. Aluminium-organic matter complexes are not readily extractable with molar KC1, a fact that accounts for the decrease in KC1 extractable Al as organic matter of soils increases, or when organic manure is added to the soil. Thus, exchangeable Al is generally lower in surface, organic matter-rich horizons than in lower horizons of the same profile (Sanyal, 1995). On the other hand, the translocation of Al^{3+} in presence of strong organic ligands from its immobile soil exchange site to the mobile organic complex releases cation exchange sites for possible occupation by other available cations. Furthermore, based on pure solution experiments, short-chain carboxylic acids were grouped as strong (citric, oxalic, tartaric), moderate (malic, malonic, salicylic), and weak (succinic, lactic, formic, acetic, phthalic) Al-complexing agents (Hue *et al.*, 1986). The Al-detoxifying capacity of these acids was positively correlated with the relative position of OH/COOH groups on their main C-chain, positions that favoured the formation of stable 5- or 6-bond ring structures with Al (Hue *et al.*, 1986). It is, thus, apparent that Al (and also Fe behaving in a similar way) activity in soil solutions is controlled, *not* only by the mineral phases present, but also by the presence of complexing organic ligands. However, little is known about the nature of these organic ligands, and even less about the stability constants of their possible Al and Fe complexes, or the phytotoxicity of such soluble organo-Al or Fe complexes (Haynes, 1984).

4.6. Other Causes of Soil Acidity

4.6.1. Soluble Salts

The presence of salts in soil solution leads to an increase in soil acidity due to displacement, followed, by hydrolysis, of adsorbed Al^{3+}, Mn^{2+}, and Fe^{3+} ions (and also H^+ ions) by the cation of the salt. Divalent cations have a greater effect than monovalent cations.

The addition of a fertilizer in a band would lead to generation of high 'local' concentration of soluble salts, leading to enhanced acidity. Thus, out of two wheat rotations, spanning over 17 years, with one rotation receiving only ammonium phosphate, while the other ammonium phosphate plus ammonium nitrate at an average rate of N of 35 kg ha^{-1} yr^{-1}, the soil pH was lowered in both the cases, the latter having it more (by 0.5 unit pH); soil pH seemed inversely related to soil NO_3-N (Campbell and Zentner, 1984). Especially, in the soils, marginal in soil pH, further acidification through such an avenue may be detrimental to plant growth, and recommendations for more extensive crop rotations should have provision for eventual need to lime the soil to sustain the production in course of intensive cultivation practices, followed round the year which necessitates the increased use of, for instance, nitrogenous fertilizers (Sanyal, 1995).

4.6.2. Carbon Dioxide

The pH of a soil, containing free $CaCO_3$ and in equilibrium with CO_2 at 1 atm pressure of the above-ground air is 8.5. An increase of 0.02 atm in partial pressure of CO_2 of the soil atmosphere leads to a fall of pH by one unit. The carbonic acid formed by the reaction between soil water and CO_2 released during the decomposition of organic residues in soil will contribute to soil acidity through its weak dissociation. Root activity and metabolism also contribute CO_2 to soil atmosphere (Tisdale *et al.*, 1995).

4.6.3. Acid Rain

On a world scale, the composition of acid rain is variable, but it always contains H$^+$ ions at a concentration greater than 2.0 µM, a value for an aqueous solution in equilibrium with atmospheric CO_2. Cations other than H$^+$ ion like NH_4^+ are present in rain water. The positive charge is balanced by a variety of anions, generally sulphate and nitrate, usually in the molar ratio between 3: 1 and 1: 1 (Sanyal, 1995).

The effect of acid precipitation on soil processes is a function of the nature and concentration of its cations, principally H$^+$ ion, and also of the anionic composition. Nitrate as a plant nutrient is readily taken up by growing vegetation, causing a release of an equivalent quantity of hydroxyl or other weakly basic anions. These, in turn, neutralize the H$^+$ ion, thereby avoiding *net* acidification. In the absence of growing plants, however, for instance, in a clear-cut forest or fallow land, biological nitrate uptake is eliminated, and nitrate, unlike phosphate, and to a lesser extent sulphate, being mobile geochemically, moves readily into the soil solution, causing soil acidity.

4.7. Buffering Action of Acid Soils

The exchangeable H$^+$ ions and Al^{3+} ions on soil colloids are in equilibrium with the corresponding ions in soil solution. The Al^{3+} ions in solution phase on hydrolysis leads to the generation of protons. The degree of ionization of these cations from the exchange phase into soil solution determines the nature of soil acidity. The former constitutes what is known as the potential or reserve acidity of soil. The free H$^+$ ions in the solution phase determine the active acidity. The potential acidity is in equilibrium with active acidity in a given soil.

On neutralization of active acidity, and or precipitation of the soluble Al^{3+} as Al(OH)$_3$ on addition of extraneous base, the reserve acidity restores the active acidity while maintaining the above equilibrium. The soil pH is thus maintained through such buffer action.

The magnitude of potential acidity far exceeds that of active acidity, more so in a clayey soil having high organic matter content. The buffering capacity, which is inversely related to the slope of the usual pH-neutralization titration curve of an acid soil (*vide*. Figure 2.11 in Section 2.7.1), is thus greater in clayey soils than in sandy soils. Furthermore, the Ultisols and Oxisols, containing chiefly (1 : 1) type of clays, have typically smaller buffering capacity as compared to soil (such as Mollisols and Alfisols) having larger amount of (2 : 1) type of clays (Sanyal, 1995).

4.8. Determination of Active and Potential Acidity

4.8.1. Active Acidity

A suitable measure of active acidity of soil can be obtained by measuring the soil pH using a glass electrode. The soil: water ratio in the suspension is generally 1:2.5 which is purely conventional, and was used originally perhaps for the reasons of experimental convenience. In most of the agricultural conditions, cropped to plants, such a high dilution is rarely obtained. Efforts were made (unpublished work of A. Chatterjee, Calcutta University; by private correspondence) to gradually reduce the relative content of water while monitoring the pH of soil suspension. The latter, naturally, decreased (for acid soils) on gradual "drying".

In measuring soil pH, suspension effect has been noted by several workers. Explanations have been put forward in terms of (i) exchange acidity adsorbed on the soil particle surfaces, and (ii) non-zero junction potential at the KC1-salt bridge used to couple the glass electrode to the reference (Calomel) electrode.

Lime Potential

Soil pH, as stated earlier, is affected by the presence of soluble salts which could be taken care of (*i.e.*, eliminated), from the soil by repeated leaching before taking the pH measurement. This, however, is a time-consuming process. An alternative approach is to use 0.01 M $CaCl_2$ (or KC1) solution in place of water for preparing the soil suspension in which to measure the pH. Such a high salt concentration, as compared to that generally encountered in soil systems, is expected to "level out" the effect on soil pH of much smaller differences in salt content of soil systems. Furthermore, as has been argued for a Donnan system, such as soil (*see* Section 3.2.4 and Eq. 3.44a), the ionic activity ratios (on equivalent basis) of similarly charged ions, rather than a single ion activity (*e.g.* a_{H+}) are expected to remain more constant throughout the colloidal Donnan system (Sanyal *et al.*, 2009), such as soil. Thus, the Schofield's Ratio Law states that for a soil having H^+ and Ca^{2+} ions (as would happen for 0.01 M $CaCl_2$ being used to prepare the soil suspension) on the exchange complex which is in equilibrium with the solution phase, the following ratio would remain constant, *i.e.*,

$$\left[a_{H^+} / \sqrt{a_{ca^{2+}}} \right]_{exch} = \left[a_{H^+} / \sqrt{a_{ca^{2+}}} \right]_{soln} = \text{Constant}$$

$$\left(\frac{a_{H^+}}{\sqrt{a_{ca^{2+}}}}\right)_{exch} = \left(\frac{a_{H^+}}{\sqrt{a_{ca^{2+}}}}\right)_{soln} = \text{Constant}$$

Or, $- \log (a_{H^+})_{soln} + 1/2 [\log (a_{Ca}^{2+})_{soln}] = \text{constant}$

Or, $\text{pH} - 1/2 \, \text{pCa} = \text{constant}$

The expression, (pH - 1/2 pCa), is termed as the lime potential of the soil, and is found to remain fairly constant over a wide range of soil-salt levels in acid soils.

Soil pH is a useful indicator of exchangeable H^+ and Al^{3+} ions. The former is normally present in measurable quantities at pH values only below 4.0.

4.8.2. Potential Acidity: Lime Requirement of Acid Soils

Soil pH is not indicative of the reserve of potential acidity, and hence of the amount of lime (commonly adopted acid soil-reclamation method, *see* later) to be applied for raising the soil pH for optimum crop growth. For the purpose, potential acidity must be given due consideration, and also the buffering action of the given soil.

The liming materials are calcic limestone ($CaCO_3$), dolomite [double carbonate of Ca and Mg; Ca Mg(CO_3)$_2$], quick lime (CaO), hydrated (slaked) lime [Ca(OH)$_2$], coral shell lime or Marl or chalk ($CaCO_3$) and basic slags, obtained as by-products from iron and steel plants (mostly $CaSiO_3$), press-mud from sugar factories that use carbonation process, as well as miscellaneous sources of lime, such as wood ash, ground oyster shells, paper mill sludge, tanneries and by-product $CaCO_3$ from the fertilizer factories, using gypsum process, are used for reclaiming acid soils (Sanyal *et al.*, 2015a). At the same pH, a fine textured acid soil requires much larger quantity of lime than does a light textured soil and the lime should be applied based on the lime requirement of the soil. The Indian Council of Agricultural Research (ICAR) formulated a Network Project on Acid Soils for optimized liming dose for different crops in acid soil areas throughout the country. The general guideline emerging from the studies revealed that application of lime @ 0.2-0.4 t ha⁻¹in combination with 50 per cent recommended dose of NPK fertilizers results in equal or a slightly higher yield than the application of 100 per cent NPK alone. Since leaching under high rainfall conditions is quite fast, application of amendment is not a one-time process and needs to be repeated at regular intervals. Further the cost of liming can be reduced by application of lime in cropped furrows only or by applying it in split doses. With appropriate management options, the interval between liming can be stretched, while continuing with high yields (Chaitanya, *et al.*, 2016; Sanyal *et al.*, 2015a).

The reactions of these liming materials in correcting the soil acidity are schematically shown as follows:

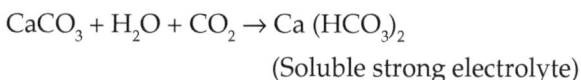

$CaCO_3 + H_2O + CO_2 \rightarrow Ca (HCO_3)_2$

(Soluble strong electrolyte)

$Ca(HCO_3)_2 \rightleftharpoons Ca^{2+} + 2HCO_3^-$

Fully dissociated   in soil solution
in soil solution

| Colloidal anion of | H^+ | | | Colloidal anion | |
| an acid soil | H^+ | $+ Ca^{2+} \rightarrow$ | | of acid soil | $Ca^{2+} + 2H^+$ |

$2H^+ + 2HCO_3^- \rightarrow 2H_2CO_3 \rightleftharpoons 2H_2O + 2CO_2 (\uparrow)$

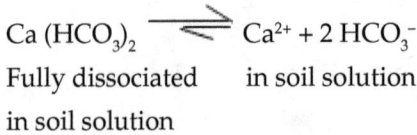

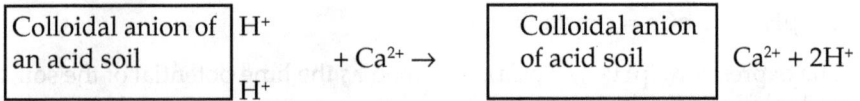

In soil solution   unstable

For application of CaO and $Ca(OH)_2$ as liming material to acid soils, the formation in soil solution of soluble calcium bicarbonate [$Ca(HCO_3)_2$], a strong electrolyte which dissociates completely in soil solution (as shown above), follows the pathways given below.

$CaO + H_2O + 2CO_2 \rightarrow Ca(HCO_3)_2$

$Ca(OH)_2 + 2CO_2 \rightarrow Ca(HCO_3)_2$

The acid neutralizing capacity of these liming materials differs considerably, and this necessitates a methodology to compare the same referred to that of some standard material. The latter is chosen to be pure calcium carbonate. The following reactions are worth considering:

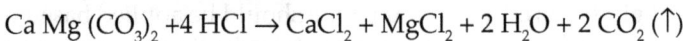

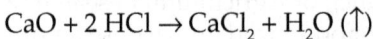

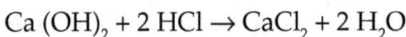

$CaCO_3 + 2HCl \rightarrow CaCl_2 + H_2O + CO_2 (\uparrow)$

$MgCO_3 + 2HCl \rightarrow MgCl_2 + H_2O + CO_2 (\uparrow)$

$CaMg(CO_3)_2 + 4HCl \rightarrow CaCl_2 + MgCl_2 + 2H_2O + 2CO_2 (\uparrow)$

$CaO + 2HCl \rightarrow CaCl_2 + H_2O (\uparrow)$

$Ca(OH)_2 + 2HCl \rightarrow CaCl_2 + 2H_2O$

$CaSiO_3 + 2HCl \rightarrow CaCl_2 + H_2O + SiO_2 (\downarrow)$

Obviously, while 100 g of $CaCO_3$ neutralizes two moles of HCl, the same amount of acid is neutralized by, respectively, 84 g of $MgCO_3$, 92 g of $CaMg(CO_3)_2$ (dolomite), 56 g of CaO, 74 g of $Ca(OH)_2$ and 116 g of $CaSiO_3$ (basic slag). Hence the neutralizing values [Calcium Carbonate Equivalent (CCE) of Liming Value] of these molecules (pure forms) are as follows:

$CaCO_3$: $(100/100) \times 100 = 100$ per cent

$MgCO_3$: $(100/84) \times 100 = 119$ per cent

$CaMg(CO_3)_2$: $(100/92) \times 100 = 109$ per cent

CaO: $(100/56) \times 100 = 179$ per cent

$Ca(OH)_2$: $(100/74) \times 100 = 135$ per cent

$CaSiO_3$: $(100/116) \times 100 = 86$ per cent

There appears to be no standard pH to decide as when to lime an acid soil; it varies from soil to soil, and with crop species. Because Al^{3+} ion is often the predominant phytotoxic, soluble and exchangeable cation in acid soils, affecting, through its adverse influence on root growth, the nutrient and water uptake by plants (Marschner, 1986; Miranda and Rowell, 1987), an index of its concentration is often used as an indication of whether lime is required, especially in tropical soils. Historically, exchangeable Al (1M KCl-extractable) was used as an index of Al toxicity in acid soils. However, both exchangeable Al and Al saturation percentage were shown to be poor predictors of Al toxicity. The concentration of soil solution Al has been considered by some to be a better measure of Al toxicity potential (Haynes, 1984).

Even though several studies have suggested that lime should be added in amounts equivalent to neutralize the exchangeable Al, such a measure often greatly underestimates the amount of lime that will react with highly weathered acid soils. Usually lime rates equivalent to 1.5-3.0 times the exchangeable Al are required to neutralize the entire Al in acid soils. Such results indicate that there are forms of reactive Al in soils that are not exchangeable with 1M KCl, but that will react with lime.

Indeed, the level of soluble and exchangeable Al in highly weathered acid soils may be partially controlled by a buffering reserve of potentially reactive Al. This includes (1) positively charged hydroxy-Al polymers of various sizes and degrees of hydration, (2) interlayered hydroxy-Al in vermiculite and montmorillonite, (3) allophane and allophane-like constituents, and (4) Al^{3+} and hydroxy-Al in the forms of non-exchangeable organic matter-Al complexes (McLean, 1976). Such a reactive pool of Al may be particularly important when chemical fertilizers are applied to acid soils, because the resulting salt effect can create a decrease in local soil pH and the release of Al^{3+} ions into the soil solution from the reserve. The latter may also be important in intensively cropped soils where the plant acts as a sink for soluble and exchangeable Al. Thus, exchangeable Al is only a short-term estimate of the potentially phytotoxic pool of soil Al, and hence only a short-term estimate of lime requirement (LR) of an acid soil.

Use of 0.5M $CaCl_2$ for extraction of reactive Al in surface soils has been suggested The strong complexing ability of Cu (II) in combination with the acidity of the $CuCl_2$ solution makes it an effective extractant for Al bound to organic matter, although it does also appear to extract some inorganic reactive Al such as that in clay interlayers (Sanyal, 1995).

Titratable acidity, as determined by $BaCl_2$ – triethanolamine, buffered at pH 8.2, can agree closely with the amount of $CaCO_3$ that reacts with highly weathered soils (Mehlich *et al.*, 1976). Titratable acidity values are a consequence of weakly dissociated acidic groups in soils which give rise to pH-dependent CEC; the buffering reserve of reactive Al greatly contributes to such groups.

The main difficulty of evolving a standard extractant for reactive Al is that the latter comprises different forms of Al in soil, the relative importance of which differs in different soils and in different soil horizons.

4.9. Buffer-Lime Requirement Methods

The lime requirement (LR) determined by most chemical methods is an estimate of the amount of liming material required by a given volume of soil to attain a desired pH or base saturation. Most LR determinations are conducted by quick-test methods involving buffer solutions (Shoemaker *et al.*, 1961; Tisdale *et al.*, 1995). These buffer solutions measure the exchangeable Al plus a variable portion of the non-exchangeable reactive pool of Al, depending upon time of reaction and the particular soil studied. In fact, these methods are based on measurement of depression of pH of standard buffered solutions [such as in the Shoemaker, McLean and Pratt (SMP) buffer methods] when a given quantity of the acid soil under study is added to a definite quantity of the former solution. These pH depressions, which are functions of the original pH and the buffering action of the experimental soil, are calibrated to obtain the amount of lime required to bring the given acid soil to a desired and prescribed higher pH.

The drawback of such methods is that they are generally aimed at adjusting soils to a given pH, usually 6.5. In highly weathered soils, such as Ultisols of the warm and humid climates, liming to pH greater than 5.6 to 5.7 and up to pH 6.5 can induce deficiencies of many nutrients including Mg, K, Mn, Zn, B, Fe, and particularly P (Sanyal and De Datta, 1991). Nevertheless, in some acid soils, where Mn toxicity and/or Ca deficiency is limiting crop growth, a pH of 6.0-6.5 may be required for optimal production. Indeed, Noble *et al.* (1988) studied the alleviation of Al toxicity in nutrient solution culture by $CaSO_4$, and such alleviation was reported to have been partly due to an increase in the formation of the less phytotoxic $[Al(SO_4)]^+$ species.

Extent of fertilizer applications can also be an important consideration. Thus, when an acid producing nitrogen fertiliser, *e.g.*, ammonium sulphate, is used, even at modest rates, it can neutralize an entire lime application within 2-3 years (Sanyal, 1995). Fox (1980) also pointed out that in variable-charge soils, the pH will fluctuate adjacent to roots widely unless the concentrations of basic cations in soil solution are closely matched to plant uptake.

4.10. Subsoil Acidity

In many highly weathered soils such as Ultisols, subsoil acidity limits crop yields even when the plough layer is adequately limed. Root proliferation may be severely restricted in subsoil due to Al toxicity, and consequently, the water stored in the subsoil is rendered unavailable to crops.

In order to avoid exposure of the infertile subsoil for deep incorporation of lime, recourse is made to leaching of surface-applied amendments to correct for the subsoil acidity. It has been demonstrated that the soil pH and the exchangeable Ca in the subsoil can be modified in 2-5 years by the leaching of surface-applied lime (Sanyal, 1995). To some extent, heavy fertilization with ammonium and potassium would, over a period of time, effect considerable downward movement of Ca by cation exchange. Shainberg *et al.* (1989) further suggested the use of gypsum ($CaSO_4$) as an ameliorant for subsoil acidity since Ca from this source leached downward at much faster rates than did that from $Ca(OH)_2$. Hern *et al.* (1988) suggested leaching of lime and EDTA, applied together at the surface, is responsible for the subsequent chelation and removal of exchangeable Al from acid subsoil. The results obtained supported the hypothesis that organic ligands can have a marked effect on the migration and subsequent ionic reactions of metal ions within the soil exchange matrix.

Although much research has centered on complex laboratory methods of estimating the lime requirements of soils, the practical problem of correcting the subsoil acidity, an important limiting factor in many acid soils, has lagged well behind. The problem certainly deserves further research attention.

4.11. Acid Sulphate Soils

4.11.1. Introduction

Millions of hectares of land on coastal plains would be suitable for rice cultivation but for the presence of potential and actual acid sulphate soils. Potential acid sulphate soils (Sulfaquents) occur in tidal swamps. They have high levels of pyrite, low levels of bases and produce strongly acid sulphate soils when pyrite oxidizes to sulphuric acid after drainage. Pyrite formation is favoured in brackish and saline mangrove swamps, dissected by many tidal creeks, where coastal aggradation is slow. Where salinity is low, Sulfaquents can give good rice yields (van Breemen and Pons, 1978).

Productivity of crops, *e.g.* rice is higher on the deeper-developed and more-leached acid sulphate soils (Sulfic Tropaquepts). The dominant adverse soil factors are iron toxicity and sometimes phosphate deficiency in Sulfic Tropaquepts. Aluminum toxicity may be a problem, especially in deep water areas where rice is broadcast on the unflooded land at the start of the wet season.

Liming (Sulfaquepts), applying phosphate (in Sulfic Tropaquepts) and the use of suitable varieties are important in increasing soil productivity. But the key to improving crop productivity lies mainly in good water management, especially in tidal areas with Sulfaquents, and in swamps with highly acidic Sulfaquepts. By keeping the water table above the pyritic subsoil and by preventing prolonged drying of the surface soil, the pH, which controls all other adverse factors, is kept in a safe range. In this respect, areas of acid sulphate soils with a continuously wet climate are more favourable for a rice crop than those with a monsoon climate (van Breemen and Pons, 1978).

Millions of hectares of low-lying coastal land in the tropics are either poorly-productive or unsuitable for agriculture because the soils are acid sulphate or potentially acid sulphate.

Acid sulphate soils are soils that have, somewhere within a 50-cm depth, a pH below 3.5 (for Entisols) or 4.0 (for Inceptisols) that is directly or indirectly caused by sulphuric acid, formed by oxidation of pyrite (cubic FeS) or rarely of other reduced sulphur compounds. Potential acid sulphate soils are poorly drained and highly pyritic with a nearly neutral or slightly acid reaction in the field. They become acid sulphate soils if pyrite oxidizes after drainage (van Breemen and Pons, 1978).

Despite the high acidity and its associated adverse effects on the plant growth, acid sulphate soils have characteristics favourable, especially for submerged rice crop. They are commonly supplied with plant nutrients from appreciable amounts of 2:1 clay minerals and organic matter. Moreover, the increase in pH caused by reduction of soil upon flooding is a favourable factor. Therefore, it is not surprising that moderately acid sulphate soils are often used for rice growing.

4.11.2. Occurrence, Genesis, Properties and Classification

Of the actual and potential acid sulphate soils (5 million ha) in Southeast and East Asia soils, about two-thirds are found in Indonesia, Thailand, India and Vietnam. Millions of hectares of shallow peat land in Indonesia are also underlain by potentially acid sediments. Data on the potential acid sulphate soils (Sulfaquents) in tidal marshes in Bangladesh, Burma and India tend to be tentative. The total area of actual and potential acid sulphate soils in West Africa may be as large as 6.6 million ha (Beye et al.,1975) and smaller areas are found in coastal areas of East Africa, Australia and tropical South America, especially in the Guianas and Venezuela (Orinoco Delta) (van Breemen and Pons, 1978).

Indeed, the formation of potential acid sulphate soils is favoured in mangrove swamps dissected by many creeks where the tidal amplitude is large, and where the supply of new sediment is too small to cause rapid coastal accretion but high enough to prevent the formation of pure mangrove peats. Such conditions are frequently seen near the mouth of rivers, supplying small amounts of sediment. Even if the supply of sediment is large, however, coastal accretion can be slow if there is a relative rise in the sea level.

Genesis of Acid Sulphate Soils

Appreciable aeration of potential acid sulphate soils and subsequent acidification start only after the water table stays below the upper part of the highly pyritic zone for several weeks. A perquisite for such drainage is decreased tidal influence. This is brought about either gradually by natural processes (coastal accretion or a relative decrease in sea level, e.g. by tectonic uplift) or more abruptly, by empoldering. But in many of the mangrove areas is Southeast Asia still under the tidal influence, acidification takes place in material from the subsoil brought to the surface by the mound building mud lobster Thalassina anomale (van Breemen and Pons, 1978)).

Oxidation of Pyrite

After drainage, the physically unique soil will crack, and air will penetrate. In well-aerated but moist soil, the fine-grained pyrite, typical of mangrove sediments, is first oxidized to dissolved ferrous sulphate and sulphuric acid. The intermediate reaction steps are not completely understood (Bloomfield and Coulter, 1973; van Breemen, 1973), but the overall reaction can be expressed as:

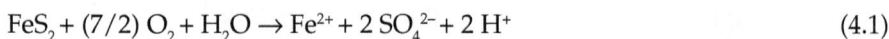

$$FeS_2 + (7/2) O_2 + H_2O \rightarrow Fe^{2+} + 2 SO_4^{2-} + 2 H^+ \tag{4.1}$$

Chemical oxidation of pyrite is quite slow but the oxidation rate increases greatly in the presence of microbes of the genus Thiobacillus. These autotrophs, some of which can function below pH 2, derive energy from the oxidation of reduced sulphur compounds and, in the case of *T. ferrooxidans,* from the oxidation of Fe^{2+} to Fe^{3+}. Dissolved Fe^{3+} rapidly oxidizes pyrite according to (Breemen and Pons, 1978) the following reaction.

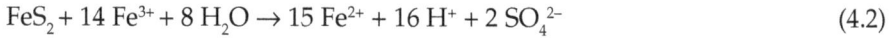

$$FeS_2 + 14\ Fe^{3+} + 8\ H_2O \rightarrow 15\ Fe^{2+} + 16\ H^+ + 2\ SO_4^{2-} \tag{4.2}$$

By this process, pyrite quickly removes Fe^{3+} which would soon be depleted if *T. ferrooxidans* did not catalyze the oxidation of Fe^{2+} to Fe^{3+}, a normally slow reaction at low pH (van Breemen and Pons, 1978). This plus the low pH required for high levels of dissolved Fe^{3+} explains the frequently reported positive effect of high acidity and the presence of *T. ferrooxidans* on pyrite oxidation. Equation (4.2) also explains why at low pH, ferric oxides and the basic ferric sulphate, jarosite, $[KFe^{III}_3(SO_4)_2(OH)_6]$, which is responsible for the typical yellow mottles found in most acid sulphate soils, cannot persist in the vicinity of pyrite; the ferric iron in these minerals would be reduced to Fe^{2+} ion. So the ferrous sulphate resulting from reaction (4.1) can only be oxidized at some distance from the pyrite, after being transported by diffusion or mass flow. Acid is generated further by this process (van Breemen and Pons, 1978):

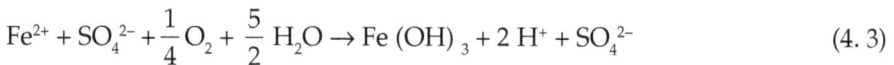

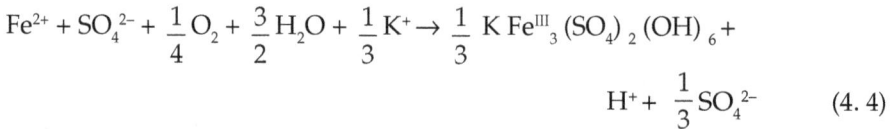

$$Fe^{2+} + SO_4^{2-} + \frac{1}{4}O_2 + \frac{5}{2}H_2O \rightarrow Fe\,(OH)_3 + 2\,H^+ + SO_4^{2-} \tag{4.3}$$

$$Fe^{2+} + SO_4^{2-} + \frac{1}{4}O_2 + \frac{3}{2}H_2O + \frac{1}{3}K^+ \rightarrow \frac{1}{3}K\,Fe^{III}_3\,(SO_4)_2\,(OH)_6\ +$$

$$H^+ + \frac{1}{3}SO_4^{2-} \tag{4.4}$$

When potentially acid soil samples are exposed to air in moist conditions, strong acidification often occurs within a few weeks. Within a few months, the pH falls below 3 or 4 (and often below 2) and pyrite contents in the order of 3 to 6 per cent may be halved in 2 or 3 months (van Breemen and Pons, 1978). However, except when pyritic soil is directly exposed by excavation, pyrite oxidation in the field is much slower, probably because the supply of oxygen by rather slow diffusion through wet soils is the rate-limiting step (Richmond *et al.*, 1975). In mangrove marshes where pyrite still occurs at shallow depth, drainage may cause appreciable acidification within a few months. By contrast, pyrite oxidation during a dry season is hardly noticeable in well-developed acid sulphate soils where the pyritic substratum is found in 1 or 2 m below the soil surface (van Breemen and Pons, 1978).

4.11.3. Soil Management, Reclamation and Improvement

It is clear that different problems are posed by *potential* acid sulphate soils, young acid sulphate soils and old acid sulphate soils. The management options, reclamation and improvement of these different categories of acid sulphate soils are discussed hereunder (van Breemen and Pons, 1978):

(1) Tidal marshes with potential acid sulphate soils (Sulfaquents), physically unripe.

    (2) Empoldered, artificially drained land in tidal areas with young acid sulphate soils (Sulfaquents), physically half-ripe near the surface.

    (3) Nontidal swamps with very acid Sulfaquepts, having potentially acid soil at shallow depth, physically ripe to half-ripe near the surface; and Nontidal, periodically flooded coastal plain areas with deeply developed and moderately acid Sulfic Tropaquepts or Sulfic Haplaquepts, physically ripe to at least 50-cm depth.

In reclaiming land of the first three categories, two approaches are possible in addition to applying chemical amendments, namely to limit pyrite oxidation and try to inactivate the existing acidity by maintaining a high water table, or else to drain intensively to achieve maximum oxidation of pyrite and try to remove the acidity by leaching (van Breemen and Pons, 1978).

Under favourable hydrologic conditions, as in soil piled on ridges for dryland crops, the second method can give productive soil within a few years. But in paddies, this process is slow and the first approach is normally preferred. The prospects for such reclamation strongly depend on the possibilities of fresh water supply, as well as climate. It is generally easier to keep the pyritic substratum water saturated in continuously wet equatorial regions than in areas that have one or two pronounced dry seasons. However, physically unripe Sulfaquents need to be drained to some extent to increase the bearing capacity, and thus complete prevention of pyrite oxidation is not always practical (van Breemen and Pons, 1978).

4.12. Salt Affected Soils

4.12.1. Salinity and Sodicity

Soil salinization, a generic term for soil salinization and sodification, is the process of accumulation of salts in the root zone to an extent that these adversely impact the plant growth (Gupta and Gupta, 2014; Sanyal *et al.*, 2015a). These soils are commonly encountered in arid and semiarid regions of generally low to moderate rainfall where evaporation exceeds the rainfall, but are also found in the coastal regions (Gupta and Gupta, 2014). In saline soils (Figure 4.3), increased osmotic pressure of soil solution in the crop rhizosphere restricts the entry of water into the plant body, thereby affecting the crop growth. This aids in plant desiccation. The harmful effects of excess salts in the crop root zone also leads to preferential absorption of one ion that might retard the absorption of other essential plant nutrients, as well as excess uptake of some of the salt constituents to cause toxicity of specific ions in the plants (Gupta and Gupta, 2014; Sanyal *et al.*, 2015a).

The accumulation of excess salts in the soil results from the parent material, and also from irrigation of soils using the groundwater rich in salts, without making adequate provision for optimum drainage, thereby leading to capillary rise and accumulation of salts in the soil, subjected to evapo-transpiration losses when the water evaporates, leaving behind the salts on the surface soil and the crop rhizosphere. Further, the seepage from the salt-rich canal or irrigation channels also causes salt accumulation in soils. In several occasions under arid to semi-arid region, salt incrustations are formed on the soil surface, thereby severely restricting the soil fertility and plant growth. Obviously under such environmental conditions of less rainfall and the water shortage, the leaching of the salts down the soil profile is very much restricted.

The soluble salts accumulating in saline soils are generally chlorides and sulphates of Na^+, Ca^{2+} and Mg^{2+} ions, while the K^+, carbonate and bicarbonate concentrations are generally less. A measure of such salt accumulation in saline soils is obtained from the experimentally determined value of the electrical conductivity (specific conductance) of the saturation extract of the given soil, denoted by EC_e. For a saline soil, the value of EC_e is more than 4.00 dS m^{-1} at 25°C. The pH of these soils is generally less than 8.5. The exchangeable sodium percentage [ESP (per cent) = (Exchangeable sodium on the exchange complex of the soil in cmol (p$^+$) kg^{-1}/cation exchange capacity of the soil in cmol (p$^+$) kg^{-1}) x 100] of the saline soils is less than 15 per cent. Yet another parameter is known as the sodium absorption ratio (SAR) which refers to the irrigation water quality as well as the soil solution characteristic. This is given by the ratio,

$$SAR = (K^+)/\sqrt{[(Ca^{2+}) + (Mg^{2+})]/2}$$

Where ( ) terms denote the concentrations of the respective cations in irrigation water or the soil solution, expressed in the unit of cmol.L^{-1}. Obviously the SAR values take care of the moderating influence of Ca^{2+} and Mg^{2+} ions on the deteriorating effect of excess Na^+ ions on soil properties (*see* later). With increased evapotranspiration under arid to semi-arid climate (that favours the formation of salt-affected soils), the concentration of the soil solution constituents increases. As a result, the relatively sparingly or lowly soluble components, namely calcium sulphate and carbonate and magnesium carbonate (unlike sodium chloride, sulphate or carbonate or the corresponding potassium salts) get precipitated out of the soil solution, thereby leading to an increase in SAR values of the soil solution. The latter favours the increased sodium saturation of the soil exchange complex and hence facilitates the formation of sodic or alkali soil (*see* later).

The above mentioned EC_e values are also used to calculate the total salt loading in a saline soil by using the following formulations:

For soils with EC_e ranging from 0.1 to 5 dS m^{-1}, total dissolved salts in the soil = EC_e (dS m^{-1}) x 640, and

Sum of soluble cations or anions (mmol L^{-1}) = EC_e (dS m^{-1}) x 10

For soils with EC_e values ranging from 3.0 to 30 dS m^{-1}, the osmotic potential of the soil solution = EC_e (dS m^{-1}) x (-0.36). The more negative is such osmotic potential, the higher is the osmotic stress experienced by the crops growing in saline soils, and hence more adverse is the effect on crop growth.

Another category of salt affected soils are known as the Sodic soils. These are also termed alkali soils (Figure 4.3). The formation of sodic soils is preceded by accumulation of relatively excess amount of sodium salts in soil solution which is in a dynamic equilibrium with the exchange complex. That is, salinity development in soil is the precursor to the development of sodic soils. Such excess Na$^+$ ion accumulation in soil solution leads through ion exchange process to the gradual Na$^+$ ion enrichment of the soil exchange complex, and hence an increased degree of exchangeable sodium (Na$^+$) saturation of the said soil exchange complex. This, in turn, increases the zeta (ξ) potential of the diffuse double layer of the soil colloidal fraction (*see* Section 2.6.1 and Eq. 2.4). The latter causes dispersion of the fine soil colloidal particles which block the soil pores, leading to increased soil dispersion, destruction of the soil structure and formation of crust on the soil surface that hinders air/water movement in the soil (Gupta and Gupta, 2014; Sanyal *et al.*, 2015a). Besides, toxic effects of Na and mismatch between the plant uptake of chemical constituents also affect the crop growth. Sodic soils have pH more than 8.5, EC$_e$ less than 4.00 dS m^{-1} at 25°C and ESP greater than 15 per cent. Majority of these sodic soils in India contain calcium carbonate at a depth about a meter below the surface.

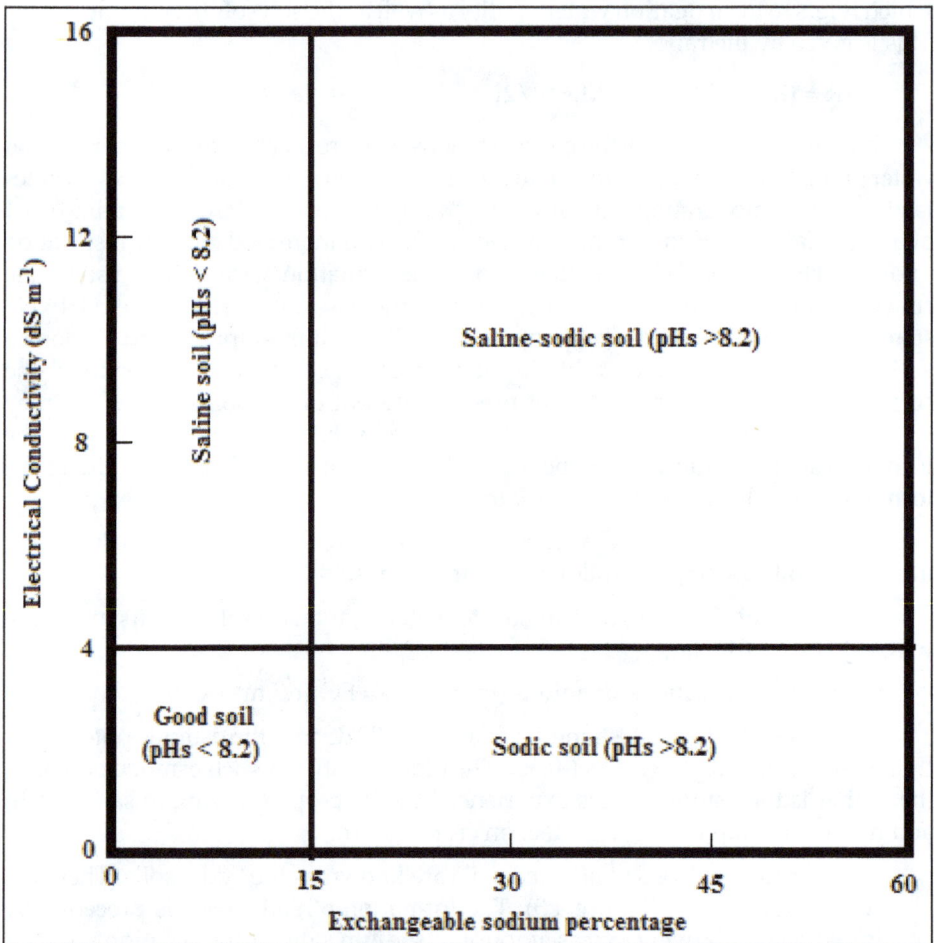

Figure 4.3. Categorization of Salt-affected Soils.
Source: Gupta and Gupta (2014) and Sanyal *et al.* (2015a).

Yet another type of salt affected soils is saline-sodic soils which have both the excess salts and increased content of exchangeable sodium percentage (Figure 4.3). These are characterized by pH more than 8.5, EC_e more than 4.00 dS m^{-1} at 25°C and ESP greater than 15 per cent. Besides the adverse impacts on chemical and physical processes, soil salinization/sodification has a direct negative effect on soil biology that also aids in reducing the crop productivity (Gupta and Gupta, 2014; Sanyal *et al.*, 2015a). An area of 6.73 Mha in India is saline/sodic in nature, 2.96 Mha being saline and 3.77 Mha sodic (ICAR and NAAS, 2010).

Extreme events, climatic aberrations and anthropogenic interventions are likely to further aggravate the spatial extent of these soils (CSSRI, 2006; Sanyal *et al.*, 2015a).

Amelioration

Prevention and reclamation of salt affected soils require an integrated approach, comprising of physical/mechanical processes, hydraulic processes, chemical processes, biological processes, and reuse and/or safe disposal of drainage water, wherever necessary (CSSRI, 2006; Gupta and Gupta, 2014; Sanyal *et al.*, 2015a).

Leaching of soluble salts from the root zone is practiced with relatively good quality irrigation water, *not* rich in salts (*e.g.* harvested excess rainwater may be used). This would be effective when adequate drainage system (*e.g.* surface and/or sub-surface) is in place or else has to be installed. Such leaching process is especially effective in heavy textured soils, mostly characterized by excessive salt accumulation at the soil surface. In this context, the **Leaching Requirement (LR)** is obtained by taking the ratio of equivalent depth of the drainage water to the depth of the applied water, that is, the fraction of water that must be leached through the root zone to control the soil salinity at a specified level (Biswas and Mukherjee, 1994). However, for obtaining the overall beneficial effect, the *environmentally safe disposal* of the drainage water, rich in leached salts, must be ensured (CSSRI, 2006).

Ideally, LR is given as, $LR = EC_{applied\ water} / EC_{drainage\ water}$

Further, the on-farm land development including dyking, leveling, tillage to conserve rainfall and/or increase infiltration rate and deep tillage is also beneficial for reclamation of the saline soils (Gupta and Gupta, 2014; Sanyal *et al.*, 2015a).

The chemical amendments used to reclaim the sodic soils (characterized by high ESP > 15 per cent , and high soil pH > 8.5) are mostly soluble (or sparingly) salts of calcium, namely gypsum (relatively less costly and available), $CaSO_4.2H_2O$ (sparingly soluble); calcium chloride, $CaCl_2$; and phospho-gypsum, available as industrial by-product; acid formers, such as elemental sulphur, pyrites (FeS_2), iron and aluminium sulphates, etc. (Bajwa and Swarup, 2009). However, a non-calcareous soil does not respond to the acidifying amendments (Sanyal *et al.*, 2015a).

The mode of action of the most commonly used amendment for sodic soils, namely gypsum is shown below. As an amendment, it has to bring down the ESP and pH of a sodic soil. In this context, it is worth recapitulating the conjugate acid-base theory for aqueous acids acid bases, *i.e.*, the conjugate acid of a strong base is a weak acid, and *vice versa*. Furthermore, water is amphoteric in nature and behaves as a weak acid or a weak base, depending on whether it reacts (for instance, during the hydrolysis reaction) with a stronger base or a stronger acid. It is common knowledge that NaOH and KOH are strong bases, stronger than $Ca(OH)_2$ (pK_b of NaOH, KOH and $Ca(OH)_2$ at 25°C are, respectively, 0.2, 0.5, and 2.43). Hence the conjugate acids, namely Na^+ and K^+ ions act as much weaker acids than does Ca^{2+} ion. Similarly NH_4^+ being the conjugate acid of a weak base, NH_4OH, it is a strong acid. Thus, whereas hydrolysis of Na^+ and K^+ ions in aqueous solutions (*e.g.* soil solution) does *not* take place, that of Ca^{2+} ion occurs to a considerable extent. Furthermore, the SO_4^{2-} ion is the weak conjugate base of the corresponding strong parent acid, namely aqueous H_2SO_4. As a result, while Ca^{2+} ion undergoes hydrolysis in soil solution (of a, for instance, sodic soil), SO_4^{2-} ion does *not*. On the other hand, gypsum being sparingly soluble in water, a small amount of it passes into the solution phase prepared in water, forming a saturated solution, while most of it remains undissolved (solid) in dynamic equilibrium with such saturated solution. However, being a salt and thus a strong electrolyte, the dissolved gypsum dissociates fully into its constituent ions (Ca^{2+} and SO_4^{2-} ions) in its saturated solution. Therefore, when gypsum is added to a sodic soil, the following chemical reactions take place, bringing down the ESP and the pH of the soil.

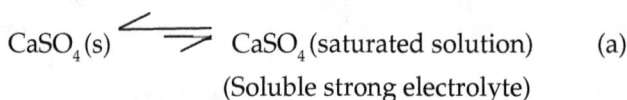

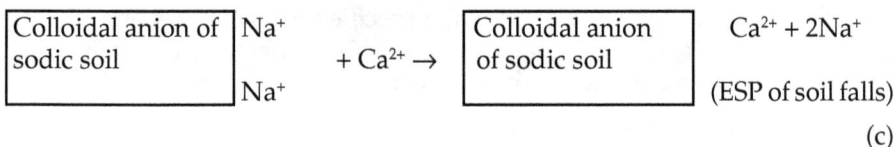

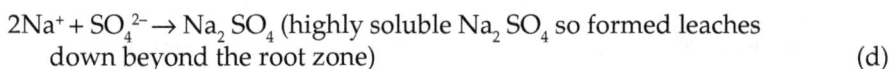

$$CaSO_4(s) \rightleftharpoons CaSO_4 \text{ (saturated solution)} \qquad (a)$$
$$\text{(Soluble strong electrolyte)}$$

$$\underset{\text{Fully dissociated in soil solution}}{CaSO_4 \text{ (saturated solution)}} \rightleftharpoons \underset{\text{in soil solution}}{Ca^{2+} + SO_4^{2-}} \qquad (b)$$

$$\boxed{\begin{array}{l}\text{Colloidal anion of } Na^+ \\ \text{sodic soil} \qquad\qquad Na^+\end{array}} + Ca^{2+} \rightarrow \boxed{\begin{array}{l}\text{Colloidal anion} \\ \text{of sodic soil}\end{array}} \quad Ca^{2+} + 2Na^+$$
$$\text{(ESP of soil falls)} \qquad (c)$$

$$2Na^+ + SO_4^{2-} \rightarrow Na_2SO_4 \text{ (highly soluble } Na_2SO_4 \text{ so formed leaches}$$
$$\text{down beyond the root zone)} \qquad (d)$$

In soil solution

$$Ca^{2+} + 2H_2O \rightarrow Ca(OH)^+ + H_3O^+ \qquad (e)$$
$$\text{In soil solution}$$

$$H_3O^+ + OH^- \rightarrow H_2O \text{ (Soil pH drops)} \qquad (f)$$
$$\text{In soil solution}$$

On such utilization of Ca^{2+} ions as above through pathways (c) and (e), the equilibria represented by the Eqs. (a) and (b), are disturbed, and hence more of solid gypsum passes into the soil solution forming the saturated solution [*i.e.*, Eq. (a)], and the fresh gypsum so dissolved in soil solution undergoes complete dissociation [*i.e.*, Eq. (b)] to provide for the fresh supply of Ca^{2+} ions which continues to amend the soil sodicity through the pathways (c), (e) and (f). In this manner, the said amendment process is sustained.

Magnesium sulphate ($MgSO_4 . 7H_2O$), on the other hand, is not used in place of gypsum (calcium sulphate dihydrate, $CaSO_4 . 2H_2O$) because the former is highly soluble in water (about 300 times more soluble than gypsum), and hence the salt will completely dissolve in soil solution on application, and thus will not be able to maintain a sustained supply of the ameliorating ion (*e.g.* Mg^{2+}), unlike gypsum.

In presence of carbonates in sodic soil, elemental sulphur, when applied, forms sulphuric acid on microbial oxidation, mediated by the *Thiobacillus* sp. The acid so formed *in situ*, reacts with $CaCO_3$ to form calcium sulphate that amends the sodic soil. Iron pyrite, on application, also undergoes oxidation to generate sulphuric acid *in situ*, which in turn, reacts with $CaCO_3$ to form calcium sulphate that amends the sodic soil. Thus,

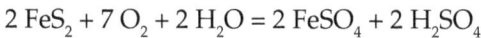

$$2 FeS_2 + 7 O_2 + 2 H_2O = 2 FeSO_4 + 2 H_2SO_4$$

$$CaCO_3 + H_2SO_4 = CaSO_4 + H_2O + CO_2$$

On simple considerations, the Gypsum equivalents of some of the amendments described above are given as follows:

Gypsum: 1.00; calcium chloride dihydrate: 0.85; sulphur: 0.18; iron pyrite: 0.63; iron sulphate ($FeSO_4 . 7H_2O$): 1.62; limestone ($CaCO_3$): 0.58.

Biological interventions such as green manuring, application of FYM or other organic amendments also provide for rapid improvements in soil physical conditions of the sodic soils and serve as a means of integrated nutrient management (Gupta and Gupta, 2014; Sanyal *et al.*, 2015a).

References

Bajwa, M. S. and Swarup, A. (2009). Soil salinity and alkalinity. **In**: *Fundamentals of Soil Science* (N. N. Goswami, R. K. Rattan, G. Dev, G. Narayanasamy, D. K. Das, S. K. Sanyal, D. K. Pal and D. L. N. Rao, Eds.) Second Edition, Indian Society of Soil Science, New Delhi, pp.329-347.

Beye, G., Toure, M. and Arial, G. (1975). Acid sulphate soils of West Africa: Problems of their management for agriculture use. *Proc. Int. Rice Res. Conf.*, International Rice Research Institute, Los Baños, Laguna, Philippines, p.10.

Biswas, T. D. and Mukherjee, S. K. (1994). *Textbook of Soil Science*, Second Edition, Tata McGraw-Hill Publishing Co. Ltd., Bew Delhi, pp. 404-405.

Bloomfield, C. and Coulter, J. K. (1973). Genesis and management of acid sulphate soils. *Adv. Agron.*, **25**, 265-326.

Campbell, C. A. and Zentner, R. P. (1984). Effect of fertilizer on soil pH after 17 years of continuous cropping in South-Western Saskatchewan. *Can. J. Soil Sci.*, **64**, 705-710.

Chaitanya, A. K., Badole, S., Gupta, A. K. and Pal, B. (2016). Prevention, reclamation and management of acid and acid sulphate soils. **In**: *Emerging Issues and Technologies in Salinity Management for Agriculture* (S. K. Gupta and M. R. Goyal, Eds.). Apple Academic Press Inc., Ontario, Canada (in Press).

CSSRI. (2006). *CSSRI-A Journey to Excellence*. Central Soil Salinity Research Institute, Karnal. pp. 156.

Fox, R.L. (1980). Soils with variable charge: Agronomic and fertility aspects. **In:** *Soils with Variable Charge* (B. K. G. Theng, Ed.), New Zealand Soc. Soil Sci., Lower Hutt, p. 195-224.

Gupta, S. K. and Gupta, I. C. (2014). *Salt Affected Soils: Reclamation and Management*. Scientific Publishers (India), Jodhpur.

Haynes, R. J. (1984). Lime and phosphate in the soil-plant system. *Adv. Agron.*, **37**, 249-315.

Hern, J. L., Menser, H. A., Sidle, R. C. and Staley, T. E. (1988). Effects of surface-applied lime and EDTA on subsoil acidity and aluminum. *Soil Sci.*, **145**, 52.

Hue, N. V., Craddock, G. R. and Adam, F. (1986). Effect of organic acids on aluminum toxicity in subsoils. *Soil Sci. Soc. Am. J.*, **50**, 28-34.

ICAR and NAAS (2010). Degraded and Wastelands of India–Status and Spatial Distribution. Indian Council of Agricultural Research, New Delhi, p. 158.

Jackson, M. L. (1963). Aluminum bonding in soils: a unifying principle in soil science, *Soil Sci. Soc. Am. Proc.*, **27**, 1-10.

Majumder, R. N. and Mukherjee, S. K. (1979). Degradation characteristics of hydrogen montmorillonites. *J. Indian Soc. Soil Sci.*, **27**, 26-37.

Marschner, H. (1986). *Mineral Nutrition of Higher Plants*. Academic Press, New York, p. 486.

McLean, E. O. (1976). Chemistry of soil aluminum. *Commun. Soil Sci. Pl. Anal.*, **7**, 619-636.

Mehlich, A., Bowling, S. S. and Hatfield, A. L. (1976). Buffer pH acidity in relation to nature of soil acidity and expression of lime requirement. *Commun. Soil Sci. Pl. Anal.*, **7**, 253-263.

Miranda, L. N. de and Rowell, D. L. (1987). The effects of lime and phosphorus on the function of wheat roots in acid top soils and subsoils. *Pl. Soil*, 104, 253-262.

Mitra, R. P. and Kapoor, B. S. (1969). Acid character of montmorillonite: Titration curves in water and some non-aqueous solvents. *Soil Sci.*, **108**, 11-23.

Mukherjee, S. K. (1974). Electrochemistry of clays and clay minerals. **In:** *Mineralogy of Soil Clays and Clay Minerals*. Bulletin No. 9 (S. K. Mukherjee and T. D. Biswas, Eds), Indian Society of Soil Science, New Delhi, pp. 87-102.

Noble, A. D., Sumner, M. E. and Alva, A. K. (1988). The pH dependency of aluminum phytotoxicity: Alleviation by calcium sulfate. *Soil Sci. Soc. Am. J.*, **52**, 1398-1402.

Oades, J. M. (1979). Interaction of metal ion species with clays. **In:** *Colloids in Soils - Principles and Practice (D. E. Yates, Ed.)*. pp. 6.1–6.34.

Panda, N. and Koshy, M. M. (1982). Chemistry of acid soils. In: *Review of Soil Research* (N. S. Randhawa, N. N. Goswami, I. P. Abrol, G. S. R. Krishna Murti, A. B. Ghosh, S. S. Prihar, R. S. Murthy, Sant Singh, T. G. Sastry and G. Narayanasamy,

Eds.), Part 1, 12ᵗʰ International Congress of Soil Science, February 8-16, 1982, New Delhi, Indian Society of Soil Science, New Delhi, pp. 160-168.

Rengasemy, P. and Oades, J. M. (1978). Interaction of monomeric and polymeric species of metal ions with clay surfaces. III. Aluminium (III) and chromium (III). *Aust. J. Soil Res.,* **16**, 53 – 66.

Richmond, T. De A., Williams, J., and Datt, U. (1975). Influence of drainage on pH, sulphate content and mechanical strength of a potential acid sulphate soil in Fiji. *Trop. Agric.* (Trinidad), **52**, 325-334.

Sanyal, S. K. (1995). Ionic environment of acid soils. **In**: *Acid Soil Management* (M. A. Mohsin, A. K. Sarkar and B. S. Mathur, Eds.), Kalyani Publishers, Ludhiana, New Delhi, pp. 31-47.

Sanyal, S. K. and De Datta, S. K. (1991). Chemistry of phosphorus transformations in soil. *Adv. Soil Sci.,* **91**, 1-120.

Sanyal, S. K., Gupta, S. K., Kukal, S. S. and Jeevanrao, K. (2015a). Soil degradation, pollution and amelioration. **In**: *State of Indian Agriculture - Soil* (H. Pathak, S. K. Sanyal and P. N. Takkar, Eds.), National Academy of Agricultural Sciences, New Delhi, pp. 234-266.

Sanyal, S.K., Poonia, S.R. and Baruah, T.C. (2009). Soil colloids and ion exchange in soil. **In**: *Fundamentals of Soil Science.* (N.N Goswami, R.K. Rattan, G. Dev, G. Narayanasamy, D.K. Das, S.K. Sanyal, D.K. Pal and D.L.N. Rao, Eds.), Second edition, Indian Society of Soil Science, New Delhi.

Shainberg, I., Sumner, M. E., Miller, W. P., Farina, M. P. W., Pavan, M. A. and V. Fey, M. (1989). Use of gypsum on soils: A review. *Adv. Soil Sci.,* **9**, 1-111.

Shoemaker, H. E. McLean, E. O. and Pratt, P. F. (1961). Buffer methods for determining lime requirement of soils with appreciable amounts of extractable aluminum. *Soil Sci. Soc. Am. Proc.,* **25**, 274-277.

Tisdale, S. L., Nelson, W. L., Beaton, J. D. and Havlin, J. L. (1995). *Soil Fertility and Fertilizers.* Fifth Edition, Prentice Hall of India, New Delhi

van Breemen, N. (1973). Soil forming processes in acid sulphate soils. **In**: *Acid Sulphate Soils* (H. Dost, Ed.), *Proc. Int. Symp.,* ILRI Publ., 18, Vol. I, Wageningen, The Netherlands, pp. 66-130.

van Breemen, N. and Pons, L. J. (1978). Acid sulphate soils and rice. **In**: *Soils and Rice.* International Rice Research Institute, Los Baños, Laguna, Philippines, pp. 739-761.

Soil Organic Matter

5.1. Introduction and Definitions

Soil organic matter (SOM) consists of a mixture of plant and animal residues in various stages of decomposition, substances synthesized microbiologically, and/or chemically, from the breakdown products, and the bodies of live and dead microorganisms and their decomposing remains (Schnitzer, 2000).

Although SOM is largely of plant origin (Stevenson, 1994), such organic matter is not recognizable under a light microscope as possessing the cellular organization of plant material, and is termed Humus. Flaig *et al.* (1975), among others, discussed the origin of humus. These humic substances (HS) are best described as a series of acidic yellow/brown to black poly-electrolytes with variable molecular weights (Stevenson, 1994; Schnitzer, 2000). Indeed, the indication is that one is dealing here with a complex and difficult group of compounds, where some of the normal concepts are difficult to apply (Swift, 1999). Thus the naturally occurring soil HS are *not* defined in terms of their chemical composition or typical functional group contents. As stated earlier briefly (*see* Section 1.1.2), they are defined instead operationally, in terms of solubility, *i.e.*, their solubility behaviour in aqueous solution at different pH ranges (Swift, 1999). Soil humic acids (HAs) are generally defined as being soluble in alkali and insoluble in acidic solution (pH = 1 to 2). Such a definition lacks specificity since a large number of organic compounds, insoluble in acids, but soluble in alkali, is not necessarily HAs or soil HS. The fulvic acid (FA) fraction of SOM, on the other hand, is defined as that which is soluble in both acid and alkali, while the organic matter, that is not solubilized by alkali or acid, is referred to as the humin fraction, and usually makes up about 20 per cent of SOM (Orlov, 1985). Besides a potential source of confusion and uncertainty, inherent in the above mentioned definitions, there is yet some degree of additional confusion as to the synonymy (or otherwise) of a number of terms often used interchangeably by the soil scientists, namely SOM, humus and HS. Stevenson (1994) identified SOM with humus, but according to Schnitzer (2000), total HS is synonymous with SOM and humus as long as losses occurring during the extraction and separation procedures are held to a minimum. Schnitzer (2000) further went on to define HS, synonymous with SOM, as the sum of humic acid (HA), fulvic acid (FA) and humin (Sanyal, 2002b).

A number of objections have been raised by many workers against the use of alkaline extracts for SOM. According to Stevenson (1994), alkali (i) encourages contamination of SOM extract by promoting silica dissolution from mineral matrix, (ii) allows auto-oxidation by air of some organic components, during prolong periods of standing in contact with air, and (iii) causes condensation reaction between amino acids and carbonyl group of reducing sugars or quinones to form Maillard reaction products. Schnitzer (2000) adds to the list the difficulty of laborious and time-consuming nature of the given extraction and fractionation procedure of SOM. Schnitzer (2000) suggest that a new approach, independent of wet chemical methods, will be useful, and proceeds on to discuss the ^{13}C NMR spectroscopic and pyrolysis field ionization mass spectrometric method as tool for direct analysis of SOM (*see* later).

However, Krosshavn *et al.* (1992), while studying the influence of wet chemical fractionation procedure on the chemical composition of SOM by using the solid-state-^{13}C NMR spectroscopic technique, detected no measurable structural charges in the humus fractions on chemical treatment with aqueous NaOH and HCl. Thus, the relative contents of functional groups in the humus fractions, when totaled, became very close to values of the functional group content of the whole soil (Figure 5.1).

5.2. Formation of Humus

The transformation of biotic debris accumulating in soil is believed to occur through initial degradative processes that lead to the formation of humic substances. The more recalcitrant lignin components are selectively preserved in this stage. Subsequently the substrates and the preserved products of the earlier stage are subjected to further transformations through synthetic processes to lead to humus formation (Stevenson, 1994).

Humus formation is essentially a biochemical process. The degradative reactions involving organic residues are catalyzed by microbial enzymes. The subsequent synthesis of humus can be mediated by both enzymes and abiotic catalysts (such as clays and hydrous oxides of Fe and Al) present in soil (Bollag *et al.*, 1998). Soil enzymes are mainly of microbial origin, and to a lesser extent, of plant origin (Sanyal, 2002b).

Notwithstanding the fact that the synthesis of humus in soil environment has been dealt by many workers, and as a result, many different theories have been proposed (Flaig *et al.*, 1975; Stevenson 1994; Schnitzer, 2000), the exact pathways of formation of humus are yet to be fully established. These theories may be classed into three major types. One of the major oldest theories proposed minor biochemical

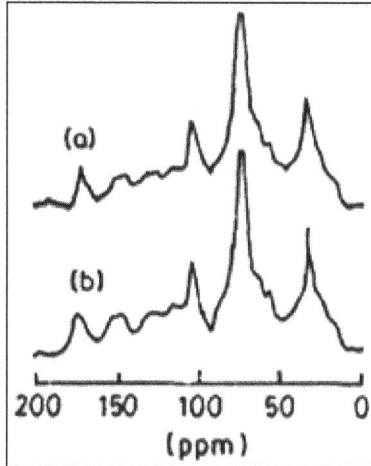

Figure 5.1: ¹³C NMR Spectra of (a) Whole Soil and (b) Sum of Humus Fractions.
Source: **Krosshavn *et al.* (1992).**

modifications, microbial transformation and degradations of lignin or other resistant plant materials under soil conditions (Waksman and Iyer, 1932). Another pathway suggests that plant materials serve exclusively as a source of energy for certain bacteria and fungi which in the process synthesize high molecular weight polymers, releasing them into soil environment upon their death (Bollag *et al.*, 1998). A yet third theory limits the role of the microorganisms to the production of various amino compounds, peptides, polysaccharides and phenols that, once discharged into soil, polymerize to form humus (Sanyal, 2002b). The essential difference between the first pathway, on one hand, and the second and the third, on the other, is that the former is based on oxidative degradation of the existing resistant polymers of plant origin, while the latter involves the synthesis of new macromolecules which undergo oxidative degradation over time (Swift, 1999).

While reviewing these theories, Stevenson (1994) emphasized the important role of lignin in humus formation in soils; however, he (1994) also pointed out that the phenolic moieties of many HAs may not have arisen from lignin precursors, but must have been formed *in situ* by microbial synthesis instead. It thus appears that all the pathways, mentioned above, and possibly others (Swift, 1999) contribute to humus formation in most soils.

5.3. Structural Aspects

The structural chemistry of humic substances is controversial, and is likely to remain so, because it seems that no two molecules are alike, with substantial amounts of these random polymers remaining till date unidentified. Despite the importance of the concept of biogenesis of HAs (Stevenson, 1994), the chemical route of humic acid synthesis has received attention. Attempts have been made to synthesize humic substances by polycondensation and polymerisation of different phenols, carbohydrates, etc., with or without nitrogen-containing compounds (Bollag *et al.*, 1998; Schnitzer, 2000; Datta *et al.*, 2001). Traditionally, phenolic compounds derived from lignin, such as catechol, hydroquinone, pyrogallol, syringic acid, have been used as substrates for polymerisation reactions. Such polycondensation reaction was shown to be catalyzed by clays (Schnitzer and Khan, 1972; Sanyal, 2002b). The synthetic polymers so produced are quite similar to the natural humus, particularly with respect to elemental analysis, exchange capacity, total acidity, functional group make-up, resistance to microbial degradation and identity of phenols recovered upon sodium amalgam degradation (Kononova, 1966; Schnitzer, 2000; Datta *et al.*, 2001).

Apart from establishing the role of polymerisation reactions in humus synthesis, the synthetic analogues of HAs, the so-called synthetic HAs, have also provided useful starting materials for studying much more complex naturally occurring soil humic substances. Indeed, the main difficulty of studying the properties of humic acids, and their interaction with other soil components, appears to be linked to the molecular complexities of the HS. Thus, unlike other macromolecules such as proteins and polysaccharides, these substances cannot be split into definite fractions by the usual degradation techniques (Datta *et al.*, 2001).

The basic structure of soil humic substances is believed (Stevenson, 1994) to be an aromatic ring of the di- or trihydroxy phenyl type, bridged together by -O-, $-CH_2-$, -NH-, and other groups (including aliphatic), and also containing carboxylic groups, attached both directly to the ring, and on the aliphatic side chains. Some oxygen may also exist as quinone/carbonyl (> C = O) group. In the natural state, humic substances contain attached proteineous and carbohydrate residues, which are also capable of forming complexes with trace and heavy metal ions. Such binding with cations occurs at a continuum of reactive sites, ranging from weak forces of attraction to the formation of strong coordinate links (Datta *et al.*, 2001).

Schnitzer (2000), on the basis of findings from the analytical methods, involving pyrolysis field ionization mass spectrometric, ^{13}C NMR spectroscopic and GC/MS techniques, suggested that alkyl aromatics are important building blocks of HAs and other humic substance. This author (2000) further proposed a structural skeleton shown in Figure 5.2. The proposed structure (Figure 5.2) has oxygen present as carboxyl, phenolic (OH), alcoholic (OH), esters, ethers, and ketonic groups, while nitrogen as nitrile and heterocycle structure. The number of structural voids, inherent in this skeleton, implies high flexibility and micro-porosity, and hence the ability of HA to entrap organic and inorganic macromolecules (such as lipid and biocides, clay minerals and hydrous oxides), as well as water. In addition, the voids and such flexible nature of the polymers could explain the ability of humic substances to bind cations that influence the stereochemical arrangement of functional groups to form stable complexes.

The elemental composition of HA, according to the proposed structure (Figure 5.2), is $C_{308} H_{334} O_{90} N_5$, with molecular weight of 5540 Dalton. The proposed structure (Figure 5.2) is consistent with the chemical, oxidative and reductive degradative, colloid-chemical, electron microscopic and ^{13}C NMR and X-ray data obtained in course of many years (Schnitzer, 2000). Further, the carbohydrates and the

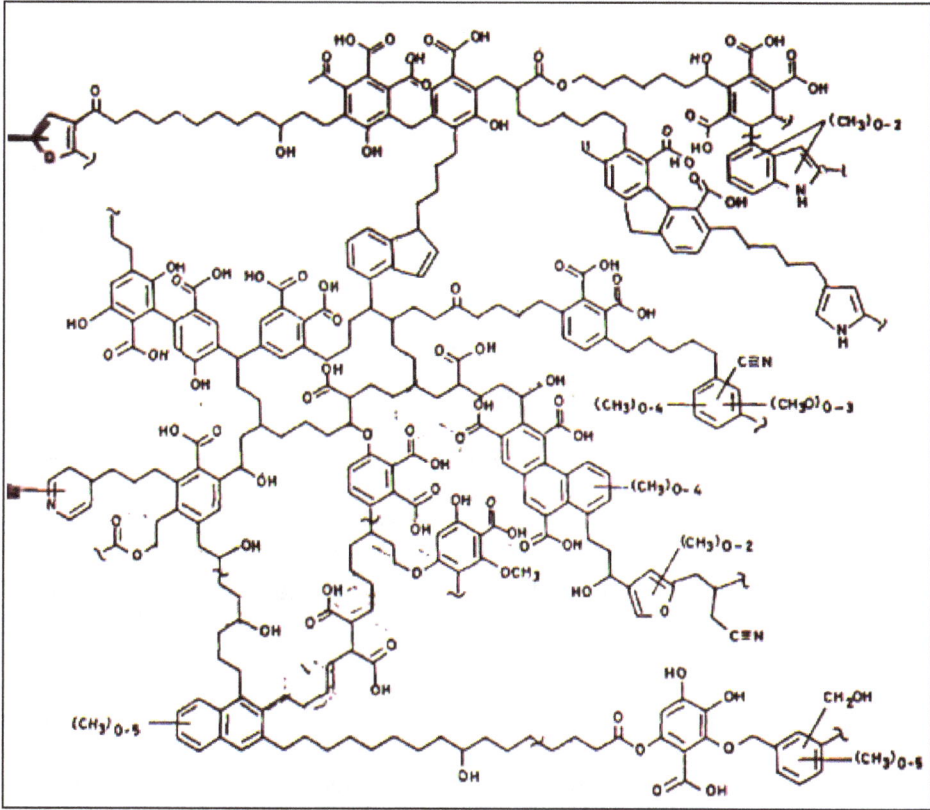

Figure 5.2: Two-dimensional Humic Acid Model Structure.
Source: Schnitzer (2000).

proteineous materials are adsorbed on external surfaces and in internal voids with hydrogen-bonds playing a major role immobilizing them within the humic structure (Schnitzer and Schulten, 1995).

Earlier, Stevenson (1994) had proposed a type of structure of HA on the basis of survey of spectral data for natural humus. This is shown in Figure 5.3. One would notice an extensive degree of commonalities between these two structures (Figs. 5.2 and 5.3), which is considered satisfactory.

The findings on humic structure, enriched by increasing use of ^{13}C NMR spectroscopy, has been summarized (Tate, 2000) with the suggestion that aromatic rings comprise a smaller portion of HA molecule than was previously believed.

5.4. Shape and Size of Humic Molecules

There is a wide array of views as to the shape and size of the humic molecules. Based on the early work on electron microscopic investigation, HAs were thought to be spherical in shape, but the structure viewed was a highly condensed result of the vacuum and electron beam (Orlov, 1985). On the other hand, viscosity measurements

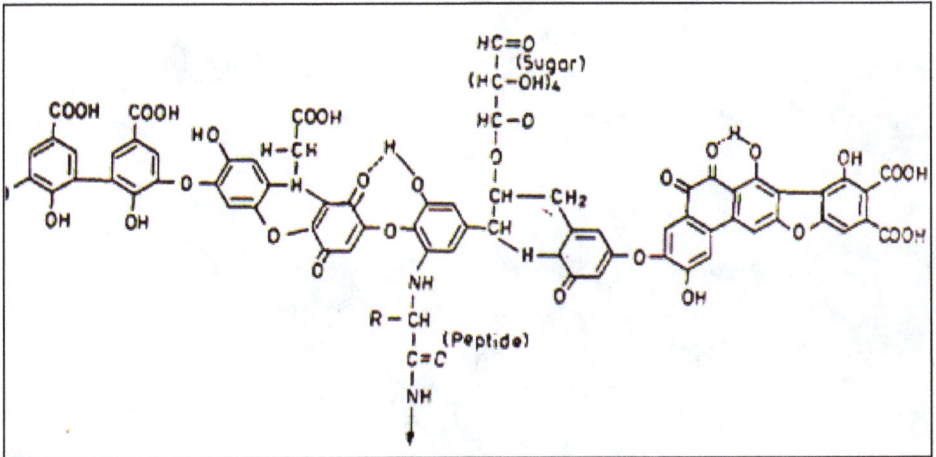

Figure 5.3: Structural Type of Humic Acid.
Source: Stevenson (1994).

of low molecular weight humic polymers were interpreted as indicating ellipsoid or rod-like polymers with axial ratios of about 10 (Relan *et al.*, 1984; Mandal and Sanyal, 1984; Tomar *et al.*, 1992a; Schnitzer, 2000). Tomar *et al.* (1992a) proposed from viscosity studies that a rod-like configuration for humic substances is more probable than a disc one. Chen and Schnitzer (1976) and Cameron *et al.* (1972) argued that the polymers behaved as flexible linear polyelectrolytes, at least at pH values at which they naturally occur in soil. From a study of sedimentation values and the ratios of the frictional coefficients of humic polymers with solvent, Cameron *et al.* (1972) concluded that except for high molecular weight-humic polymers (with molecular weight exceeding 10^5), the remaining molecules were of linear configuration. Later work by Swift (1999) used the frictional ratio (f/f_0) of the molecules studied, where f is the frictional coefficient of the molecule under consideration, while f_0, that of a hypothetical molecule of the same volume, but having a condensed spherical conformation. The value of this ratio (f/f_0) is helpful to infer about the shape of the macromolecule concerned. Using this approach, in particular, the relationship obtained between (f/f_0) and molecular weight (MW) for wide range of molecular weight of the polymers was very close to the one derived theoretically for a random coil structure of the macromolecule, namely $f/f_0 = 0.3 \, MW^{0.167}$. This is illustrated in Figure 5.4 (Swift, 1999). Some deviations observed at the higher MW side may have arisen from the corresponding chain branching and cross-linking, leading to a more condensed core. Presence of a small amount of dispersed colloidal mineral matter in such high molecular weight samples may also be partly responsible for such deviation (Swift, 1999).

Visser (1985), from viscosity measurements, suggested that the higher molecular weight HAs had more elongated structure than their lower molecular weight-counterparts, such as FAs, and that on hydration, the humic molecules become more elongated. However, Ghosh and Mukherjee (1971) had earlier shown that the HA molecule was a linear colloid in aqueous medium, but a spherocolloid in

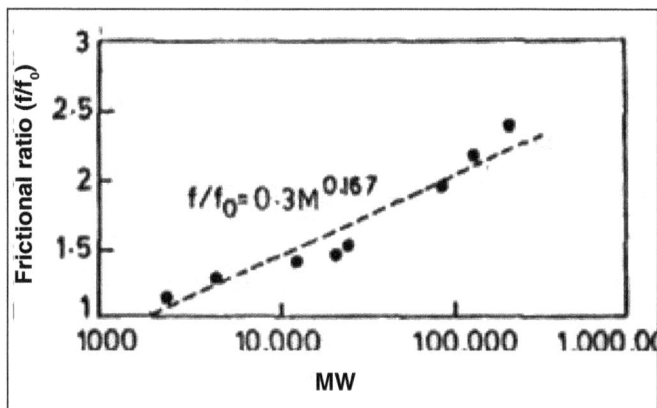

Figure 5.4: Plot of the Experimental Relationship between the Frictional Ratio and Molecular Weight (•) Obtained for Humic Substances and the Theoretical Line for the same Relationship for Molecules with Random Coil Structure (- - - - - -).
Source: **Swift (1999).**

presence of large amounts of neutral electrolytes, whereas Khan (1971) considered HAs as mixtures of spherical and linear particles. The earlier studies on shape and dimensions of humic substances have been reviewed by Stevenson (1994), and HA shapes have been projected as more of linear type rather than three dimensional network structure. The ^{13}C NMR studies suggested a predominantly aromatic character of these linear coils (Stevenson, 1994).

Schnitzer (2000) summarized the information on molecular size and shape of HAs and FAs, obtained by his research school, using viscosity and surface tension data at varying concentrations of HAs and FAs and neutral salts, as well as at different pH. According to Ghosh and Schnitzer (1980), the humic concentration, pH and ionic strength of the medium control the molecular characteristics of HAs and FAs. Thus, in an attempt to resolve the controversy (discussed above) as regards the shape and the size of humic molecules, Ghosh and Schnitzer (1980) proposed that HAs and FAs behaved like rigid, uncharged spherocolloid at (i) high concentration (> 3.5 to 5.0 g/liter), (ii) low pH (< 6.5 for HA and < 3.5 for FA), and (iii) electrolyte concentrations < 0.05 M. These conclusions about molecular configurations of HA and FA are presented in Figure 5.5 (Ghosh and Schnitzer, 1980; Schnitzer, 2000). Thus, in soil solutions, where normally both humic and salt concentrations would be rather low, HA (at pH > 6.5) and FA (at pH > 3.5) would be expected to occur as flexible linear polyelectrolytes (Schnitzer, 2000). Additional support for such inference was provided by the transmission electron micrograph of HA (Figure 5.6) which demonstrated a linear chain-like structure (Schnitzer, 2000).

5.5. Polyectolytic Behaviour from Viscosity Studies

An examination of the polyelectrolyte behaviour of HAs and its variations with HA concentration would be in order (Sanyal, 2002b). The viscosity characteristics of polyelectrolytes are distinct compared to those of uncharged polymers. Thus for the latter, the reduced viscosity, η_{red} [$\eta_{red} = \eta_{sp}/C = \{(\eta/\eta_0)-1\}/C$], is virtually

Sample	FA									
Sample conc. ↓	Electrolyte(NaCl) conc. in M					pH				
	0.001	0.005	0.010	0.050	0.100	2.0	3.5	6.5		9.5
Low conc.										
High conc.										

Sample	HA									
Sample conc. ↓	Electrolyte(NaCl) conc. in M					pH				
	0.001	0.005	0.010	0.050	0.100			6.5	8.0	9.5
Low conc.										
High conc.										

Figure 5.5: Macromolecular HA and FA Configurations at Different pH Values and Electrolyte Concentrations.
Source: Schnitzer (2000).

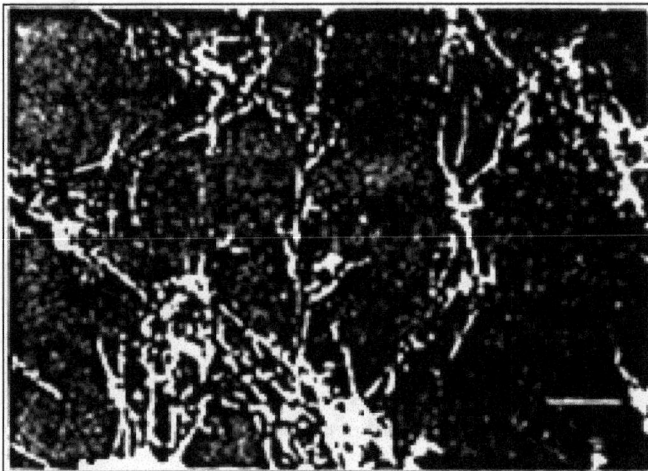

Figure 5.6: Transmission Electron Micrograph of a 0.01 per cent HA Solution.
Source: Schnitzer (1991).

independent of concentration but for polyelectrolytes, such concentration dependence of η_{red} is expected to be of varying nature in the lower and the higher concentration ranges of the polymer concerned. On the lower concentration side, η_{red} increases with dilution in response to molecular dilation of the polyelectrolyte due to electrostatic repulsion of similarly charged groups which dissociate upon dilution. The periphery of humic molecules is often considered to be open and flexible, probably maintained in swollen state by repulsion between the negatively charged groups. The above mentioned dilation will cause a sharp rise in η_{sp} following the Einstein's equation, namely $\eta_{sp} = (\eta/\eta_0) - 1 = \gamma\phi$, where ϕ is the volume fraction of the colloidal disperse phase, and the factor, $\gamma > 1.0$ ($\gamma = 2.5$ for spherocolloid) (*see* Section 2.11.1 and Eq. 2,8). At infinite dilution, the humic molecule is expected to assume an extended configuration accompanied by the maximum possible charge separation (Visser, 1985). At higher concentrations, the counter ions may be expected to neutralize, through "ion-pair" formation, some of the charged sites, thereby bringing down the ionic character of the polyelectrolyte. The same would be expected at high concentration of neutral electrolytes (*see* later). This will cause a partial coiling up of the molecule, due to hydrophobic hydration effect, forming the so-called "hydrophobic bonds" (Sanyal, 1984). At high enough concentration, the humic molecule will behave like an uncharged polymer. Ghosh and Schnitzer (1980) found this limiting concentration to be 3.5 to 5.0 g/liter for HA as mentioned above. It is thus apparent that such concentration dependence of η_{red} will lead to a

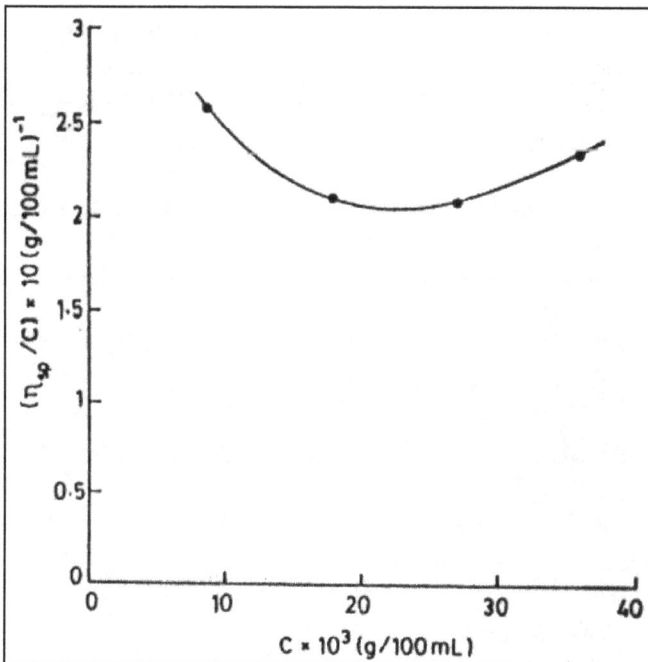

Figure 5.7: A Representative Plot of η_{red} (= η_{sp}/C) *vs.* C for a Synthetic Humic Acid Sample.
Source: Datta *et al.* (2001).

hyperbolic nature of variation of η_{red} ($=\eta_{sp}/C$) with concentration, as demonstrated in Figure 5.7 (Datta *et al.*, 2001).

Furthermore, an increase of ionic strength of the medium due to increased presence of neutral salts would cause compression of the electrical double layer, and a lowering of the corresponding zeta potential (ξ potential) around the charged sites of the humic molecule, based on the simple Helmholtz considerations [*see* Section 2.6.1 and Eq.2.4, namely $\xi = (4\pi ed)/D$, where e is the numerical value of charge on either layer, d, the thickness of the electrical double layer, both of which would fall on partial charge neutralization of the colloidal particle, and D is the dielectric constant of the medium], thereby bringing down the electrostatic repulsion between HA (and FA) moieties. This will facilitate the conglomeration of the colloidal particles at a high neutral salt concentration in a manner as to produce spherocolloid similar to the shape of uncharged polymers, *i.e.*, the effect one would expect at high concentration of HA molecules themselves (Sanyal, 2002b).

Essentially similar effect for HA would be expected at pH values in the acidic range when the high concentration of free H_3O^+ ions would render dissociation of weakly acidic groups of HA difficult (and less so in FA where total acidity is much higher; *see* later), and hence reduce the mutual repulsion between similarly charged species. Tombacz (1999), based on theoretical considerations of model HA molecules, also arrived at similar conclusions: that repulsion between the charged part (either inter- or intra-particular) of HA molecule decreases with both decreasing pH and increasing ionic strength. In parallel with this, the particle-size also decreases. This author (1999), from his own findings, suggested that the humic colloidal stability increased with increasing pH and decreasing salt concentrations, findings which are in agreement with those of Schnitzer (2000). Further, the sensitivity of HA fraction to salt was shown to be a function of size and acidic group density of HA, being greater for larger-sized macromolecules (Tombacz, 1999).

An examination of data and findings of Ghosh and Schnitzer (1980), Schnitzer (2000), Visser (1985) and Tombacz (1999) in the light of the above discussion would help one to appreciate the essential conclusions, arrived at by these authors, on shapes of humic and fulvic colloids discussed earlier (Figure 5.5). In support, the scanning electron micrographs (SEM) of humic substances were found to show gradual decreases in particle orientation as well as increases in fibre thickness as the neutral salt concentration was increased (Ghosh and Schnitzer, 1982). In this regard, one may possibly wish to draw an analogy from the classical theory of orientation of clay colloidal particles, for instance, through edge-to-face (EF), edge-to-edge (EE) and face-to-face (FF) interactions (Van Olphen, 1963; Sanyal, 2002a). In presence of high concentration of neutral electrolytes, the edge and face double layers are compressed which facilitates FF associations through van der Waals interactions. The latter causes thickening of the colloidal particles at the expense of EF and EE interactions (Sanyal, 2002a).

5.5.1. Viscosity B Coefficient

The viscosity B coefficient of the Jones-Dole's equation (or, Fuoss-Strauss equation) has been used to derive information on micro-interactions of HAs with neighbouring water in soil solution (Chen and Schnitzer, 1976; Mandal and Sanyal, 1984). The Jones-Dole's viscosity B coefficient was approximated by Chen and Schnitzer (1976) to the intrinsic viscosity, [η], of HAs that leads to the evaluation of the number-average molecular weight as follows:

$$[\eta] = \left[\left(\frac{\eta}{\eta_0} - 1 \right) / C \right]_{C \to 0}$$

(5.1)

Where η and η_0 denote the respective viscosity coefficient at the same temperature of an aqueous colloidal system of concentration C, and water. The Jones-Dole's equation (or Fuoss-Stauss equation) reads,

$$\frac{\eta}{\eta_0} = 1 + A\sqrt{C} + BC + \text{higher-order terms in } C$$

$$\text{Or,} \eta_{sp}/\sqrt{C} = A + B\sqrt{C}, \text{neglecting higher-order terms in } C$$

(5.2)

Thus, B, which is sensitive to micro-interactions between the disperse phase and dispersion medium, is obtained as a slope of the best-fit linear plot of

$$\left[\frac{\eta}{\eta_0} - 1 \right] / \sqrt{C}$$

versus √C, and in fact, is identical with [η] given by Eq. (5.1) over a low concentration range. Such intrinsic viscosity is given by the modified Staudinger's equation as, [η] = K (MW)$^\alpha$, where MW is the number-average molecular weight of the humic polymer and K and α are constants.

The sign and magnitude of the temperature coefficient, (dB/dT), rather than the B coefficient itself, is a more prominent criterion for determining the nature of solute-solvent micro-interactions for large amphiphilic electrolytes. The findings of Sanyal and Mandal (1983) on viscosity behaviour of a series of alkyl carboxylates of progressively increasing chain length were in agreement with the above, and was explained in terms of on the "local" structure of water (Sanyal, 1984). Similar considerations apply to the temperature coefficient of the viscosity B coefficient (dB/dT) of humic systems as well (Sanyal, 2002b).

Mandal and Sanyal (1984) proposed a model in which micro-viscosity measurement of aqueous humic solutions, *i.e.*, the viscosity B coefficient, was used to derive the semi-quantitative information on shape of natural, as well as synthetic HAs. The findings of these studies suggested long cylindrical chains for humic shape, rather than three-dimensional structure, or spherical type. The results were in agreement with those of Visser (1985), discussed earlier.

5.5.2. Surface Tension Measurements

The hydrophobic moieties such as long alkyl side-chains of fatty acid residue impart to humic molecules the amphiphilic character, and consequently surface-active properties (Tombacz, 1999). Thus, humic molecules have been shown to accumulate in the interfacial layers, causing a fall of surface tension of their aqueous solutions (Chen and Schnitzer, 1978; Ghosh and Schnitzer, 1980; Tombacz, 1999; Saha and Sanyal, 1988; Schnitzer, 2000; Datta *et al.*, 2001), as well as their interfacial tension at water-oil interfaces (Tombacz, 1999). Tombacz (1999) discussed the formation of self-assembly systems, caused by hydrophobic interactions, leading to the formation of oriented bilayers in membrane-like structures when humic substances form coatings on the mineral surfaces in soils or as micellar solutions that can stabilize otherwise water-insoluble organic compounds (Tombacz, 1999). Such hydrophobic interaction in aqueous system of macromolecules of mixed organic nature was discussed earlier by Sanyal (1984) from rigorous thermodynamic considerations in terms of free energy changes accompanying humic accumulation in soil solution. Such falling surface tension of the HAs and FAs with concentration leads to a positive surface excess (τ) of the humic molecules as evident from the Gibbs adsorption equation, given below.

$$\tau = - (1/RT). (d\gamma/d \ln C) \tag{5.3}$$

Where γ is the surface tension of humic solution at absolute temperature T and R is the Universal gas constant. The τ value was found to increase with the increasing hydrophobicity of the HA molecules (Datta *et al.*, 2001) as one would expect, for the higher the hydrophobicity, the greater will be the tendency of the humic molecule to minimize its contact surface area with water, under "thermodynamic compulsion" (Sanyal, 1984). The latter ensures the accommodation of a relatively large number of molecules per unit surface area of water. A sharp rise in τ was observed for humic samples in presence of neutral salts (*e.g.* 0.1 M NaCl) which obviously results from a higher degree of "exclusion" from the aqueous system of the predominantly hydrophobic HA in presence of strongly hydrated electrolytes. A representative diagram, demonstrating the concentration dependence of surface tension of two humic acid molecules, is given in Figure 5.8 (Datta *et al.*, 2001).

The surface pressure, π, which is defined as the lowering of surface tension of a liquid (such as soil solution) due to formation of a monolayer of a, say, largely hydrophobic substance (such as humic molecules) on the surface, is given by,

$$\pi = \gamma_0 - \gamma \tag{5.4}$$

Where γ_0 and γ are the respective surface tension of the clean interface (*e.g.*, soil water/air interface) and the interface plus monolayer, *e.g.* aqua humic acid system of soil. The variation of surface pressure with the area available to spread the material is represented by a π-A (force-area) curve. Such π-A curves may be regarded as two-dimensional equivalent of p-V curves for a three-dimensional system. The π-A curves of HAs were interpreted as indicating monolayer formation (by HA molecules) at the air-water interface.

5.6. Humic Substances and Water Retention: Influence on Soil Physical Properties

It is clear that the surface activity of humic substances arises from the balance of hydrophobic and hydrophilic surface sites. The surface activity of humic

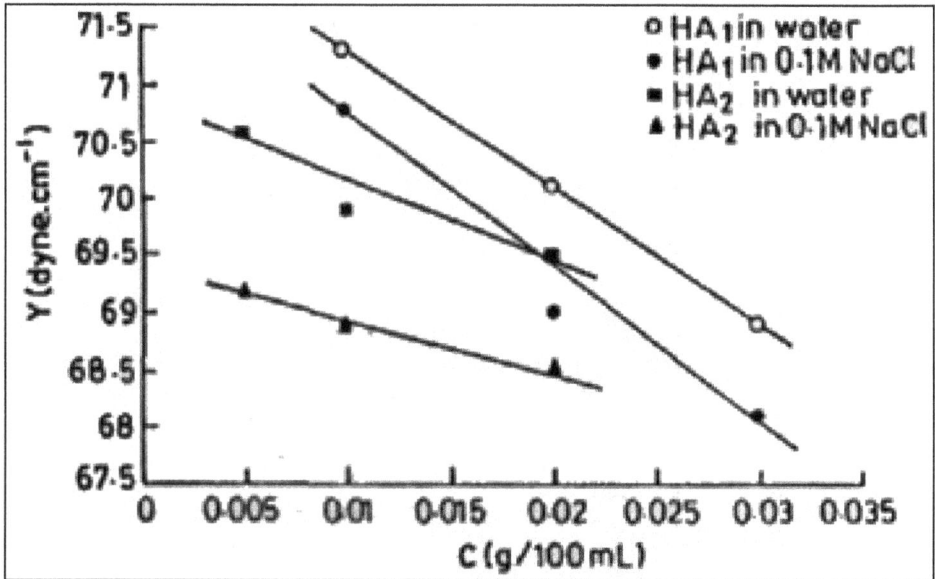

Figure 5.8: Representative Plots of Surface Tension (γ) *vs.* C for Two Synthetic Humic Acid Samples (HA₁ and HA₂) in Absence and Presence of Aqueous Electrolyte.
Source: **Datta *et al.* (2001).**

substances was found to affect the capillary control and seepage loss in soil (Sanyal, 2002b), whereas the hydrophilic character of humic substances was suggested to lead to their ability to adsorb quite substantial amounts of water vapour (Chen and Schnitzer, 1978; Schnitzer, 2000). Further, both FA and HA were reported to be surface tension-reducing agents for water, or as materials which can improve the wetting of silicate surface by the solutions (Chen and Schnitzer, 1978). Thus, the humic substances were suggested to act as water repelling agents in neutral or acid soils, but not at pH greater than 6.5 (Chen and Schnitzer, 1978). Figure 5.9 illustrates the water vapour sorption isotherms of HA and FA at 40°C (Schnitzer, 2000). These isotherms were found to follow similar course up to about 60 per cent relative humidity (RH), but beyond that, the FA isotherm rose sharply. The data were also fitted to the BET isotherm to compute the weight of water adsorbed by 1 g of HA or FA to form a monolayer and the corresponding heat of adsorption. As shown in Table 5.1, these amounts of water at 35 per cent RH were rather similar for HA and FA, but at 60 per cent RH, when the two isotherms became nonlinear and divergent (Figure 5.9), FA adsorbed larger amount than did HA. Finally at 90 per cent RH, the amount of water vapour adsorbed by FA was about double of that by HA. Schnitzer (2000) tentatively suggested that clustering of water molecules around the carboxylic groups, whose content was considerably larger in FA than in HA (*see* later), could have been responsible for such wide difference in water retention by HA and FA at high RH.

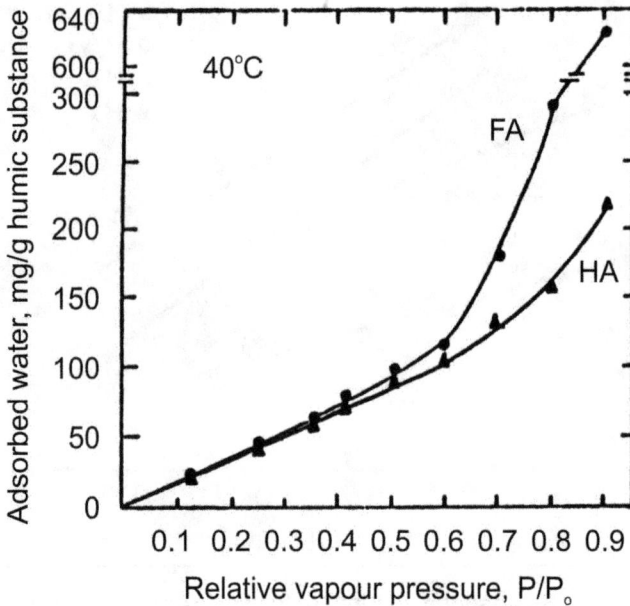

Figure 5.9: Isotherm for Adsorption of Water Vapour by HA and FA at 40° C.
***Source*: Schnitzer (2000).**

Table 5.1: Water Adsorbed by 1.0 g of HA and 1.0 g of FA at different Relative Humidity (RHs)

RH (per cent)	H_2O (mg) Adsorbed by 1.0 g HA/FA	
	HA	FA
35	58	60
60	110	120
90	225	508

Source: Schnitzer (2000).

Tarchitzky *et al.* (2000) suggested that the presence of humic substances affects soil wettability in a manner which seems to be a function of their macromolecular size, shape degree of polymerisation, as well as the drying process employed on the soil. It was further argued that when the soil is wet, the hydrophilic functional groups of humic polymers face the outside of the soil aggregate, whereas the hydrophobic sites face the mineral surfaces inside the aggregates (Tarchitzky *et al.*, 2000). During the drying process, such orientation reverses itself, thereby facilitating the hydrophobic properties of the aggregates, and hence their stability. The earlier findings (Sanyal, 2002b) that the polyvalent cations such as Al^{3+} and Fe^{3+} ions enhance the hydrophobic effect of soil humic polymers, has been coupled with those of Tarchitzky *et al.* (2000) to infer that some of the organic and inorganic components in soil form a coating on mineral surfaces, thereby reducing their hydrophilic reactivity, in addition to their important role in soil aggregate formation. This activates water-repellence on a micro-scale near aggregate surfaces, and thus improves the aggregate stability. Tarchitzky *et al.* (2000) demonstrated the ratio of the levels of various fractions of humic substances to that of clay in each of the aggregate size-fractions in a soil, wherein the highest surface coating with either humic acid, fulvic acid or humic substances was evident for 20-50 mm fraction aggregates.

5.7. Chemical Characterization

The UV-visible and fluorescence spectroscopic characterization of the ultrafiltered HA fractions showed a significant diversity in the chemical compositions of the fractions. The humic substances contain relatively high concentration of oxygen-containing functional groups (carbonyl, phenolic-OH and carboxyl) per unit weight (Schnitzer, 2000). Between HA and FA, the total acidity, carboxyl and phenolic acidity and oxygen-containing functional groups are higher in the latter, due, obviously, to lesser degree of molecular condensation of FA molecule, compared to that in HA molecules. The alcoholic (OH) group content is also generally higher in FAs than in HAs, signifying the dominance of polysaccharides in FAs. Representative data on functional group contents of HA and FA, extracted from forest soils at different altitudes of Sikkim, are shown in Table 5.2 (Martin *et al.*, 1998). These authors also explored the effect of altitude on C/H, C/N and O/H ratios in HAs and FAs. There was a close correlation between (O/H) ratio of HAs extracted from these forest soils and temperature, indicating a higher degree of oxidation and increased humification in soils of lower warmer sites. The HAs also had higher molecular weight than the corresponding FA fractions (Table 5.2), due, obviously, to a greater degree of polycondensation of aromatic rings in HAs.

Martin *et al.* (1998) also observed that the cultivation decreased the (C/H) and the (C/N) ratios, but in general, increased the (O/H) ratios. Such decrease in (C/N) ratio was the consequence of accelerated polysaccharide mineralization, whereas the increase in (O/H) ratio indicated accelerated oxidation and humification. The decrease in (C/H) ratio of humic substances under cultivation suggested accelerated mineralization of lignin (Martin *et at.*, 1998).

Being led by the criterion developed by Sturrock (1968), several workers have concluded that the HAs are essentially polyprotic in nature (Banerjee and Mukherjee, 1975; Stevenson, 1994; Orlov, 1985; Saha and Sanyal, 1988; Datta *et al.*, 2001). The carboxylic and phenolic (OH) group contents of HAs and FAs are obtained by potentiometric (and conductometric) titrations. A fall in final pH of HA solution on standing for several hours after titration with alkali was observed (Datta *et al.*, 2001). The same trend was reported for the synthetic HAs (Saha and Sanyal, 1988). A coiled nature of aqueous HAs, under "thermodynamic compulsion" (Sanyal, 1984), was the probable reason for such behaviour. A partial unfolding of HA is believed to take place at high pH due to concomitant larger enthalpic compensation for the entropic deficiencies of largely hydrophobic aqueous HA system (which tends to render the free energy change accompanying the solution of HA at constant temperature and pressure negative), leading to release of "locked-up" acidity in the coil. The latter lowers the solution pH. The time and excess fresh alkali needed to attain the stable

Table 5.2: Functional Group Analysis, Molecular Weight and (E₄/E₆) of Humic and Fulvic Acids in Cultivated and Natural Forest Soils of Sikkim

Site No. (alt, m)	Cultivation/Dominant Forest sp.	Acidity/meq. g^{-1}						Mol. Weight		E_4/E_6 Ratio	
		Total		(COOH)		Phenolic (OH)					
		HA	FA	HA	FA	HA	FA	HA	FA	HA	FA
1 (600)	Cultivated	5.2	9.0	3.6	2.0	1.6	7.0	1550	528	4.62	7.05
	Shorea sp.	5.3	6.7	3.1	2.3	2.2	4.5	1561	720	4.12	5.76
2 (700)	Cultivated	5.2	5.9	1.8	1.9	3.4	4.0	1316	662	4.94	5.21
	Pinus sp.	4.3	6.8	1.9	1.0	2.5	5.7	1411	680	4.87	5.04
3 (2100)	Cultivated	5.9	6.6	3.2	4.9	2.7	1.7	800	496	7.05	7.23
	Castonopsis sp.	3.7	5.9	3.3	1.9	0.5	3.9	1127	617	5.21	5.31
4 (2300)	Cultivated	4.7	6.7	2.2	1.6	2.6	5.2	976	485	5.74	9.38
	Rhododendron sp.	4.3	8.0	3.1	1.9	1.1	6.1	1056	540	5.59	6.21
5 (2700)	Cultivated	5.9	7.1	3.0	2.2	2.8	4.2	1010	500	5.74	16.16
	Abies sp.	4.6	7.1	3.2	3.0	1.3	4.1	1033	526	5.23	11.04

Source: Martin et al. (1998).

higher pH (after initial titration of the HAs, and then allowed to stand) was found to be related to the extent of initial coiling in the HAs (Datta *et al.*, 2001).

5.8. Spectroscopic Properties

The absorption spectra of HA in the UV-visible region are generally featureless, exhibiting no extremum with the optical density (O.D.) falling as the wavelength increases (Banerjee and Mukherjee, 1972; Tomar *et al*, 1992b). The ratio of O.D. of dilute aqueous HA or FA solutions at 465 and 665 nm (E_4/E_6), is taken to provide a measure of the degree of condensation of aromatic rings present in these materials, *i.e.*, the relative preponderance of aliphatic and aromatic groups (Schnitzer and Khan, 1972; Tomar *et al.*, 1992b; Sarmah and Bordoloi, 1993; Lahiri and Chakravarti, 1995; Rivero *et al.*, 1998; Datta *et al.*, 2001). The high O.D. values at longer wavelengths are believed to be associated with an increased mobility of the delocalized π-electron clouds over aromatic carbon nuclei, and hence HAs may be expected to register higher E_6 (and lower E_4) values than the corresponding FAs. The ratio, (E_4/E_6), of soil humic substances is primarily governed by the particle size, molecular weight and degree of aromaticity (Tomar *et al.*, 1992b; Datta *et al.*, 2001)). The fulvic colloids, being less condensed and aromatic, possess typically higher (E_4/E_6) ratio than do their corresponding HA fractions. Representative values of (E_4/E_6) ratio for HAs and FAs are given in Table 2.7 (given earlier in Section 2.9.2).

The (E_4/E_6) values of HAs extracted from Inceptisol, Alfisol, Entisols, Vertisol and Oxisols were quite similar, while the Spodosol HA exhibited a higher (E_4/E_6) (Rivero *et al.*, 1998). This was consistent with a relatively low particle-size and molecular weight of the latter, thereby suggesting that the Spodosol HA was characterized by a lower degree of condensation and has less aromatic structure, as compared to the other HAs. The (E_4/E_6) ratio of HAs extracted from manures and sewage/sludge was found to be close to those of the soil humic acids (Deiana *et al.*, 1990).

A lowering of (E_4/E_6) ratio was observed for a number of synthetic (and also soil) HAs with rise in pH (Datta *et al.*, 2001). This is possibly linked with the aforesaid partial decoiling of the coiled structure of HA, at a higher pH, which renders the release of "locked up" hydrophobic parts inside the HA coil. The latter being largely aromatic in character, this leads to a lowering of (E_4/E_6) value.

Of the spectroscopic methods generally used for structural investigations, the infrared (IR) spectroscopic techniques have been used for functional group analysis in soil humic substances (Saha and Sanyal 1988; Tomar *et al.*, 1992b; Stevenson 1994; Rivero *et al.*, 1998; Martin *et al.*, 1998; Schnitzer, 2000). The fingerprint region (7 to 11μm) and beyond of the IR spectrum of the humic molecules is taken advantage of. The IR spectroscopic technique has also been used to detect structural alterations in humic substances ensuing from oxidation, pyrolysis, acetylation and esterification. This technique is useful for detecting changes in the absorption frequencies of functional groups of the polymer, resulting from their interaction with pesticides, vitamins and other adsorptives (Sanyal, 2002b).

Two developments in IR spectroscopy in fairly recent times are expected to influence studies of organo-mineral interactions in soils, namely Fourier-transform technique (FTIR) and the reflected IR spectra (which are obtained by resorting to diffuse-reflectance IR techniques) from the first few layers of atoms in a molecule. The latter can be applied directly to quartz grains for obtaining spectra of organic molecules (*e.g.* HAs) on surfaces with no chemical pretreatment (Schnitzer, 2000).

The IR and FTIR spectra of humic substances generally show bands at 3400 cm^{-1} (H-bonded OH), 2000 cm^{-1} (aliphatic C-H stretch), 1725 cm^{-1} (C= O of COOH, C = O stretch of ketonic C = O), 1630 cm^{-1} (COO$^-$, C = O of carbonyl and quinone), 1450 cm^{-1} (aliphatic C-H), 1400 cm^{-1} (C-O stretch or OH deformation of COOH), and 1050 cm^{-1} (Si - O of silicates) (Schnitzer, 2000).

The different functional groups present in humic and fulvic acids of soil humic colloids, as revealed by the representative IR spectra, are summarized in Table 2.8 (given in Section 2.9.3) (Sarmah and Bordoloi, 1993).

The concentration of (COOH) groups in HAs was determined directly from FTIR spectra by totaling absorbance at 1720-1710 cm^{-1} (COOH) and 1620-1600 cm^{-1} (COO$^-$) (Schnitzer, 2000). Such carboxyl content was in good agreement with that determined by wet chemical method and by ^{13}C NMR spectroscopy. However, the IR techniques, important though as they are for functional group identification, are not able to enlighten us much about the chemical structure of humic substances (Sanyal, 2002b).

The electron spin resonance (ESR) spectroscopic data for the free radical contents of humic substances were also obtained (Hayes and Swift, 1978). The number of these free radicals generally decreases in the order:

Humin > Humic acid > Fulvic acid

These organic free radicals (such as semi-quinone radicals) enable the soil humic substances to participate in a wide variety of organic-organic and organic-inorganic interactions. The wide scan-range (5000 Gauss) ESR spectra of all the HAs, examined by Rivero *et al.* (1998), featured a resonance signal at g = 4.3 - 4.2, typically attributed to high spin Fe^{3+} ions held in the inner-sphere complexes of tetrahedral and/or octahedral coordination by oxygenated functional groups of HA macromolecules (Schnitzer, 2000).

The nuclear magnetic resonance (NMR) spectroscopy is one of the most promising spectroscopic techniques for structural elucidation of the humic materials. Some of the NMR data indicate that humic materials are less aromatic than thought earlier from degradation techniques, although there are variations also to this trend (Hayes, 1984; Krosshavn *et al.*, 1992; Stevenson 1994; Schnitzer, 2000). The characterization of humic acids, extracted from manures and sewage/sludge, by the ^{13}C NMR technique demonstrated that the manure HAs were more aromatic than the sewage/sludge HA, with greater functional group contents, and were quite similar to soil HAs (Deiana *et al.*, 1990). The sewage/sludge HA was, on the other hand, richer in carbohydrates and proteineous material, more aliphatic (than manure HA) and resembled the aquatic humic substances. Solid-state NMR is also a powerful tool for comparison of changes in organic matter during decomposition of litter, and for comparison of sequence of soils (Schnitzer, 2000), as well as for monitoring the structural changes of humic components of forest soils, subjected to cultivation (Martin *et al.*, 1998). The advantage of NMR spectroscopy is that it does *not* alter the chemical composition (being a non-de-structive technique), and can be used for studying the nature of organo-mineral complexes. A comparative distribution of carbon percentage in a Haploboroll HA and a Spodosol FA, as determined by the ^{13}C NMR technique, is shown in Table 5.3 (Schnitzer, 2000). The main structural features, along with degree of aromatic and aliphatic character, of HA and FA are seen to be quite similar, which tends to suggest similar chemical structure for these two components. The HA, however, is shown to be more aromatic than FA (Sanyal, 2002b).

Table 5.3: Distribution of C (Per cent) in a Haploboroll HA and a Spodosol FA as Determined by ^{13}C NMR Spectroscopy

Chemical Shift Range (ppm)	Per cent of C	
	HA	FA
0-40	24.0	15.6
41-60	12.5	12.8
61-105	13.5	19.3
106-150	35.0	30.3
151-170	4.5	3.7
171-190	10.5	18.3
Aliphatic C (0-105 ppm)	50.0	47.7
Aromatic C (106-150 ppm)	35.0	30.3
Phenolic C (l51-170 ppm)	4.5	3.7
Aromaticity[a]	44.1	41.6

[a][(Aromatic C + phenolic C)/(Aromatic C + Phenolic C + Aliphatic C)] x 100.

Source: Schnitzer (2000).

The pyrolysis-mass spectrometric techniques have been used for obtaining structural information and molecular weights of humic substances and the information generated has been summarized (Schnitzer, 2000). While this technique is yet to yield any new information on soil humic substances, small untreated samples are required and the method is a rapid one. This enables one to have a rather quick overall chemical characterization of soil organic fraction. By this method and the solid-state ^{13}C NMR spectroscopic technique, it is also possible to obtain important information on the chemical composition of soil organic matter in whole soils *in situ*, *i.e.*, without the need of extracting the soil humic substances and fractionating them by classical methods (Schnitzer, 2000). Further, the information on structural voids in HAs, that emerge from such investigations, may throw light on the entrapment of organic macromolecules by the humic substances (Schnitzer, 2000).

5.9. Metal-Humic Interactions

The binding of metal ions by humic colloids is a widely researched subject. Such metal-humic interactions would be expected to occur first at those sites which lead to the formation of coordinate to ring structures. Binding at the weaker sites becomes progressively important as the stronger sites are saturated. Some investigations also emphasized the formation of chelate rings (Stevenson, 1994; McBride, 1994).

The increasing input of toxic heavy metals as well as organic chemicals to agricultural production system through the use of untreated sewage/sludge, industrial effluents and chemical pesticides tends to exert a long-term impact on soil-crop environment. The question as to how soil humic substances may exert moderating influence (if any) in controlling the activity, behaviour and survival rate of these metal ions/chemicals is important from the viewpoint of soil ecosystem, and hence the significance of studies which examined the metal-humic interactions (Datta *et al.*, 2001; Sanyal, 2002b).

The functional groups of humic substances active in retaining/complexing metal ions are carboxylic, phenolic (OH), amine, and thiol (sulfhydryl) groups (Stevenson, 1994). Although (COOH) and phenolic (OH) groups are most abundant in humic substances, the role of organic sulphur in complexation of heavy and trace metals is potentially important since organic sulphur accounts for more than 90 per cent of total sulphur, for instance, in temperate soils (Sanyal, 2002b). Reduced sulphur-containing functional groups, rather than oxygen-containing functional groups, play a significant role in binding of soft cations, *e.g.* Hg^{2+} ions (Stevenson, 1994).

The IR spectroscopic studies of metal-humic complexes suggested the participation of phenolic (OH) and carboxylic groups in binding of metal ions, followed by $> C= O$ and $> NH$ groups (Stevenson, 1994), although considerable controversy exists as to the extent to which the carboxylate linkages are covalent or ionic. Structures similar to salicylic acid have been proposed (Stevenson, 1994); others opined that the structures are similar to those of phthalic acid, while there has also been report opposing the possibility of phthalic-type of structures for metal-humic complexes (Stevenson, 1994; Sanyal, 2002b).

Stevenson (1994), from potentiometric titration results, suggested that (2:1) complexes are formed, simultaneously with (1:1) complexes between bivalent metal ions and HAs. The proposed reactions, leading to the formation of phthalate type of structures, are shown in Figure 5.10. Stevenson (1994) also observed that both (2:1) and (1:1) complexes are formed in reactions between Cu^{2+} ions and humic substances.

The results of ESR studies have been rather inconclusive with the opinion divided on whether the electron donation (coordination) to metal ion from the humic ligand occurs through N-containing groups or the carboxylates (McBride, 1994; Stevenson, 1994). By using the electron paramagnetic resonance (EPR) spectroscopy, it was noted that HAs form two types of complexes with Cu^{2+} ions, one with oxygen-containing group, while the other with nitrogen-containing groups, but that FA forms such complexes only with oxygen-containing functional groups

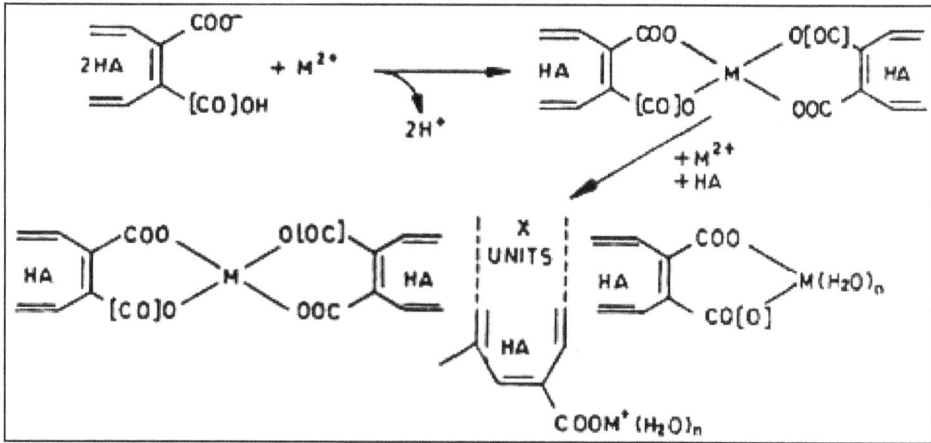

Figure 5.10: Reactions Leading to the Binding of Bivalent Metal Ions by Humic Acids. The metal ion is shown to be held: (a) as a (2:1) chelate complex holding two molecules together as a chain, (b) as a (1:1) complex at the end of a chain, and (c) in a salt-like linkage with an isolated (COOH) group.
Source: **Stevenson (1994).**

(Neto *et al.*, 1991). McBride (1994) concluded, on the other hand, that only a small proportion of the acidic groups of FA were involved in the formation of inner-sphere complexes of Mn^{2+} ion. The ESR spectra of improved resolution were analyzed by Senesi *et al.* (1985). At a Cu^{2+} ion/FA molar ratio of 0.20, the ESR spectra provided evidence for three different coordination environments for bound Cu^{2+} ions. The finding clearly suggested the formation of mixed complexes, in agreement with the trend shown in Figure 5.10.

While studying the Cu (II) activity in aged suspension of organic matter and goethite (α-FeOOH), McBride *et al.* (1998) concluded that under slightly to strongly acidic soil conditions, organic matter is more likely to limit Cu^{2+} ionic activity and phytotoxicity than iron oxides. In general, HA interaction with metal ions in soil solution increased with pH, humic acid concentration, but decreased with metal ion concentration (Spark *et al.*, 1997).

Significant amounts of Li^+, Na^+ and K^+ ions were observed to associate with soil humic matter between pH 3 and 8 (Bonn and Fish, 1993). These authors (1993) could describe their findings in terms of a diffuse-double layer model of hydrated HA in which the alkali metal cations neutralize the humic charge. The association of Na^+ and K^+ ions with HA at pH 1.0 was successfully described by a Langmuir adsorption model.

The stability constants of metal-humic complexes are expected to elucidate the tendency of the humic substances to retain the metal ions in their matrix. These stability constants appear to be generally lower than those for the complexes of the same metal ions with well-known chelating agents, such as EDTA, and also many naturally occurring biochemical compounds (Norvell, 1991). However, the nature of the complexing humic colloid is also important in this regard. Thus, insoluble

Fe-humate complexes, rich in aromatic carbon, are less susceptible to release iron to natural chelators present in soil solution than complexes formed with humic substances, rich in aliphatic carbon (Piccolo *et al.*, 1993).

Datta *et al.* (2001) investigated the Cd^{2+} complexation reaction with one natural (soil) and several synthetic humic acids. The resulting stability constants (log K) were positively correlated with phenolic (OH) acidity (r = 0.952*), $\Delta (E_4/E_6)$ (r = 0.804), and surface excess, τ, in water (r = 0.828) of the given HA samples. These correlations are shown in Figure 5.11 (Datta *et al.*, 2001). A linear multiple regression equation (Eq. 5.5) was also developed describing log K (of Cd^{2+}-HA complexes) as a function of phenolic (OH), τ in water and $\Delta (E_4/E_6)$ values of the humic samples. Thus,

$$\log K = 26.2 + 3.35** \text{ Phenolic (OH) } + 7.7 \, (\tau \text{ in water}) + 0.134 \, \Delta (E_4/E_6) \qquad (5.5)$$

Adjusted $R^2 = 0.86$; Multiple R = 0.96

Equation (5.5) strongly suggests the participation of the phenolic (OH) in the Cd^+-complexation process. This is in agreement with earlier observations (Sanyal, 2002b). Indeed, an analogy may perhaps be drawn with metal-oxine (8-hydroxy quinoline) complexation of Cd^{2+} ion.

As suggested by Figure 5.11, the hydrophobic moiety of the humic acid samples may provide a cage (or clathrate)-type conformation around Cd^{2+} ion because of hydrophobic hydration effect (Sanyal, 1984), thereby imparting the desired stability to the complex. This is supported by a direct relationship between log K and $\Delta (E/E_6)$. Furthermore, a greater degree of hydrophobicity of HA is expected to render the phenolic (OH) group more readily accessible to complex Cd^{2+} ion in the aqueous phase. A positive relationship between log K and τ (in water) of the given HAs tends to subscribe to this hypothesis.

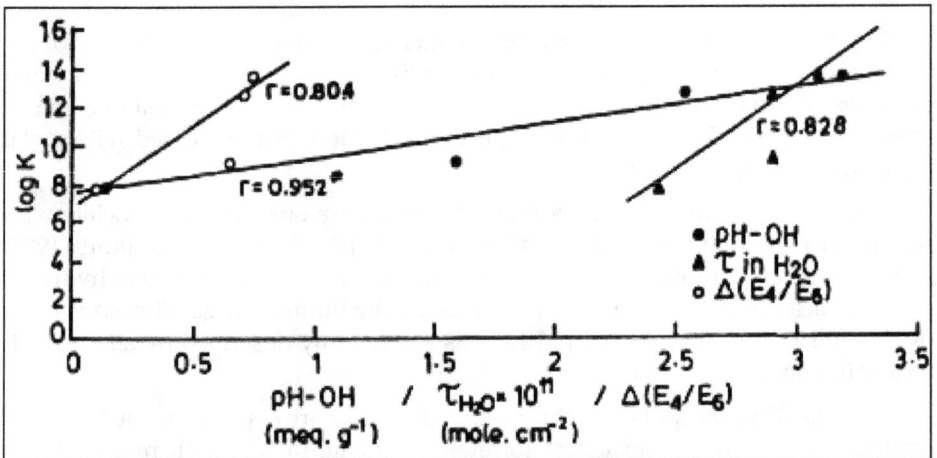

Figure 5.11: Stability Constant of Cd^{2+}-Humic Complexes as a Function of Phenolic (OH) Content, Surface Excess (τ) in Water and $\Delta (E_4/E_6)$ of Four Synthetic Humic Acid Samples.
Source: Datta *et al.* (2001).

The kinetics of iron fulvate and other metal-humic complex formations have been investigated by a number of workers (Dyanand and Sinha, 1979, 1980; Sanyal, 2002b). A number of kinetic equations, covering zero-order, first-order, second-order, parabolic-diffusion, and Elovich kinetic equations, have been used by different workers and have been reviewed extensively (Sanyal, 2002b).

Modelling approach is a fairly recent trend of probing metal-humic interaction and complexation behaviour (Stevenson, 1994; Manunza *et al.*, 1995). The normal distribution model has been successfully applied to the study of proton and metal-humate associations (Deiana *et al.*, 1995). The bimodal Gaussian model was applied to the description of complexes of Cu^{2+}, Pb^{2+}, Mn^{2+} and Cd^{2+}ions with a soil humus matrix (Manunza *et al.*, 1995).

Due to experimental difficulties to take care of a variety of cation-humate interactions possible in the natural environment, the use of computer modelling was resorted to by combining sophisticated mixture model to represent HA and a general geochemical speciation code to simulate the interactions between HA and inorganic components (Mountney and Williams, 1992). After an initial investigation of Cd^{2+}-HA and Cu^{2+}-FA interactions, this coupled code approach was modified to include a sensitivity analysis which identified the ligands that were statistically significant in any system.

Mineral-bound humic substances exert variable effect on metal-ion binding, depending on the structure and chemical properties of the humic substances. Zachara *et al.* (1994) concluded that mineral-bound humic substances exhibit an additive effect on metal sorption, commensurate with their aqueous phase complexation behaviour.

Sufficient research has been conducted to allow generalization of the potential effect of HAs on metal ion sorption by Fe and Al oxides. This is shown in Figure 5.12 (Murphy and Zachara, 1995). The adsorption of metal ions increases typically with increasing pH, while humic substances show anion-like adsorption behaviour, increasing with decreasing pH. Adsorbed HAs tend to increase the adsorption of metal ions at lower pH where their own surface density is greatest, and where the surface complexation of the metal ions by inorganic surface hydroxyls (of Fe and Al hydroxides having relatively high ZPC) is low. The complexation process may be visualized as a competitive one between metal ions, electrolyte cations and H^+ ions for free carboxylate groups on mineral-bound humic substances. At pH values higher than the intersection point of their respective pH-adsorption edges (*i.e.*, pH 7.5 in Figure 5.12), metal ion adsorption may decrease as the HA is stabilized (dispersed) in aqueous phase and forms weakly sorbing metal-humate complexes.

The sorption of heavy metal ions is enhanced in the combined mineral-humic acid systems for the minerals goethite and silica due to secondary reactions in which metal-humic acid complexes are adsorbed by the minerals. Such sorption of the metal-humic acid complex in the combined systems for α-alumina and kaolinite is not enhanced, possibly due to competing reactions associated with the sorption of the HA by these minerals (Spark *et al.*, 1997).

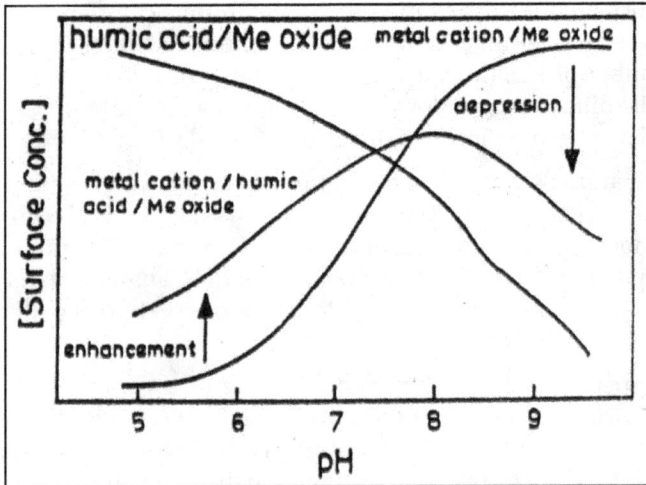

**Figure 5.12: Generalized Effect of Humic Acids on
Metal Ion Sorption by Fe and Al Oxides.**
Source: **Murphy and Zachara (1995).**

Kendorff and Schnitzer (1980) reported the efficiency of adsorption of metal ions on HA, from a solution containing equimolar concentration of 11 different metal ions, to increase with pH and HA concentration and falling metal ion concentration. At pH 2.4, the order of adsorption was: $Hg^{2+}>Fe^{3+}> Pb^{2+}> Cu^{2+} = Al^{3+}> Ni^{2+}> Cr^{3+} = Zn^{2+}= Cd^{2+} = Co^{2+} = Mn^{2+}$. At pH 4.7, the order got modified to $Hg^{2+} = Fe^{3+} = Pb^{2+} = Al^{3+} = Cr^{3+} > Cd^{2+}> Ni^{2+} = Zn^{2+} > Co^{2+}> Mn^{2+}$.

5.10. Interaction of Humic Substances with Pesticides

Schnitzer (2000) discussed the persistence, degradation, mobility, leachability, volatility and hence bioavailability of pesticides in the light of the nature and extent of their interactions with soil humic colloids. For charged pesticide species, the interaction could involve, among others, ion exchange or charge-transfer processes, while for adsorption of nonionic pesticides, the lyophobic (hydrophobic) sites would be expected to be more active than the hydrophilic sites of HA, through hydrophobic bonding and/or van der Waals forces. Among the different SOM fractions tested, HA had the highest affinity for metolachlor and hydrophobic bonding was the suggested mechanism (Kozak *et al.*, 1983). The diphenyl ether herbicide, namely acifluorfen, was reported to be apparently adsorbed unchanged on the external surfaces of HA and in the internal voids of the proposed model HA structure (Celi *et al.*, 1997; Schnitzer, 2000).

Adsorption of alachlor, a non-ionisable chloracetanilide herbicide, by HAs was extensively studied (Senesi *et al.*, 1994). The spectroscopic and quantitative data suggested that multifunctional hydrogen bonds and charge-transfer bonds may be responsible for preferential adsorption of alachlor at low concentrations, especially by the well humified soil HA, rich in aromatic structures and oxygen-containing functional groups. The soil humic substances can also promote the nonbiological degradation of pesticides and form strong linkages with the corresponding metabolites and residues (Stevenson, 1994). Indeed such knowledge of pesticide-humic interactions is important in relation to detoxification modes of the pesticide residues, and hence environmental decontamination.

5.11. Humic-Mineral Interactions: Clay-Humus Complex

Ghosh and Ghosh (1998) have reviewed extensively the organo-mineral interactions in soil. The extent of the interactions between soil minerals, including clays and HA, have been found to depend on the nature of the mineral, pH and, in some cases, the concentration of the background electrolyte (Yuan *et al.*, 2000). The said interaction influences the solubility of the HA as well as the surface charge density and electrophoretic mobility of the mineral. While examining the role of soil mineral composition and cation suite in rendering the soil organic matter (SOM) of the tropical soils much younger than that of the temperate zones, kaolinitic was found much less active in sorption reaction than were the expanding clays, dominant in soils of temperate zones. The activity of the oxides in SOM stabilization was, on the other hand, determined not by their total amounts, but by their crystallinity, particle-size, and association with kaolinite and mica. This greatly reduced the amount of oxides, effectively participating in SOM stabilization (Shang and Tiessen, 1998).

The functional groups in humic substances involved in binding metal ions are quite similar to those involved in clay-humic acid (HA)/fulvic acid (FA) associations. These include enolate (OH), amino, imino, azo, heterocyclic N (aromatic ring N), carboxyl/carboxylate, phenolic (OH), ether group, sulfhydryl (-SH), carbonyl groups.

Stable soil aggregates are formed by reactions between clay minerals and humic substances. Clay-humus complex formation retards the microbial degradation of humus. In mineral soils, nearly entire humus occurs in combination with clay. However, the mechanisms of clay-humus complexation process are far from being well understood.

In chemical association between clay and humus, the metal ion tends to form a bridge, linking the two components, *e.g.* Clay-Ca^{2+}-HA. In montmorillonite, bonding takes place at the basal surface of the mineral, while both basal surface and edges of illite plates are involved in complexation. For kaolinite, humus is bonded to the edges of the mineral.

It was noted that hematite, among goethite, hematite, gibbsite and boehmite, showed the greatest HA fixation at different oxide: HA ratios at all the pH values higher than 7.0. A gradual reduction in HA/FA fixation from pH 2.0 to 10.0 was reported for all the aforesaid minerals, except gibbsite, which exhibited a sharp fall at pH > 7.0, and a maximum at pH 5.0. The extent of fixation was independent of specific surface area or ZPC of the given oxide minerals (Varadachari *et al.*, 1997).

An examination of the humic-fixation/complexation process from the viewpoint of theoretical analysis of crystal surface structure of the oxide minerals was also made (Varadachari *et al.*, 2000). For this, the intrinsic charges on the surfaces and the edges of iron and aluminium oxides and sites for coordinate bond formation were derived. These authors (2000) inferred that the HA bonding would be strongest in hematite, in agreement with earlier experimental observation (Varadachari *et al.*, 1997). The IR spectra rendered further evidence for the participation of (OH) groups of boehmite in strong HA linkages as opposed to gibbsite, involving very weak or no (OH) group involvement. The analysis of crystal structures at the broken surfaces and edges of oxide minerals is thought to provide information on the factors primarily controlling oxide-HA complexation (Varadachari *et al.*, 2000).

The influence of crystal edges of the common aluminosilicate clays (kaolinite, illite and montmorillonite) on clay-humus complexation was examined (Varadachari *et al.*, 1995) by way of blocking these edges with hexametaphosphate (HMP) and triethanolamine (TEA). The latter led to reduction of HA retention in the dried complexes for all except HMP-Ca-montmorillonite complex. These authors (1995) suggested that (i) edge bonding *via* exchangeable cations is the primary mode of interaction of kaolinite with HA, (ii) cations at the cleavage planes provide the major bonding sites for HA on illite, with crystal edges playing only a minor role, and (iii) increased availability of basal surfaces in montmorillonite by dispersion or swelling has strong influence on montmorillonite-HA complexation. Interaction of HA with montmorillonite causes disruption of its stacking arrangement due to prying open of the interlayer of HA molecules (Varadachari *et al.*, 1991). Indeed, HA fixation by montmorillonite with reduced charge was decreased substantially owing to loss of swelling capacity of the clay, while the exchangeable cations on the illite surface were found to form stronger bonds, compared with montmorillonite or kaolinite (Varadachari *et al.*, 1994).

There have been a number of other studies as well reported in the literature on reactions of the crystalline clays with humic substances (reviewed by Sanyal, 2002b), but relatively little is known about the reactivity towards humic substances of allophane, a clay-size alumino-silicate mineral, characterized by short-range order, occurring widely in Spodosols and Andosols (Sanyal, 2002b).

Yuan *et al.* (2000) investigated the interaction of allophane with HA and cations. The findings were interpreted in terms of ligand exchange between the carboxylate groups of HA and (OH) groups, associated with allophanic aluminium exposed on defect sites at the surface of allophane particles. This reaction generated a surface negative charge, requiring the co-sorption or chemical binding of extraneous cation of the background electrolytes used, namely aqueous $CaCl_2$ or NaCl. Much more HA was sorbed in presence of $CaCl_2$ than NaCl of identical ionic strength, due, presumably, to the possibility that Ca^{2+} ion, besides compensating the negative charge generated (like Na^+ ion also did), may have been bound specifically to the surface complex, and was also more effective (being bivalent) in screening the negative charge on HA than was Na^+ ion.

The generation of surface negative charge on allophane through interaction with humic substance (Figure 5.13) could enhance the capacity of allophanic soils to retain and immobilize positively charged species (as well as organic contaminants), such as nutrient cations and also heavy metal ions over the pH range of most soils (Yuan *et al.*, 2000). Liming of such soils, along with organic manure incorporation,

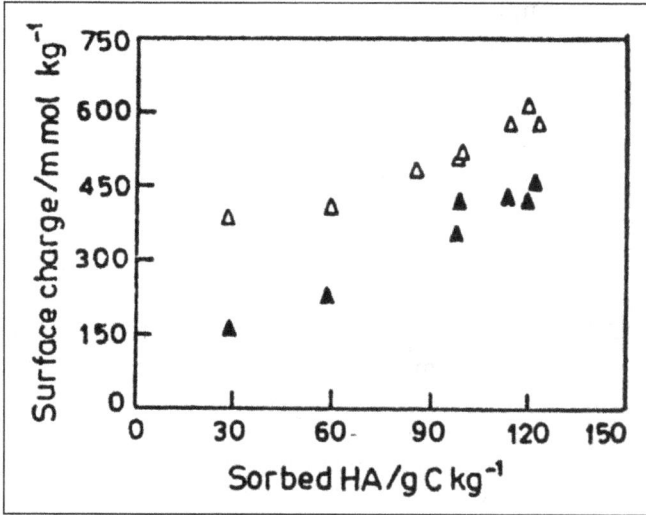

Figure 5.13: Relation between Surface Negative Charge Developed by Allophane-HA Complexes and HA Sorption in 2 mM CaCl$_2$ Solution at pH 7.0 (Δ, r = 0.96) and unadjusted pH (▲, r = 0.95).
Source: **Yuan *et al.* (2000).**

would facilitate the humic-mineral matter interaction (through the added Ca^{2+} ions), and thus help stabilize organic matter in allophanic soils (Sanyal, 2002b).

The fixation of FA by clay minerals (at pH 2.0) was found to fall in the order: illite > kaolinite > montmorillonite (Varadachari *et al.*, 1994) The more numerous negative charges of FA (than on the corresponding HA fraction) and its more hydrophilic character were cited to account for the poorer complexation of FA than that of HA by the given clay minerals (Varadachari *et al.*, 1994).

As mentioned earlier (Section 2.13), abiotic catalysts play a vital role in the transformation of phenolic compounds to humic substances. These catalysts include primary (soil) minerals, layer-lattice aluminosilicates, metal oxides, hydroxides, and oxyhydroxide, as well as poorly crystalline aluminosilicates. Huang and coworkers (reviewed by Sanyal, 2002b) have studied, since the early 1980s, the sequence of catalytic power of layer-lattice silicates and their reaction sites in the polymerisation of phenolic compounds leading to formation of the humic substances. Wang and Huang (1994) reported the Fe (III) in the octahedral sheet of nontronite to serve as a Lewis acid while catalyzing the oxidative polymerisation of hydroquinone, pyrogallol and catechol, by way of cleaving the corresponding aromatic ring structure to liberate CO_2. This is illustrated in Table 2.10 of Section 2.13 (Bollag *et al.*, 1998). The ease of cleavage decreased in the order: pyrogallol > catechol > hydroquinone, the proximity of phenolic (OH) groups apparently favouring the said cleavage.

Studies on naturally occurring clay-humus complexes have been fewer than those conducted with synthetic complexes. One reason, among others, for this trend could be related to the difficulty of the extraction and fractionation of these naturally occurring complexes *without* introducing artifacts (Sanyal, 2002b). Morra *et al.* (1991) suggested ultrasonic dispersion technique with application of no more than 3 to 5 kJ of the sonification energy to 10 g soil in 50 mL of aqueous suspension. Use of sedimentation and ultra-centrifugation techniques has also been resorted to for the purpose. Manoj Kumar (1995) reported from thermal analysis data that soil clay-organic complexes were associated with three-stage exothermic peaks, around 340°, 435° and 510° C, which was reported to ensue from the oxidation of organic matter, complexed to clay mineral surface, sesquioxides surface and allophane, respectively.

5.12. Impact on Soil Properties

The importance of soil humic substances in influencing soil quality or soil health hardly requires any over-emphasis. This is all the more so in the context of sustainable agricultural production system, the need of the day, especially in tropical regions like ours, where not only the soils are inherently poor in native organic matter status, but also the rate of its decay is fast. Notwithstanding this, there is often a tendency to treat soils as a mechanical medium for anchoring plant roots, which can then be bathed in nutrient and growth regulator solutions. The obvious drawback of such management approach lies in the neglect of the need of replenishment of organic matter and maintenance of complex biological communities in soil (Sanyal, 2002b).

It is well known that soil organic matter (SOM), the backbone of soil, contributes significantly to the improvement of soil physical tilth, particularly for maintenance of adequate and stable soil structure by way of acting as a binding agent, thereby facilitating aggregate formation. The latter, in turn, ensures optimum drainage, water-holding capacity and aeration for crop growth, while minimizing soil erosion, and hence loss of fertile top soil and consequent silting of the drainage channels (Sanyal, 2002b). The SOM also contributes tremendously to cation exchange capacity that enables the soil to buffer nutrient concentrations in soil solution, apart from helping storage of the plant nutrients (such as N, P, S and micronutrients, etc.) in soil. The detoxification of pesticide residues is also aided by the SOM. The organic matter in soil serves as the substrate for macro- and microorganisms in soil. The role of these soil microorganisms as key architect in nutrient turnover, organic matter transformations, including synthesis of humic substances, and physical engineering of soil structure is well recognized. Thus, the building of soil organic matter, especially for maintenance of optimum soil biological status, is of vital importance in soil health regeneration (Sanyal, 2002b).

References

Banerjee, S. K. and Mukherjee, S. K. (1972). Studies on the infrared spectra of some divalent transition metal humates. *J. Indian Soc. Soil Sci.*, **20**, 91-94.

Banerjee, S. K. and Mukherjee, S. K. (1975). Electrometric studies of humic and fulvic acids components of the humic fraction. *J. Indian Soc. Soil Sci.*, **23**, 310-318.

Bollag, J. M., Dec, J. and Huang, P. M. (1998). Formation mechanisms of complex organic structures in soil habitats. *Adv. Agron.*, **63**, 237-266.

Bonn, B. A. and Fish, W. (1993). Measurement of electrostatic and site-specific associations of alkali metal cations with humic acid. *J. Soil Sci.*, **44**, 335-345.

Cameron, R. S., Thornton, B. K., Swift, R. S. and Posner, A. M. (1972). Molecular weight and shape of humic acid from sedimentation and diffusion measurements on fractionated extracts. *J. Soil Sci.*, **23**, 394-408.

Celi, L., Schnitzer, M., Khan, S. U. and Schnitzer, M. (1997). Mechanisms of Acifluorfen interaction with humic acid. *Soil Sci. Soc. Am. J.*, **61**, 1659-1665.

Chen, Y. and Schnitzer, M. (1978). Surface tension of aqueous soil humic substances. *Soil Sci.*, **125**, 7-15.

Chen, Y. and Schnitzer, M. (1976). Viscosity measurements on soil humic substances. *Soil Sci. Soc. Am. J.*, **40**, 866-872.

Datta, A., Sanyal, S. K. and Saha, S. (2001). A study on natural and synthetic humic acids and their complexing ability towards cadmium. *Pl. Soil*, **235**, 115-125.

Deiana, S., Gessa, C., Manunza, B., Rausa, R. and Seebar, R. (1990). Analytical and spectroscopic characterization of humic acids extracted from sewage sludge, manure, and worm composts. *Soil Sci.*, **150**, 419-424.

Deiana, S., Gessa, C., Manunza, B., Rausa, R. and Solinas, V. (1995). Iron (III) reduction by natural humic acids: a potentiometric and spectroscopic study. *Eur. J. Soil Sci.*, **46**, 103-108.

Dyanand, S. and Sinha, M. K. (1979). Kinetics of FeEDTA reactions in calcareous soils. *Soil Sci.*, **127**, 202-210.

Dyanand, S. and Sinha, M. K. (1980). Kinetics of reaction of Fe-EDTA and Fe-Fulvate in calcareous soils. *J. Indian Sec. Soil Sci.*, **28**, 429-433.

Flaig, W., Beutelspacher, H. and Rieu. E. (1975). Chemical composition and physical properties of humic substances. **In:** *Soil Components*, Vol.1, Organic Components. Springer-Verlag, New York, pp. 1-211.

Ghosh, K and Schnitzer, M. (1982). A scanning electron microscopic study of effects of adding neutral electrolytes to solutions of humic substances. *Geoderma*, **28**, 53-56.

Ghosh, K. and Mukherjee, S. K. (1971). Hymatomelanic acids as polyelectrolytes. I. Viscometric and osmometric studies. *J. Appl. Polym. Sci.*, **15**, 2073-2077.

Ghosh, K. and Schnitzer, M. (1980). Macromolecular structures of humic substances. *Soil Sci.*, **129**, 266-276.

Ghosh, S. K. and Ghosh, K. (1998). Organo-mineral complexation in soils. *Bull. Indian Soc. Soil Sci.*, New Delhi, No. **19**, pp. 68-79.

Hayes, M. H. B. (1984). Structures of humic substances. **In:** *Organic Matter and Rice.* 1RRI, Los Baños, Philippines, pp. 93-115.

Hayes. M. H. B. and Swift, R. S. (1978). The chemistry of soil organic colloids. **In:** *The Chemistry of Soil Constituents.* John Wiley and Sons. New York, pp. 179-320.

Kendorff, H. and Schnitzer, M. (1980). Sorption of metals on humic acid. *Geochim. Cosmochim. Acta*, **44**, 1701-1708.

Khan, S. U. (1971). Distribution and characteristics of organic matter extracted from the black solonetzic and black chernozemic soils of Alberta. The humic acid fraction. *Soil Sci.*, **112**, 401-409.

Kononova, M. M. (1966). *Soil Organic Matter*, Pergamon, New York.

Kozak, J., Weber, J. B. and Sheets, T. J. (1983). Adsorption of prometryn and metolachlor by selected soil organic matter fractions. *Soil Sci.*, **136**, 94-101.

Krosshavn, M., Kogel-Knabner, I., Southon, T. E. and Steinnes, E. (1992). The influence of humus fractionation on the chemical composition of soil organic matter studied by solid-state ^{13}C NMR. *J. Soil Sci.*, **43**, 473-483.

Lahiri, T. C. and Chakravarti, S. K. (1995). Distribution and nature of organic matter in some hill soils of West Bengal at various altitudes in the Eastern Himalayan region. *J. Indian Soc. Soil Sci.*, **43**, 464-466.

Mandal, S. K. and Sanyal, S. K. (1984). The shape of humic acid – natural polymer-form viscometry, *J. Indian Soc. Soil Sci.*, **32**, 224-229.

Manoj Kumar (1995). Ph.D. Thesis, IARI, New Delhi; cited in Ghosh and Ghosh (1998).

Manunza, B., Deiana, S., Maddau, V., Gessa, C. and Seeber, R. (1995). Stability constants of metal-humate complexes: Titration data analyzed by bimodal Gaussian distribution. *Soil Sci. Soc. Am. J.*, **59**, 1570-1574.

Martin. D., Srivastava, P. C., Ghosh, D. and Zech, W. (1998). Characteristics of humic substances in cultivated and natural forest soils of Sikkim. *Geoderma*, **84**, 345-362.

McBride, M. B. (1994). *Environmental Chemistry of Soils*. Oxford Univ. Press, New York.

McBride, M. B., Martinez, C. E. and Sauvé, S. (1998). Copper (II) activity in aged suspensions of goethite and organic matter. *Soil Sci. Soc. Am. J.*, **62**, 1542-1548.

Morra, M. J., Blank, R. R., Freeborn, L. L. and Shafili, B. (1991). Size fraction of soil organo-mineral complexes using ultrasonic dispersions. *Soil Sci.*, **152**, 294-303.

Mountney, A.W. and Williams, D.R. (1992). Computer simulation of metal ion–humic and fulvic acid interactions. *J. Soil Sci.*, **43**, 679-688.

Murphy, E.M. and Zachara, J.M. (1995). The role of sorbed humic substances on the distribution of organic and inorganic contaminants in groundwater. *Geoderma*, **67**, 103-124.

Neto, L. M., Nascimento, O. R., Talamoni, T. and Popp, N. R. (1991). EPR of micronutrient-humic substance complexes extracted from a Brazilian soil. *Soil Sci.*, **151**, 369-376.

Norvell, W. A. (1991). Reactions of metal chelates in soils and nutrient solutions. **In**: *Micronutrients in Agriculture* (J. J. Mortvedt, F. R. Fox, L. M. Shuman and

R. M. Welch, Eds.) Second Edition, Soil Sci. Soc. Am., Madison, Wisconsin, pp. 187-227.

Orlov, D. S. (1985). *Humus Acids of Soils*, Oxonian Press Pvt. Ltd., New Delhi, India.

Piccolo, A., Pietramellara, G. and Celano, G. (1993). Iron extractability from iron-humate complexes by a siderophore and a mixture of organic acids. *Can. J Soil Sci.*, **73**, 293-298.

Relan, P.G., Girdhar, K.K. and Khanna, S.S. (1984). Molecular configuration of compost's humic acids by viscometric studies. *Pl. Soil*, **81**, 203-208.

Rivero, C., Senesi, N., Paolin, J. and D'Orazio, V. (1998). Characteristics of humic acids of some Venezuelan soils. *Geoderma*, **81**, 227-239.

Saha, P. B. and Sanyal, S. K. (1988). Synthetic humic acids– Their constitution, shape and dimension, *J. Indian Soc. Soil Sci.*, **36**, 35-42.

Sanyal, S. K. (1984). Structure of water in solution of organics-Hydrophobic hydration. *Chem. Edn.* (UGC), **1** (No.2), 14-18.

Sanyal, S. K. (2002a). Soil Colloids. **In:** *"Fundamentals of Soil Science"* (G. S. Sekhon, P. K. Chhonkar, D. K. Das, N. N. Goswami, G. Narayanasamy, S. R. Poonia, R. K. Rattan and J. L. Sehgal, Eds.), Indian Society of Soil Science, New Delhi, pp. 229-259.

Sanyal, S. K. (2002b). Colloid chemical properties of soil humic substances: A Relook. **In:** *Bull. Indian Soc. Soil Sci.*, New Delhi, No. **21**, pp. 278-307.

Sanyal, S.K. and Mandal, S.K. (1983). Viscosity B coefficients of alkyl carboxylates, *Electrochim. Acta*, (*London*), **28**, 1875-1876.

Sarmah, A. C. and Bordoloi, P. K. (1993). Characterization of humic and fulvic acids extracted from two major soil groups of Assam. *J. Indian Soc. Soil Sci.*, **41**, 642-648.

Schnitzer, M. (1991). Soil organic matter – the next 75 years. *Soil Sci.*, **151**, 41-58.

Schnitzer, M. (2000). A lifetime perspective on the chemistry of soil organic matter. *Adv. Agron.*, **68**, 1-58.

Schnitzer, M. and Khan, S. U. (1972). *Humic Substance in the Environment*, Marcel Dekker. New York.

Schnitzer, M. and Schulten, H. R. (1995). Analysis of organic matter in soil extracts and whole soils by pyrolysis-mass spectrometry. *Adv. Agron.*, **55**, 168-198.

Senesi, N., Bocian, D. F. and Sposito, G. (1985). Electron spin resonance investigation of copper (II) complexation by soil fulvic acid. *Soil Sci. Soc. Am. J.*, **49**, 114-119.

Senesi, N., Brunetti, G., Cava, P. L. and Miano, T. M. (1994). Adsorption of Alachlor by humic acid from sewage sludge and amended and non-amended soils. *Soil Sci.*, **157**, 176-184.

Shang, C. and Tiessen, H. (1998). Organic matter stabilization in two semiarid tropical soils: Size, density, and magnetic separations. *Soil Sci. Soc. Am. J.*, **62**, 1247-1257.

Spark, K.M., Wells, J. D. and Johnson, B. B. (1997). Sorption of heavy metals by mineral/humic acid systems. *Aust. J. Soil Res.*, **35**, 113–122.

Stevenson, F. J. (1994). *Humus Chemistry: Genesis, Composition, Reactions.* Second Edition, John Wiley and Sons, New York.

Sturrock, P.E. (1968). Is a weak acid monoprotic? A new look at titration curves. *J. Chem Edn.*, **45**, 258.

Swift, R.S. (1999). Macromolecular properties of soil humic substances: Fact, fiction, and opinion. *Soil Sci.*, **164**, 790-802.

Tarchitzky, J., Hatcher, P.G. and Chen, Y. (2000). Properties and distribution of humic substances and inorganic structure-stabilizing components in particle-size fractions of cultivated Mediterranean soils. *Soil Sci.*, **165**, 328-342.

Tate, R.L. III (2000). *Soil Microbiology,* Second Edition, John Wiley and Sons, New York.

Tomar, N. K., Yadav, K. P. and Relan, P.S. (1992b). Characterization of humic and fulvic acids extracted with NaOH and NaOHNapyrophosphate mixture from soils of arid and sub-humid regions 2. Spectroscopic properties. *Arid Soil Res. Rehab.*, **6**, 187-200.

Tomar, N. K., Yadav, K. P. and Relan, P.S. (1992a). Characterization of humic and fulvic acids extracted with NaOH and NaOHNapyrophosphate mixture from soils of arid and subhumid regions 1. analytical characteristics. *Arid Soil Res. Rehab.*, **6**, 177-185.

Tombacz, E. (1999). Colloidal properties of humic acids and spontaneous changes of their colloidal state under variable solution conditions. *Soil Sci.*, **164**, 814-824.

Van Olphen, H. (1963). *An Introduction of Clay Colloid Chemistry.* Intersci., New York.

Varadachari, C, Chattopadhyay, T and Ghosh, K (2000). The crystallo-chemistry of oxide-humus complexes. *Aust. J. Soil Res.*, **38**, 789-806.

Varadachari, C., Chattopadhyay, T. and Ghosh, K. (1997). Complexation of humic substances with oxides of iron and aluminium. *Soil Sci.*, **162**, 28-34.

Varadachari, C., Mandal, A. H. and Ghosh, K. (1991). Some aspects of clay-humus complexation: effect of exchangeable cation and lattice charge. *Soil Sci.*, **151**, 220-227.

Varadachari, C., Mondal, A. H. and Ghosh, K. (1995). The influence of crystal edges on clay-humus complexation. *Soil Sci.*, **159**, 185-190.

Varadachari, C., Mondal, A.H., Nayak, D.C. and Ghosh, K. (1994). Clay-humus complexation: Effect of pH and the nature of bonding. *Soil Biol. Biochem.*, **26**, 1145-1149.

Visser, S. A. (1985). Viscometric studies on molecular weight fractions of fulvic and humic acids of aquatic, terrestrial and microbial origin. *Pl. Soil*, **87**, 209-221.

Wagner, G.H. and Stevenson, F. J. (1965). Structural arrangement of functional groups in soil humic acid as revealed by infrared analysis. *Soil Sci. Soc. Am. Proc.*, **29**, 43-48.

Waksman, S. A. and Iyer, K. R. N. (1932). Contribution to our knowledge of the chemical nature and origin of humus: I. On the synthesis of the "Humus Nucleus". *Soil Sci.*, **34**, 43-70.

Wang, M. C. and Huang, P. M. (1994). Structural role of polyphenols in influencing the ring cleavage and related chemical reactions as catalyzed by nontronite. **In:** *Humic Substances in the Global Environment and Implications in Human Health* (N. Senesi and T.M. Miano, Eds.). Elsevier, Amsterdam, The Netherlands, pp. 173–180.

Yuan, G., Theng, B. K. G., Parfitt, R. L. and Percival. H. J. (2000). Interactions of allophane with humic acid and cations. *Eur. J. Soil Sci.*, **51**, 35-41.

Zachara, J. M., Resch, C. T. and Smith, S. C. (1994). Influence of humic substances on Co^{2+} sorption by a subsurface mineral separate and its mineralogic components. *Geochim. Cosmochim. Acta*, **58**, 553-566.

Chemistry of Phosphorus, Waterlogged Soils and Micronutrients in Soil

6A. Chemistry of Phosphorus in Soil

6A.1. Phosphorus Cycle in Nature and Soils

The continued cycling of phosphorus (P) from soil to crop and *via* crop to animals and humans, as well as its return to soil through the decomposition of plant residues and animal and human wastes are well recognized phenomena (Figure 6A.1). Phosphorus in soil is derived from the weathering of primary minerals, notably the apatites of igneous rocks, and progressively transferred to plants and animals (Majumdar and Sanyal, 2015). Since only a very small fraction of P is present in plant-available form in soil solution at any point of time, the plant available pool of P in soil solution has to be enriched through the addition of chemical fertilizers as well as organic manures for crop production (Figure 6A.1). Much of the added (and native) P in soil solution gets transformed through adsorption and precipitation processes into P reaction products of varying degrees of solubility through interactions with soil components. The solubility of these reaction products decreases continuously with passage of time. However, a fraction of such immobilized P gets slowly mineralized, mediated by the soil micro-organisms, back to the soil solution for plant uptake over a period of time. The temporary tie-up of a portion of the mineralized P in course of this process within the microbial biomass is also subject to gradual release in soil solution through the subsequent microbial

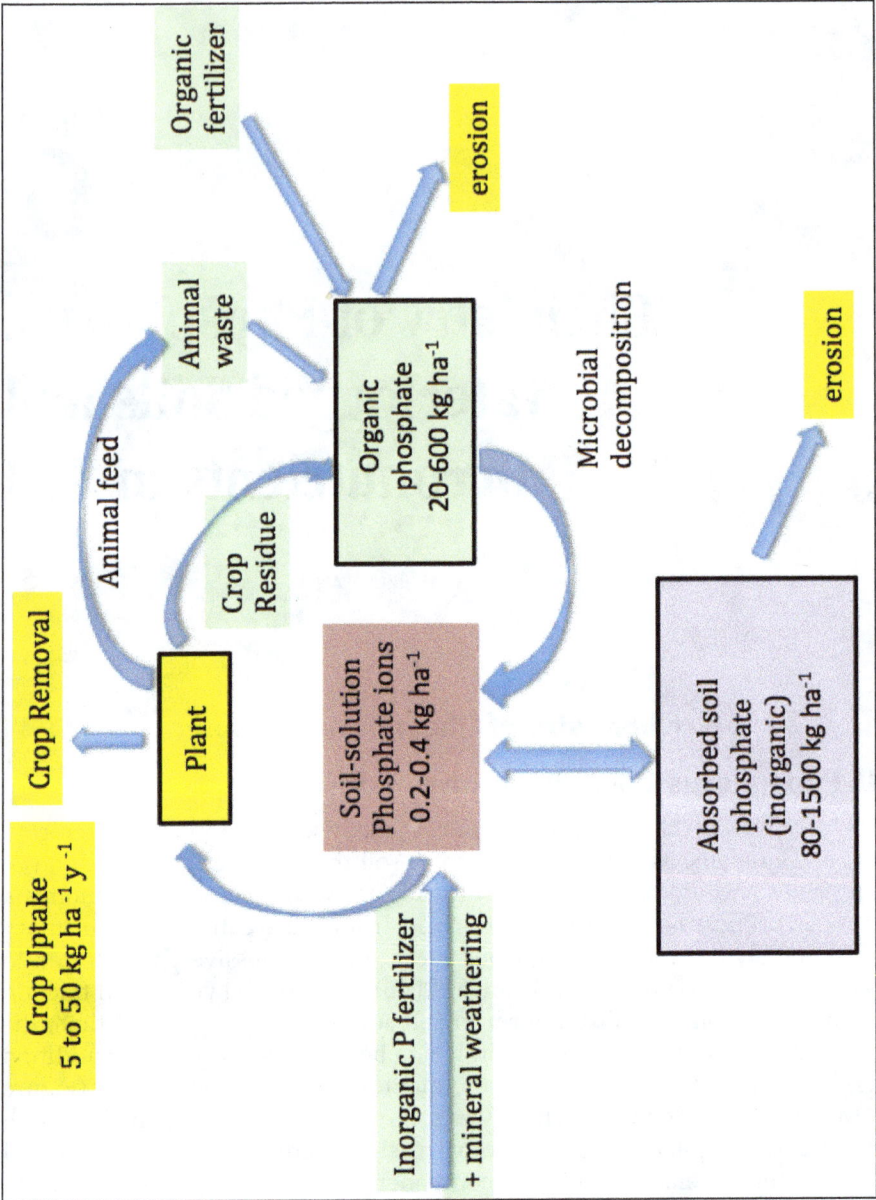

Figure 6A.1. The Phosphorus Cycle.
Source: Majumdar and Sanyal (2015).

decay processes. A portion of the soil P is lost through the erosion process as well (Sanyal and De Datta, 1991; Majumdar and Sanyal, 2015).

6A.2. Phosphorus Sources, Forms and Fractions in Soil

Phosphorus is the tenth most abundant element in the earth's crust. About 200 minerals, occurring in nature, have been reported to contain P, with apatite group of minerals being the most important. Indeed, fluorapatite (*see* later) and the related minerals act as the source material for bulk of the phosphatic fertilizers. The surface plough layer of soil (0-0.15 m) contains P in the range of 200 to 2000 kg P. ha^{-1}, the average being 1000 kg P. ha^{-1}. Total P content of some representative Indian soils ranges from 580 to 2900 kg P. ha^{-1} (Majumdar and Sanyal, 2015). Generally, P is present in soil predominantly in the inorganic forms, with organic P content varying from 20 to 80 per cent of total P in the soil. The inorganic phosphates in soil comprise the (i) Fe and Al phosphates (in acidic soils), and (ii) Ca phosphates (in neutral and alkaline soils). The Al-P and Fe-P contents are generally higher in the fertilized soils. Relatively less important inorganic P fractions in soil include the occluded (*i.e.*, closed or enveloped) or the reductant soluble P. The occluded phosphates remain occluded within the oxide matrices.

The organic P in mineral soils normally declines sharply down the soil profile, and its content is higher in clayey soils than in coarse-textured soils. Poor drainage characteristics, high soil pH and cultivation practices adversely affect the organic P content in soil. The organic P is mostly located in the fulvic, and then in the humic acid fractions of the soil humus. The soil organic P compounds may generally be classified into three groups, namely (i) inositol phosphates, comprising up to 60 per cent of soil organic P, (ii) nucleic acids, and (3) phospholipids. The presence of phosphoproteins, sugar phosphates, glycerophopsphates has also been reported in soil organic P (Sanyal and De Datta, 1991). Research on organic P suggests that the readily decomposable or soluble fractions of soil organic P is often important in supplying P to plants in highly weathered soils (*e.g.* Ultisols and Alfisols). Mineralization of organic P in soils is influenced by factors such as temperature, moisture, tillage, etc. Thus one would take note of the following transformations occurring in soil (Majumdar and Sanyal, 2015):

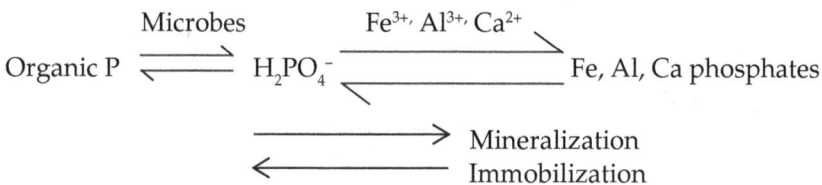

$$\text{Organic P} \underset{\xleftarrow{\hspace{1cm}}}{\overset{\text{Microbes}}{\rightleftharpoons}} H_2PO_4^- \underset{\xleftarrow{\hspace{2cm}}}{\overset{Fe^{3+}, Al^{3+}, Ca^{2+}}{\xrightarrow{\hspace{2cm}}}} \text{Fe, Al, Ca phosphates}$$

$$\xrightarrow{\hspace{2cm}} \text{Mineralization}$$
$$\xleftarrow{\hspace{2cm}} \text{Immobilization}$$

The *net* mineralization of organic P is likely if the organic residues added to soil have a (C/P) ratio below 200:1, while *net* immobilization of soluble P is likely to occur at the (C/P) ratio above 300:1 (Majumdar and Sanyal, 2015).

Of the total P content in soil, only a very small fraction is present in plant-available form in soil solution, which generally does not exceed 0.1-0.2 mg P. L^{-1} in unfertilized soils, with the higher-end values noted in the submerged paddy soils.

6A.3. Chemistry of Phosphorus in Soil: *Historical Aspects and New Understandings*

It has long been recognized that phosphorus (P) applied to soils is removed in large proportion from the soil solution. This process was called "P fixation" or "P retention" The reaction was usually attributed to specific soil components, *e.g.* calcium carbonate in calcareous soils, and hydrous iron and aluminium oxides in acid soils. Scanty attention was paid to the subsequent bio-availability of fixed or retained phosphorus during this early phase of "P fixation" research. The need to apply more P than was removed in the harvested crop raised the question as to what happened to the residual phosphate. It had commonly been assumed that the failure of a crop to respond to fertilizer P was because of the rapid fixation of P by the soil. Pioneering work in Australia and elsewhere in the late 1960s and 1970s (Barrow, 1978) on P adsorption in soils and its reversibility (desorption) led to better understanding of P reactions in soils (Sanyal and De Datta, 1991). It is now generally agreed that changes in the extractability of soil and fertilizer P, and the decrease in plant availability of the added P with time, can be explained reasonably well by the current concepts relating to P equilibria in soils. These primarily involve adsorption and absorption reactions, which may be largely reversible with time. Syers *et al.* (2008) summarized that soil P exists in four different pools on the basis of its accessibility and extractability, such as: (a) Soil solution P; (b) Surface-adsorbed P; (c) Strongly-bonded or absorbed P; and (d) Very strongly–bonded or inaccessible or mineral or precipitated P. In the soil solution, P is immediately available for uptake by plant roots. The second pool represents readily extractable P held on the surface of soil components. This pool is considered to be in equilibrium with P in the soil solution, and can be transferred readily to the soil solution as the concentration of P in the latter is lowered by P uptake by plant roots. Phosphorus in the third pool is less readily extractable as it is more strongly bonded to soil components or is present within the matrices of soil components as absorbed P (*i.e.,* P adsorbed on internal surfaces) but can become plant-available over time. The P in the fourth pool has a low or very low extractability. This is because the P is very strongly bonded to soil components, or it has been precipitated as sparingly soluble P compounds, or it is part of the soil mineral complex, or it is unavailable because of its position within the soil matrix. Whatever the reason, such P is only very slowly available (often over periods of many years) for plant uptake. The amount of P in each of the four pools is related to differences in bonding energy for P between the sites, both on the surfaces and within the soil constituents, able to retain P, and variations in the proportion of such sites within the soil matrix. For P in the less readily available pool, the underlying reasons could be related to the fact that there can be other reactions of P with soil constituents (Syers *et al.*, 2008). The existence of different pools of P in the soil and their accessibility to crops and cropping are also recognized by researchers in India (Sanyal *et al.*, 2015b). As stated above, P fixation in soil involves both adsorption and precipitation reactions, although the former appears to be dominant over a short-range period. The phosphorus availability to crop plants is optimum at the neutral pH range and decreases as the soil pH shifts to either acidic or alkaline range. In acid soils the availability of Ca, Mg, P,

Mo, B and Si is low, whereas the availability of Fe, Mn, Cu and Zn is high. The low phosphorus (P) status of highly weathered acid soils is a particular problem because large amounts of P need to be applied in order to raise concentrations of available soil P to an adequate level. This is because such soils contain large quantities of Al and Fe hydrous oxides which have the ability to absorb P onto their surfaces. Thus, much of the added P is 'fixed' and is not readily available for crop use.

6A.4. Phosphorus Equilibrium and Physico-chemical Processes Operating in Soils

The dynamics of P transformations in soils and the fixation and release characteristics of P are governed primarily by (i) Adsorption-desorption and (ii) Precipitation-dissolution reactions in soil. The adsorption-desorption processes as well as the corresponding adsorption isotherms and the hysteresis phenomena involved in phosphate adsorption-desorption in soil were discussed earlier in a companion chapter on **Ion Exchange Processes in Soil** (Section 3.2.1). In general, the Freundlich adsorption isotherm gives a better fit of the equilibrium phosphate sorption data in soils than did the Langmuir adsorption isotherm, especially so at relatively higher concentrations of the phosphate. This may be because the Freundlich equation, although originally empirical, implied that the affinity term decreases exponentially with surface coverage, which is closer to reality than the assumption of a constant bonding energy of the adsorbent for the adsorbate, inherent in the Langmuir equation. The details of these aspects are skipped here for the sake of brevity and also for avoiding repetition.

6A.4.1. Processes of P Adsorption in Soils

Phosphate reacts with the surface of minerals (*e.g.* Fe and Al oxide and hydroxide minerals, aluminosilicate clays, calcite) if P concentration in soil solution is low. Adsorption reactions occur in acid, neutral, alkaline and calcareous soils.

As mentioned earlier in the companion chapter on **Soil Colloids: Structural Aspects and Properties** (Section 2.5.2), it is worth mentioning here that the exposed edge/surface hydroxyl groups at the broken edges of the variable–charge surfaces in soil (*e.g.* Fe and Al oxides and hydroxides as well as the aluminosilicate clays) can act both as an acid (donating a proton to the surrounding OH group) or a base (accepting a proton from the surrounding soil solution), depending on pH of the soil system. Such amphoteric nature of these soil colloids warrants that these exposed hydroxyls would behave as an "uncharged" group (neither a proton donor nor a proton acceptor) at an intermediate pH. The pH at which the net surface charge of the soil colloids is zero is known as the zero-point of charge or ZPC. The ZPC of a soil component depends upon its relative acidity with respect to water. Thus, ZPC of silicic acid is 2.0, that of goethite (α-$Fe^{III}OOH$) is 8.5 and that of gibbsite [$Al(OH)_3$] is more than 9.0. However, even at ZPC, the permanent charge of soil colloids (such as the aluminosilicate clays) persists unchanged (Sanyal *et al.*, 2009).

Further, anions derived from strong acids, *e.g.* Cl^-, NO_3^-, Br^-, ClO_4^-, etc., undergo anion exchange (Section 3.9) in soil and are held at the positively charged soil colloidal sites by virtue of electrostatic forces. Such retention of anions is also referred to as the non-specific anion retention. This may be described as the physical adsorption because no chemical bond is formed between the soil solid surface and the anions. Several other anions, including the organic anions, are generally specifically adsorbed to the surface of aluminosilicate clays or clay-sized primary minerals (*e.g.* Fe and Al oxides and hydroxides). These anions may be derived from both strong and weak acids, and are of mono- and polyvalent type. In this case, the anions, such as phosphate, form a chemical bond with the surface group, *i.e.*, undergoes chemisorption and is thus strongly held. This phenomenon is also known as ligand exchange (Section 3.9) because the extraneous anion (*e.g.* phosphate), acting as a ligand to the centrally located cation in the clay lattice (*e.g.* Fe^{3+} and Al^{3+} ions), substitute an existing surface ligand (*e.g.* an OH$^-$ ion or a H_2O molecule which satisfy the secondary valency or the coordination number of the central cation, *e.g.* Fe^{3+} and Al^{3+} ions) (Sanyal *et al.*, 2009).

While anion exchange takes place only at the positively charged colloid surface, *e.g.* for a variable-charge surface, namely hydrous oxides of Fe^{III}, such as goethite [α-$Fe^{III}OOH$], at a pH below the corresponding ZPC (pH 8.5), such oxides undergo ligand exchange at pH values below, above and even at its ZPC pH. The same is true for other types of soil colloid surfaces of variable-charge as well.

6A.4.2. Precipitation – Dissolution

The adsorption of P, as stated earlier, dominates in soils with low soil solution P concentration, while precipitation reactions proceed when solution P concentration is such as to cause the solubility product (K_{sp}) of the specific P containing reaction products (minerals) being exceeded. Surface adsorption and precipitation reactions are collectively known as P fixation or retention, which is of great practical significance for P management in soils (Sanyal and De Datta, 1991; Majumdar and Sanyal, 2015).

6A.4.2.1 Precipitation Reactions of P

In such processes, P reacts with Fe and Al oxide and hydroxide minerals or $CaCO_3$ forming precipitates of hydrated or hydroxyl phosphates of Fe and Al and Ca phosphate, if the P concentration in soil solution is high (Majumdar and Sanyal, 2015).

In acid soils, $H_2PO_4^-$ ions in the soil solution react with Fe and Al oxide and hydroxides through ligand-exchange mechanism, as shown below:

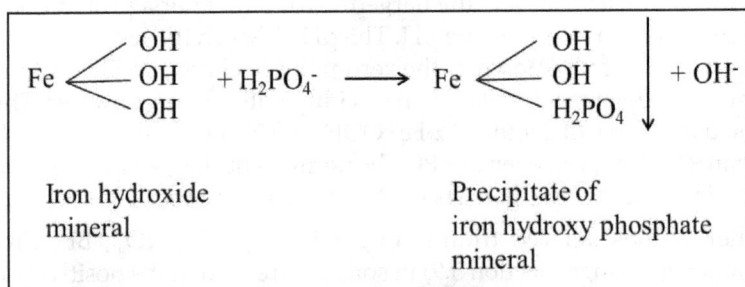

$$Fe \begin{cases} OH \\ OH \\ OH \end{cases} + H_2PO_4^- \longrightarrow Fe \begin{cases} OH \\ OH \\ H_2PO_4 \end{cases} \downarrow + OH^-$$

Iron hydroxide mineral

Precipitate of iron hydroxy phosphate mineral

In strongly acid soils, Fe^{3+} and Al^{3+} predominate in the soil solution. These react with $H_2PO_4^-$ to form hydroxy phosphates of Fe and Al.

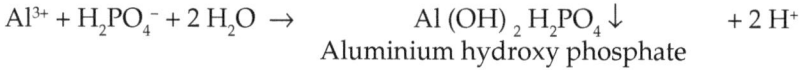

$$Al^{3+} + H_2PO_4^- + 2\,H_2O \rightarrow \qquad Al\,(OH)_2\,H_2PO_4 \downarrow \qquad + 2\,H^+$$
$$\text{Aluminium hydroxy phosphate}$$

When $H_2PO_4^-$ concentration is low, adsorption (as described earlier) instead of precipitation occurs in the same way.

In neutral, alkaline and calcareous soils, soluble Ca^{2+} ions predominate. The Ca^{2+} ions react with $H_2PO_4^-$ and HPO_4^{2-}, forming precipitates of secondary calcium phosphate minerals. Some of these reactions are shown as follows (Majumdar and Sanyal, 2015):

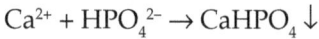

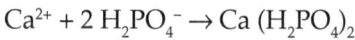

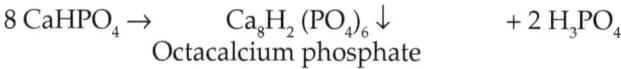

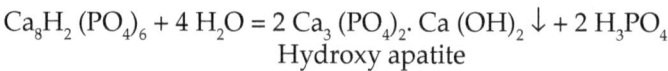

$$Ca^{2+} + HPO_4^{2-} \rightarrow CaHPO_4 \downarrow$$

$$Ca^{2+} + 2\,H_2PO_4^- \rightarrow Ca\,(H_2PO_4)_2$$

$$Ca\,(H_2PO_4)_2 \rightarrow CaHPO_4 \downarrow + H_3PO_4$$

$$8\,CaHPO_4 \rightarrow \qquad Ca_8H_2\,(PO_4)_6 \downarrow \qquad + 2\,H_3PO_4$$
$$\text{Octacalcium phosphate}$$

$$Ca_8H_2\,(PO_4)_6 + 4\,H_2O = 2\,Ca_3\,(PO_4)_2.\,Ca\,(OH)_2 \downarrow + 2\,H_3PO_4$$
$$\text{Hydroxy apatite}$$

These reactions show that in alkaline and calcareous soils, soluble P [that is, mono calcium phosphate, $Ca\,(H_2PO_4)_2$] is precipitated as dicalcium phosphate, $CaHPO_4$ (in a few weeks) and as tricalcium phosphate, $Ca_3\,(PO_4)_2$ (which precipitates in a few months). With time (on aging), fluorapatites or hydroxyapatites are formed (*see* later), which are relatively more stable than the recently formed precipitates. Besides the adsorption and precipitation reactions of P in soils, some of the phosphates previously adsorbed on the surfaces of solid particles of either hydrous oxides of Fe or Al or $CaCO_3$, with time, penetrate into these particles.

The availability of P in soils depends on the dissolution of precipitated phosphates. Dissolution is a slow process, which depends on the solubility of the compounds formed from the reactions of $H_2PO_4^-$ ions with Fe, Al or Ca compounds present in the soil. Bacteria and fungi can enhance the solubility of both Ca and Al phosphates by releasing citric and other organic acids that either dissolve the calcium phosphates or form metal complexes which release P from Fe and Al phosphates in acid soils (Majumdar and Sanyal, 2015).

6A.5. Thermodynamics of Phosphorus Adsorption in Soil

Following Sanyal *et al.* (1993), the phosphate sorption process may be regarded as a process of partition of P between the bulk soil solution phase and a surface phase. The above stated partition process is characterized by a distribution coefficient (K_0). Admittedly, the distribution coefficient (K_0) would vary with P solution concentration, and hence with the extent of soil surface coverage by the sorbed P. By way of having recourse to the appropriate extrapolation of the experimental specific phosphate adsorption data (in the experimental soils) to the infinitely dilute soil solution condition, it is possible to refer the resulting standard differential Gibbs free energy change (ΔG_0) values, accompanying the adsorption process in the limit of the infinitely dilute solution of the adsorbate (*e.g.* phosphate), to the corresponding minimum surface coverage at equilibrium (Sanyal *et al.*, 1993). Thus, the above stated concentration effect would be minimal. This would also minimize the electrostatic interactions of sorbed P with the charged surfaces of soil, or that between the sorbed P species themselves. Goldberg and Sposito (1984), using the constant capacitance model of Stumm *et at.* (1980), obtained expressions for intrinsic conditional equilibrium constants for adsorption of P by Al and Fe hydrous oxides. These equilibrium constants were identical with the conditional equilibrium constants (which ignore the electrostatic interactions involved in the given P sorption process) only at zero *net* surface charge. The two constants were, in general, related through an exponential term in surface potential, which was considered to be the solid-phase activity coefficient that corrects for the charge on the surface and of the sorbed species. Indeed one may take a measure of the latter from the extrapolated value of K_0 and thus the changes in surface charge due to P sorption is minimized, and hence the given P sorption system is considered to approach the ideal state. In so doing, the effect of charged surfaces in soil on P sorption is not ignored (and hence the calculated ΔG_0 values would be different for various soils), but the *perturbations* of such charges due to interaction of sorbed P with the soil surface is ignored. The negative magnitude of the ΔG_0 values would indicate the spontaneity of the given sorption process in the soils studied, while exhibiting significant correlations with the sorption parameters of different adsorption isotherms used to describe P-fixation by soils and soil components. Thus, the stability of P sorption reaction products in soils, relative to P in soil solution, seems to contribute towards the P-fixing characteristics of these soils. Such information may profitably be incorporated into crop management practices when planning for an appropriate P source, rate, and timing of application for a soil (*see* later), given its P sorption-desorption behaviour. Thus, for a soil with a high P sorption capacity and a strong P-binding energy, coupled with a high P sorption rate and poor P desorption characteristics, one may settle for a less soluble P source that releases P in soil solution in smaller concentrations, spread over a longer period of time. This is expected to slow down the P fixation reactions in the given soil, and maintain fertilizer P in plant-available form over a longer period

6A.6. Phosphorus Buffering Capacity

The phosphate buffering capacity (PBC) of soil characterizes the dynamic relation between labile solid phase (Quantity) and solution phase phosphate (Intensity) from which plants take their supply. It is measured from the slope of the sorption isotherm or quantity-intensity isotherm, and it varies according to the solution concentration at which isotherm slope is measured. Various indices used to express this parameter are: (i) the slope at a standard equilibrium concentration of 0.2 μg P. mL^{-1}, (ii) the slope at a standard equilibrium concentration of 0.3 μg P. mL^{-1}, (iii) the maximum slope of the isotherm as solution concentration tends to zero, (iv) the ratio between the change in quantity factor and intensity factor, designated as differential buffering capacity. A closer linear relationship exists between P-buffering capacity (PBC) and ability of soil to sorb phosphate. The PBC of the sandy soils is less than that of the fine-textured soils. The PBC of acidic and neutral soils is a function of the amount and the extent of crystallinity of hydrated oxides of Fe and Al. In calcareous soils, the amount of exchangeable calcium and $CaCO_3$ determine the PBC (Sanyal *et al.*, 2015b)

6A.7. Kinetics of Phosphorus Sorption in Soil

The study of the kinetics of phosphorus (P) sorption and release by soils is of considerable interest in soil and environmental science. The time factor is certainly of relevance for P uptake by plants. Indeed comparisons among soils on the basis of kinetic rate constant alone do not seem to be of great practical value, since amounts of P desorbing during any time interval would also depend on the reserve of the desorbable P present. On the other hand, a better understanding of the energetics of P sorption, based on kinetic studies, may help elucidate mechanisms of P adsorption-desorption processes in soils. The reaction between P and soils is rapid at first. It then becomes slow, and continues for very long time. In many cases, it is doubtful whether a true equilibrium is ever reached within a reasonable reaction period, even though an apparent equilibrium may be established. This led to several authors to use a number of adsorption isotherms to describe P adsorption and desorption by soils. As for the kinetic parameter (*e.g.* the rate constant), the latter generally decreases with increase in initial P concentration in soil solution, thereby suggesting a two-stage reaction (Barrow, 1978; Sanyal and De Datta, 1991). The commonly used kinetic models for describing the P reaction kinetics in soil are summarized in Table 6A.1.

Table 6A.1: Summary of Kinetic Equations Used for Phosphorus Reaction with Soils

Kinetic Equations	Assumptions
First-order kinetics $\log C = \log C_0 - kt$	Rate of change in concentration is proportional either to the concentration in solution or to the number of empty sites.
Second-order kinetics $(1/C_0) - (1/C) = kt$	Rate of change in concentration is proportional to both the concentration in solution and number of empty sites.
Diffusion equation $X = R \sqrt{t} + b$	Rate limiting step is the diffusion of phosphate ions either from the solution to the soil surface or from the surface to the interior of the soil particles.
Modified Langmuir equation $X = b_1 + \Delta b_1 + (b_{11} k_{11} C)/(1+k_{11}C) + (b_{111}k_{111}C)/(1+k_{111}C)$	Rate of adsorption is proportional to the concentration in the solution and the number of empty sites; with time, adsorbed phosphate redistribution in region 1.
Modified Freundlich equation $X = k C^b t^{b1}$	Phosphate reaction in soil system contained three compartments, A, B, and C and reacts according to A $\rightleftharpoons$ B $\rightleftharpoons$ C; the rate limiting step is B→C
Elovich equation $X = (1/\beta) \ln (\alpha\beta) + (1/\beta) \ln t$	Activation energy of adsorptions increases linearly with surface coverage.

Source: Adapted from Sanyal and De Datta (1991)

6A.8. Phosphorus Availability in Soils

The concentration of P in the soil solution varies from about 0.001 mg

P. L^{-1} in low-P soils to about 1 mg P. L^{-1} in high P-soils, with an average of 0.05 mg P. L^{-1}. Plant roots absorb P from soil solution as $H_2PO_4^-$ ions. In strongly acidic soils, $H_2PO_4^-$ dominates, while in alkaline soils, P is largely present as HPO_4^{2-} form. In near-neutral soils, both these anions are present. Plant uptake of HPO_4^{2-} is much slower than that of $H_2PO_4^-$. The rate of movement of P ions in soils (diffusion) is very slow. Thus, in cropped soils, phosphatic fertilizers should be applied very near to the roots. Soils are categorized as low, medium and high in available P, according to the amount of P extracted, and the fertilizer P doses are recommended accordingly. For instance, a soil analyzing less than 10 kg P. ha^{-1} (Olsen-P value) is categorized as low, between 10 and 25 kg P. ha^{-1} as medium and over 25 kg P. ha^{-1} as high in P-availability (Majumdar and Sanyal, 2015).

Some important factors affecting P availability in soils are discussed below (Majumdar and Sanyal, 2015).

(i) Iron and Aluminium Oxides and Hydroxides

Iron and aluminium oxides present in acid soils have the capacity to adsorb large amounts of solution-P, rendering it unavailable to plants. These oxides also exist as coatings on other soil particles, *e.g.* the aluminosilicate clays. The P-fixation capacity is high in soils where amorphous Fe and Al oxides contents are higher as compared to their crystalline counterpart, because of the greater specific surface area of the amorphous oxides. Thus, in highly weathered soils (rich in kaolinite clays with high Fe and Al oxides), low P availability is a major problem.

(ii) Clay Minerals

Soils containing greater amounts of clay (as in the black soils, Vertisols) will fix more P than do soils with low clay content (red and lateritic soils, Alfisols and Ultisols). Among the clay minerals, 1:1 clays (*e.g.* kaolinite) adsorb more P than do the 2:1 clays (*e.g.* montmorillonite). Cation saturation on clays also influences P adsorption. For example, clays saturated with Ca^{2+} ions retain larger amounts of P than those saturated with Na^+ ions or other monovalent cations.

(iii) pH

The soil P availability to plants is strongly dependent on soil pH that affects the dominant forms of phosphate, prevailing in soil solution at different pH. The dissociation constants of orthophosphoric acid, H_3PO_4, being, respectively, $pK_{a1}=2.12$, $pK_{a2}=7.21$, and $pK_{a3}=12.32$ at 25 °C., as shown below, the predominant phosphate ion in soil solution at low pH (acidic soil) is $H_2PO_4^-$.

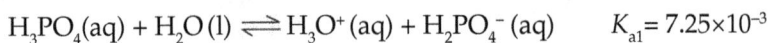

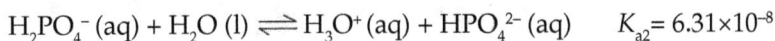

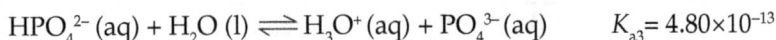

$$H_3PO_4(aq) + H_2O(l) \rightleftharpoons H_3O^+(aq) + H_2PO_4^-(aq) \qquad K_{a1} = 7.25\times10^{-3}$$

$$H_2PO_4^-(aq) + H_2O(l) \rightleftharpoons H_3O^+(aq) + HPO_4^{2-}(aq) \qquad K_{a2} = 6.31\times10^{-8}$$

$$HPO_4^{2-}(aq) + H_2O(l) \rightleftharpoons H_3O^+(aq) + PO_4^{3-}(aq) \qquad K_{a3} = 4.80\times10^{-13}$$

A mixture of HPO_4^{2-} and $H_2PO_4^-$ ions predominate at moderate soil acidity. At neutral to slightly alkaline pH, the predominant ionic species in soil solution is HPO_4^{2-}. Figure 6A.2 gives the preponderance of different phosphate ionic species in soil solution at varying pH.

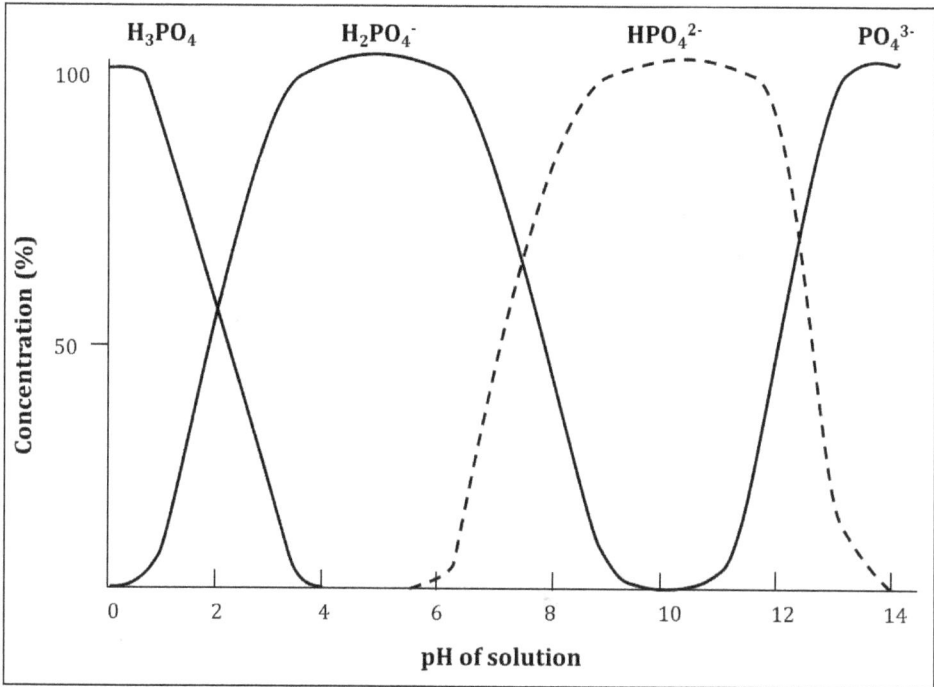

Figure 6A.2: Relationship between Soil Solution pH and the Relative Concentrations of Ionic Forms of Phosphate in Soil Solution.
Source: Majumdar and Sanyal (2015).

In acidic soils, P is fixed predominantly as Fe and Al phosphate, namely strengite ($Fe^{III}PO_4 . 2H_2O$) and variscite ($AlPO_4 . 2H_2O$). In neutral to alkaline soils, soluble P is converted gradually to dicalcium phosphate ($CaHPO_4.2H_2O$), octacalcium phosphate [$Ca_8H_2 (PO_4)_6.H_2O$], tricalcium phosphate [$Ca_3 (PO_4)_2$] and hydroxy apatite, $3Ca_3 (PO_4)_2. Ca (OH)_2$, in this order, with progressively decreasing solubility and hence plant availability of P. In general, the P availability in soil decreases in the order: Ca-P> Al-P> Fe-P (Sanyal and De Datta, 1991; Majumdar and Sanyal, 2015).

(iv) Organic Matter

Organic matter often enhances P sorption in soil and decreases P availability to plants due to association with the sesquioxides, and hence possible stabilization of the soil organic matter by the sesquioxides. In volcanic soils, P adsorption correlated with organically bound Al, and to a lesser extent, with Fe, extracted by sodium pyrophosphate. The organic matter, notably the humic colloids, competes with P for the common sorption sites in soil (*e.g.* hydrous oxides of Fe and Al). The humic and fulvic acid fractions of soil organic matter (humus) compete strongly with P for adsorption sites on goethite and gibbsite at low pH. This will cause reduced P sorption and enhanced P availability to plants. Citrate, tartrate and acetate were also effective, in this order, in reducing P sorption by soils (Sanyal and De Datta, 1991).

(v) Redox State

In submerged soil of low redox (reduction) potential, the available P increases initially because of reduction of Fe and Mn compounds. The corresponding ferrous and manganous phosphates are relatively soluble compared to the higher valent-ferric and manganic phosphate, respectively. However, drying (draining) of flooded soils increase the P sorption capacity of soil and its bonding energy for P (Ponnamperuma, 1972). Also, drying increased the amount of acid-ammonium oxalate-extractable Fe, and the Fe-bound phosphate at the expense of Al-bound phosphate. Consequently, flooding and drying were suggested to increase the activity of ferric hydrous oxides in sorbing P (by way of decrease of their crystallinity) that resulted in the added P immobilization and decreased plant-availability of P (Sanyal and De Datta, 1991; Majumdar and Sanyal, 2015).

(vi) Role of Microorganism

Microorganisms such as bacteria, fungi and actinomycetes, secrete organic acids (namely lactic, glycolic acids, etc.), and aid in the solubilization of the immobilized inorganic phosphates, thereby enhancing the P availability to plants. The bacteria oxidizing ammonium to nitric acid and sulphur to sulphuric acid also cause partial and gradual dissolution of phosphate rock and/or the insoluble calcium phosphates (*e.g.* tricalcium phosphate), thereby enhancing the P availability to plants (Majumdar and Sanyal, 2015).

6A.9. Major Phosphatic Fertilizers and their Reactions in Soils

Phosphatic fertilizers in use (Table 6A.2) are broadly classified on the basis of their solubility in water and/or 2 per cent citric acid or in 1N ammonium citrate (Sanyal and De Datta, 1991).

Table 6A.2: The Composition of Phosphatic Fertilizers

Fertilizer	Total P Per cent	Citrate Soluble P Per cent	Water Soluble P Per cent	Total N Per cent	Ammoniacal N Per cent	Nitrate N Per cent
Single superphosphate	7.09	7.09	6.88	–	–	–
Triple superphosphate	17.2	17.2	17.2	–	–	–
Dicalcium phosphate	14.6	14.6	Nil	–	–	–
Calcium metaphosphate	27.1	27.1	Nil	–	–	–
Fused magnesium phosphate	9.67	8.17	Nil	–	–	–
Phosphate rock	11.6	0.645	Nil			
Basic slag (Indian)	1.08	0.860	–	–	–	–
Monoammonium phosphate	20.6	20.6	20.6	11.0	11.0	–
Diammonium phosphate	20.0	20.0	20.0	18.0	18.0	–

Source: Raychaudhuri (1976).

Monocalcium phosphate (MCP) is the essential phosphate component in superphosphates. The latter are obtained by treating the phosphate rock with either sulphuric acid or phosphoric acid, leading to the formation of single superphosphate (which is an equi-molar mixture of MCP and gypsum) or triple superphosphate (MCP). Soil applied superphosphates undergo hydrolysis to form an acid (pH 1.48) metastable triple point solution (MTPS) of MCP, dicalcium phosphate dihydrate (DCPD), and phosphoric acid. Over a subsequent period of 7-8 days, dicalcium phosphate dihydrate gradually dissolves, and less soluble dicalcium phosphate (DCP) precipitates out. The solution, leaving the site of fertilizer granule by diffusion down the concentration gradient of soluble P, becomes highly acidic and enriched in P. This hydrolytic dissolution of MCP may be represented as (Sanyal and De Datta, 1991):

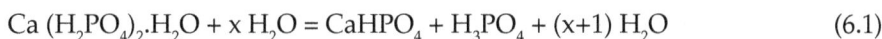

$$Ca (H_2PO_4)_2.H_2O + x H_2O = CaHPO_4 + H_3PO_4 + (x+1) H_2O \qquad (6.1)$$

The dicalcium phosphate that remains in the fertilizer granule dissolves, and the highly insoluble hydroxyapatite precipitates out and forms phosphoric acid in solution.

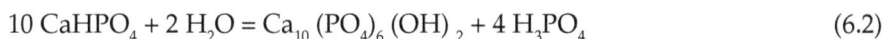

$$10 CaHPO_4 + 2 H_2O = Ca_{10} (PO_4)_6 (OH)_2 + 4 H_3PO_4 \qquad (6.2)$$

The solubility of DCP being lower than that of MCP, the diffusion gradient between the fertilizer granule site and the bulk soil solution in reaction (6.2) is much lower than that for reaction (6.1). As a consequence, P contribution to water-soluble P pool in soil from reaction (6.2) becomes very low. The concentrated phosphoric acid solution, moving out of the fertilizer granule, enters into several reactions with soil minerals. It induces the dissolution and/or exchange of appreciable quantities of cations such as Fe, Al, Mn, Ca and Mg in the soil matrix. In acid soils, where Fe and Al concentrations are high, Fe and Al phosphates are the primary reaction products, whereas in neutral and alkaline soils, Ca and Mg phosphates precipitate out, and with time, probably highly insoluble tricalcium phosphate forms in the soil (Sanyal and De Datta, 1991; Majumdar and Sanyal, 2015). The relative solubility of some of these P reaction products in soil are shown in Table 6A.3.

6A.9.1. Phosphate Rock as P Fertilizer

The apatites in igneous and metamorphic phosphate rock (PR) deposits are relatively inert, being coarse-grained with little internal surfaces. The sedimentary rocks, on the other hand, contain apatite minerals that are microcrystalline, and consist of fairly open, loosely consolidated aggregates of microcrystals with relatively large specific surfaces. Chemically these are usually carbonate fluorapatites (FA), known as francolites. The apatites in sedimentary PR form a series of minerals, with the end member formula as $Ca_{10} (PO_4)_6F_2$, fluorapatite, and carbonate fluorapatite, or francolites, namely $Ca_{10-a-b} Na_a Mg_b (PO_4)_{6-x} (CO_3)_x F_{2+y}$ ($y = 0.4 x$, generally). Indeed the isomorphous substitution of $(CO_3^{2-} + F^-)$ to PO_4^{3-} decreases the a axis dimension of the apatite crystal structure from 0.9376 nm (characteristic of pure fluorapatite), and thereby increases the specific surface area, in addition to causing charge imbalance in the apatite (Sanyal and De Datta, 1991). This leads to higher reactivity of the latter in soil.

**Table 6A.3: Solubility Products of some
Soil-phosphorus Fertilizer Reaction Products**

Compound	Formula	pK_{sp}
Variscite	$AlPO_4 \cdot 2\,H_2O$	21.5-22.5
Ammonium taranakite	$H_6\,(NH_4)_3\,Al_5\,(PO_4)_8 \cdot 18\,H_2O$	175.5
Potassium taranakite	$H_6\,K_3\,Al_5\,(PO_4)_8 \cdot 18\,H_2O$	178.7
Strengite	$FePO_4 \cdot 2H_2O$	35.3
Dicalcium phosphate	$CaHPO_4$	6.66
Dicalcium phosphate dihydrate	$CaHPO_4 \cdot 2H_2O$	6.56
Octacalcium phosphate	$Ca_8\,H_2\,(PO_4)_6 \cdot 5\,H_2O$	93.8
Hydroxyapatite	$Ca_{10}\,(PO_4)_6\,(OH)_2$	111.8
Fluorapatite	$Ca_{10}\,(PO_4)_6\,F_2$	120.8
Magnesium hydrogen phosphate trihydrate	$MgHPO_4 \cdot 3\,H_2O$	5.82
Magnesium ammonium phosphate hexahydrate	$MgNH_4PO_4 \cdot 6H_2O$	12.2

Source: Adapted from Sanyal and De Datta (1991)

The PR dissolution in soil solution (acidic) may be represented by

$$Ca_{10}\,(PO_4)_6\,F_2 + 12\,H^+ = 10\,Ca^{2+} + 6\,H_2PO_4^- + 2\,F^- \tag{6.3}$$

Thus, the Law of Mass Action favours PR dissolution in soil under conditions of low (i) soil pH, (ii) soil exchangeable Ca^{2+}, and (iii) phosphate concentration in soil solution. For dissolution of PR to continue, the soil ought to provide a source of H$^+$ ions, and also a sink of Ca^{2+} and H$_2$PO$_4^-$ ions (*vide*. Eq. 6.3). Thus, the PR should be added to acid soils months ahead of liming so that an increase in exchangeable Ca^{2+} and pH does *not* interfere with PR dissolution. The effectiveness of PR fertilizers also increases with a decrease in particle-size, owing to the increase in specific surface area of the applied PR, but the finely ground PR is also difficult to handle and spread. It is presently recommended that grinding of PR may be done to ensure that at least 80 per cent of the material passes through a 100-mesh sieve. It is, however, doubtful whether PR, including the finely ground, highly reactive materials, could be economic substitutes for the soluble P fertilizers, *e.g.* superphosphates and DAP, for annual crops (Sanyal and De Datta, 1991).

6A.10. Phosphorus Fertilizer Management Options in Different Soil and Crop Situations

Under favourable growth conditions, most agricultural crops recover 15 to 20 per cent of the applied phosphorus depending upon the growth stage of P applications. A large portion of the unused P accumulates in the soil as the soil-fertilizer reaction products, as stated earlier. Eventually the subsequent crops recover a part of the latter over time. A much smaller fraction of P is lost as runoff (as both particulate and dissolved P) or through leaching that can cause secondary off-site impacts. All the forms of P within the soil system are subjected to a variety of pathways of transport at the soil profile, hill-slope, or catchment scale. Particulate and colloid P transport is most commonly associated with soil erosion, which arises from raindrop impact and overland flow. Additionally, when fertilizer or manure application is coincident with fast or energetic water flows, this will contribute to particularly high losses (Sanyal *et al.*, 2015b; Majumdar and Sanyal, 2015).

The management of P has strong interrelation with soil characteristics, such as soil reaction, degree of weathering, nature and amount of clay minerals, organic matter content, as well as the prevailing water regime. Optimum management of P fertilizers requires ensuring the 4R Nutrient Stewardship Principles of fertilizer application (Sanyal and De Datta, 1991; IPNI, 2012). Thus,

1. Use a suitable P **source** for a given soil-crop situation to minimize reactions with soil components that render P in soil solution unavailable to crop.

2. Select a P application **rate** and **timing** that will prevent a marked rise and fall of P concentration in soil solution throughout the crop growth period.

3. Modify the soil environment or application **method** (of P fertilizer) to reduce the amount of P in the solid phase.

Use of sparingly soluble P sources (*e.g.* phosphate rock, PR) alone or in combination with soluble P sources has been considered where improved P use efficiency is difficult to attain. The appropriate time of P application is that the entire amount of the fertilizer P be added as a basal dose before planting.

Phosphorus management under specific crop-soils situations are discussed below (Majumdar and Sanyal, 2015):

Acid Soils

Liming an acid soil reduces the extent of P fixation problem. However, an increase in pH on liming decreases P adsorption by amphoteric adsorbing surfaces, whereas a high exchangeable Al content of the soil may generate *fresh* P-adsorbing surfaces in soil through the precipitation of hydroxy-Al polymers. A rise in pH also reduces the extent of P desorption. Liming appears to benefit greatly an acid soil that has a very low P status, and that is about to be fertilized with P (Sanyal and De Datta, 1991).

High P-fixing Soils

During the fixation of phosphate by clays and/or hydrous oxides of Fe/Al in clay-sized dimensions in soil by the ligand-exchange process, the *net* negative charge of the soil colloid would increase. As stated earlier in Section 3.9.1, this will raise the CEC of the soil. Indeed, in the volcanic ash soils of, for instance, the East Indies, allophanic clay minerals, being amorphous with high specific surface charge density, have very high phosphate fixation capacity. To these soils, often a massive dose of phosphatic fertilizer (*e.g.* 1000 kg P_2O_5. ha^{-1}) is added to satisfy the phosphate hunger of the soil, and in the subsequent crop seasons, moderate dose of such fertilizer is added, which is supplemented by a slow release of small amounts of phosphate from the (earlier) immobilized phosphate in soil. Furthermore, such massive application of phosphatic fertilizer causes a rise in the CEC of the soil due to ligand exchange process as explained above. In such instances, phosphate plays the role of *not only* a fertilizer, *but also* that of a soil amendment (Sanyal and De Datta, 1991; Sanyal *et al.*, 2009).

Submerged Soils

For lowland rice crop, the soluble P sources are preferred to PR in initially acidic soils because of the decrease in solubility of PR due to enhanced pH and the available soil P level in the submerged paddy soils, and also the less ability of rice to derive P from PR. Phosphorus application by surface broadcasting or incorporation of P fertilizer before transplanting is generally more effective than deep-placement of P in planting hills or between the rows of rice. For transplanted rice, the best time of P application appears to be the application of the whole amount as a basal dose at transplanting (Sanyal and De Datta, 1991; Majumdar and Sanyal, 2015).

Phosphorus Management in Cropping Systems

The variation of soil P under rotation of rice and upland crops influences the direct and the residual effects of P fertilizers. As P availability changes with alternate drying and submergence, the P applied to the upland crop may have a greater residual effect on the succeeding rice crop, than the other way around. As stated earlier, flooding a rice soil, followed by its draining for the subsequent upland crop, greatly enhances the P sorption capacity (Ponnamperuma, 1972). Thus, for the rice-wheat cropping system, it has been suggested that the total P recovery by these two crops and the total biomass yield increased significantly when all P is applied to the upland crop as compared to that when all P is applied to rice. However, exception to this trend has also been reported wherein the total P addition is divided between these two crops as per the crop requirement. Further, recycling of fertilizer P through a preceding green manure crop (*e.g. Sesbania* sp.) has been beneficial for the succeeding lowland rice crop (Sanyal and De Datta, 1991; Majumdar and Sanyal, 2015).

6A.11. Environmental Impact of Phosphorus Fertilization in Agriculture: Losses of P through Erosion and Run-off to the Adjoining Water Bodies: Pollution through Eutrophication

Although the benefits of P in agriculture are evident, phosphorus can be a pollutant as well, if it moves from the site of application in soil to streams, rivers, lakes, and eventually oceans. Phosphorus so transported from agricultural soils can promote *eutrophication* which is considered as one of the most pressing environmental problems. This is defined as the enrichment of surface water bodies with excess nutrients, notably N and P, thereby facilitating the growth of algal biomass, as well as other aquatic weeds. The latter leads to an undesirable rise in the biological oxygen demand (BOD) and the chemical oxygen demand (COD) of the water bodies, thereby affecting several forms of aquatic life and causing the undesirable changes in the aquatic ecosystems, often with serious economic consequences (Sanyal *et al.*, 2015b).

As stated earlier, all the forms of P within the soil system are subjected to a variety of pathways of transport at the soil profile, hill-slope, or catchment scale. Particulate and colloid P transport is most commonly associated with soil erosion, which arises from raindrop impact and overland flow. Additionally, when fertilizer or manure application is coincident with fast or energetic water flows, this will contribute to particularly high losses (Sanyal *et al.*, 2015b; Majumdar and Sanyal, 2015).

Phosphorus is lost from crop lands *via* erosion or run-off. Little attention has been paid to the management strategies for minimizing non-point movement of P in the landscape. As a result, non-point sources presently account for a larger share of the nation's water quality problems than ever before. The main factors influencing P movement can be divided into transport and P source factors. Transport factors include the mechanisms by which P moves within a landscape. These are rainfall- and irrigation-induced erosion and run-off. Factors which influence the source and the amount of P available to be transported are soil P content and rate and method of P applied in either mineral fertilizer or organic forms (Sanyal *et al.*, 2015b).

As run-off enters a stream channel and, ultimately, a water body, there is generally a progressive dilution of P load through water dilution and sediment deposition. Sources of particulate P in streams include eroding surface soil, plant material, stream banks, etc. As the finer-sized fractions of source material are preferentially eroded, the P content and the reactivity of the eroded particulate material is usually greater than that of the source soil.

Indeed, it now appears that much of the P transferred from agriculturally-managed land to streams, rivers and lakes is derived from specific areas ("hot spots") within a river catchment and that these are related to farming system, soil type, and hydrology. It is possible to consider these areas as: (i) critical source areas – permanent features within a catchment from which P may be lost readily; and (ii) variable source areas – temporary features, often near streams, that lead to overland water flow carrying P, often associated with mineral or organic particles. Most of the P transported from soil to water is in eroded soil particles, enriched with P or from the excessive amounts of applied P fertilizer (Syers *et al.*, 2008).

6A.11.1. Minimizing Eutrophication through Efficient P Management

Losses of P from the cultivated lands may be minimized by adopting judicious P management strategies. Continuous P fertilization over the years at rates exceeding those of the crop removal results in P build-up, often above the levels required for crop production. Once the soil test P levels become excessive, further application of P will increase the potential for P movement, while providing no further agronomic benefit. Accumulation of soil test P near the soil surface due to previous P application influences the concentration and loss of P in run-off. Highly significant linear relationships are frequently seen between the soil test P in the surface soil and the dissolved P concentration in surface run-off. Adoption of soil test-based P fertilization would, therefore, not only be economically viable, but would also avoid its excessive accumulation in soil. Assessing the impact of P management through fertilizers and manures on P losses at field as well as watershed level is important from both the agronomic and the environmental viewpoint (Sanyal *et al.,* 2015b).

References

Barrow, N. J. (1978). The description of phosphate adsorption curves. *J. Soil Sci,* **29**, 447-462.

Goldberg. S. and Sposito, G. (1984). A chemical model of phosphate adsorption by soils. I. Reference oxide minerals. *Soil Sci. Soc. Am. J.,* **48**, 772-778.

IPNI (2012). *4R Plant Nutrition Manual: A Manual for Improving the Management of Plant Nutrition* (T. W. Bruulsema, P. E. Fixen, and G. D. Sulewski, Eds.), International Plant Nutrition Institute, Norcross, GA, USA.

Majumdar, K. and Sanyal, S. K. (2015). Soil and Fertilizer Phosphorus and Potassium. **In:** *Soil Science: An Introduction* (R. K. Rattan, J. C. Katyal, B. S. Dwivedi, A. K Sarkar, Tapas Bhattacharyya, J. C. Tarafdar and S. S. Kukal, Eds.), ISBN 81-903797-7-1, Indian Society of Soil Science, New Delhi. pp. 571-600.

Ponnamperuma, F. N. (1972). The chemistry of submerged soils. *Adv. Agron.,* **24**, 29-97.

Raychaudhuri, S. P. (1976). Phosphatic and potassic fertilizers and their management. In: *Soil Fertility-Theory and Practice* (J. S. Kanwar, Ed), Indian Council of Agricultural Research, New Delhi, pp. 371-409.

Sanyal, S. K. and De Datta, S. K. (1991). Chemistry of phosphorus transformations in soil. *Adv. Soil Sci.,* **16**, 1-120.

Sanyal, S. K., De Datta, S.K. and Chan, P.Y. (1993). Phosphate sorption desorption behaviour of some acidic soils of South and Southeast Asia, *Soil Sci. Soc. Am. J.,* **57**, 937-945.

Sanyal, S. K., Dwivedi, B. S., Singh, V. K., Majumdar, K., Datta, S. C., Pattanayak, S. K. and Annapurna, K. (2015b). Phosphorus in relation to dominant cropping sequences in India: Chemistry, fertility relations and management options. *Curr. Sci.,* **108** (No. 7, April 10, 2015), 1263-1270.

Sanyal, S. K., Poonia, S. R. and Baruah, T. C. (2009). Soil colloids and ion exchange in soil. **In:** *Fundamentals of Soil Science* (N. N. Goswami, R. K. Rattan, G. Dev, G. Narayanasamy, D. K. Das, S. K. Sanyal, D. K. Pal and D. L. N. Rao, Eds.) Second Edition, Indian Society of Soil Science, New Delhi, pp. 269-315.

Stumm, W., Kummert, R. and Sigg, L. (1980). A ligand-exchange model for the adsorption of inorganic and organic ligands at hydrous oxides interfaces. *Croat. Chem. Acta,* **53**, 291-312.

Syers, J.K., Johnston, A.E. and Curtin, D. (2008). *Efficiency of Soil and Fertilizer Phosphorus Use.* FAO Fertilizer and Plant Nutrition Bulletin, No. 18, Food and Agriculture Organization of the United Nations, Rome, Italy.

6B. Chemistry of Waterlogged Soils

6B.1. Basic Concepts

When a soil is submerged, atmospheric gases can enter the soil only by molecular diffusion in the interstitial water. Oxygen movement through the flood water overlying the submerged soil, as well as the water-filled soil pores is 10,000 times slower than diffusion in gas-filled pores (Patrick and Reddy, 1978). Thus the oxygen diffusion rate suddenly decreases when a soil reaches saturation by water (Ponnamperuma, 1972). Within a few hours of soil submergence, microorganisms use up the oxygen present in the water or trapped in the soil at or slightly below the soil surface and render a submerged soil practically devoid of molecular oxygen. As a result of rapid exhaustion of the low supply of molecular oxygen, compared to its demand, two distinctly different soil layers are formed in case of waterlogged soils–an oxidized thin surface layer where oxygen is present, overlying a reduced or anaerobic layer devoid of free oxygen supply (Patrick and Reddy, 1978).

In a waterlogged soil, several electrochemical changes take place. More important ones include (i) decrease in redox potential, (ii) an increase in pH of acid soils and a decrease in pH of alkaline soils, (iii) changes in specific conductance and ionic strength, (iv) drastic shifts in mineral equilibria, (v) cation and anion exchange reactions, (vi) sorption and desorption of ions, to enlist some (Ponnamperuma, 1972).

Oxidation-reduction is a chemical reaction in which electrons are transferred from a donor to an acceptor. The electron donor loses electrons and increases its oxidation number or is oxidized; the acceptor gains electrons and decreases its oxidation number or is reduced. The source of electrons for biological reductions is organic matter (Ponnamperuma, 1972). The driving force of a chemical reaction is the tendency of the free energy (at constant temperature and pressure) of the system to decrease until, at equilibrium, the sum of the free energies of the products equals that of the remaining reactants. In a reversible oxidation-reduction reaction, this force can be measured in volts. The change in free energy, ΔG, for the given reduction reaction (shown below) at constant temperature and pressure is given by,

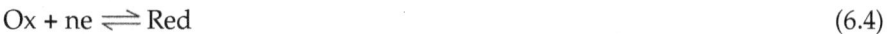

$$Ox + ne \rightleftharpoons Red \tag{6.4}$$

$$\Delta G = \Delta G^\circ + RT \ln [(Red)/(Ox)] \tag{6.5}$$

Where (Red) and (Ox) are the activities of the reduced and the oxidized species and ΔG° is the free energy change when the activities are unity, that is, when they are in their respective standard states of unit activities. Remembering that $\Delta G = -nEF$, where E is the (reduction or redox) electrode potential for the above noted reduction reaction, we have

$$E = E^\circ + (RT/nF) \ln [(Ox)/(Red)] \tag{6.6}$$

Here E° is the standard reduction or standard redox potential of the given redox system, namely Ox/Red.

It must be noted here that the single electrode potential of a redox system [*e.g.* Fe^{2+}/Fe^{3+} (Red/Ox) system] cannot be measured experimentally. It is assessed by referring to the value of the standard hydrogen electrode [Pt/H_2 (g, 1 atm)/H^+ ion in aqueous HCl with the activity of H^+ ion being unity (a_{H+} = 1) at 25°C; platinum (Pt) wire is used as a conducting material to the potential measuring potentiometer] potential at 25°C which is set equal to zero by convention. The above noted reduction (redox) potential (E), accompanying the given redox reaction (Eq. 6.4), namely

$$Ox + ne \rightleftharpoons Red,$$

is denoted by E_h. Thus,

$$E_h = E_h^0 + (RT/nF) \ln [(Ox)/(Red)] \tag{6.7}$$

Where E_h is a quantitative measure of the tendency of a given system to oxidize or reduce. The E_h is positive and high in strongly oxidizing systems, and is negative and low in strongly reducing systems (Ponnamperuma, 1972). Any chemical reaction involving exchange of electrons will be influenced by reduction (redox) potential, namely E_h. If oxidation-reduction reactions are arranged sequentially in the descending order of the E_h^0 values, as listed in Table 6B.1, a given system theoretically can oxidize the reduced form of any one below it under standard conditions. Changes in pH and activities of the reactants and resultants can, however, alter the order (Ponnamperuma, 1972). The latter results from the fact that at a given E_h, characterizing a submerged soil, the sequential reduction of the redox systems present in the soil depends more on the corresponding prevailing E_h value of each system rather than the corresponding standard redox potential (E_h^0) value of such a redox systems present in the submerged soils.

Because of lack of oxygen in a submerged soil for the reasons stated above, requirement of the electron acceptors by the aerobic, facultative and obligate anaerobic microorganisms leads to reduction of several oxidized soil components which act as the electron acceptors and get thus reduced, roughly in the sequence predicted by the corresponding standard thermodynamic redox potentials, presented in Table 6B.1. Oxygen is the first soil component to be reduced, and it gets undetectable within a day after submerging a soil. The next oxidized form is the nitrate, but nitrate reduction begins only after the oxygen concentration has dropped to an extremely low value. Indeed, some of these reduction processes partially overlap. Thus, the reduction of manganic compounds (containing Mn^{4+} forms) to the corresponding divalent manganous ion-containing compounds takes place during the reduction of oxygen and nitrate (Patrick and Reddy, 1978; Ponnamperuma, 1972). The reduction of the oxidized components of the other redox systems proceeds in a sequential manner, with a degree of partial overlap (mentioned above), as the gradual degree of the reduction processes intensifies in the submerged soils, accompanied by a progressive fall in the value of E_h of the given waterlogged soil.

Table 6B.1: Thermodynamic Sequence of Standard Redox (Reduction) Potential of Selected Redox Systems (Corrected to pH 7.0)

Redox System	E_h^o at 25°C (Volt)
$O_2 + 4 H^+ + 4 e^- = 2 H_2O$	0.814
$2 NO_3^- + 12 H^+ + 10 e^- = N_2 + 6 H_2O$	0.741
$MnO_2 + 4 H^+ + 2 e^- = Mn^{2+} + 6 H_2O$	0.401
$CH_3COCOOH + 2 H^+ + 2 e^- = CH_3CHOHCOOH$	−0.158
$Fe(OH)_3 + 3 H^+ + e^- = Fe^{2+} + 3 H_2O$	−0.185
$SO_4^{2-} + 10 H^+ + 8 e^- = H_2S + 4 H_2O$	−0.214
$CO_2 + 8 H^+ + 8 e^- = CH_4 + 2 H_2O$	−0.244
$N_2 + 8 H^+ + 6 e^- = 2 NH_4^+ + 6 H_2O$	−0.278
$NADP^+ + 2 H^+ + 2 e^- = NADPH$	−0.317
$NAD^+ + H^+ + 2 e^- = NADH$	−0.329
$2 H^+ + 2 e^- = H_2$	−0.413
Ferredoxin (Oxidized form) + e^- = Ferrodoxin (Reduced form)	−0.431

Source: Adapted from Patrick and Reddy (1978).

In particular, the behaviour of the major plant nutrient, namely phosphorus (P) in flooded lowland soils remarkably differs from that in the upland soils. Moreover the chemistry of P transformations in flooded soils has received rather scant attention, compared with those in non-flooded soils. Flooding the soils increases the availability of the native and the added P (Patrick and Mahapatra, 1968; Ponnamperuma, 1972; Sanyal and De Datta, 1991). Consequently yield responses of lowland rice to fertilizer P are generally lower than those to N or even to P for upland crops grown on the same soil (Sanyal and De Datta, 1991). Lowland rice therefore has access to soil P sources ordinarily unavailable to other crops.

Several investigators have looked into the physiochemical changes that accompany flooding and that are distinctly different from those in upland soils (Goswami and Banerjee, 1978;, Ponnamperuma, 1972, 1985; Willett, 1989; Sanyal and De Datta, 1991). These changes indirectly affect the behaviour of soil P that by itself does not participate in these redox processes.

Rice crop easily adapts to the environment. It can grow in various types of soil under a wide range of climatic and soil moisture conditions. Rice can be grown with a thin film of moisture on the soil surface, to about 10- 50 cm of standing water. It is mostly grown, however, in submerged soils with 10-30 cm standing water during most of its growth period. Rice can also be grown under continuous flooded soil or under alternate wetting and drying conditions. These changes in soil moisture conditions in the rice fields affect the changes in soil which, in turn, influence the transformation of the native and the applied P, P availability and consequently, rice nutrition and growth. Changes in P availability in alternately flooded and drained soils are also important with regard to the growth of subsequent crops in rotation with rice. In what follows, this Section will take a close look at the **chemistry of phosphorus transformations in waterlogged soils**.

6B.2. Physiochemical Changes on Flooding that Affect Phosphorus Availability

6B.2.1. Phosphorus Transformations under Continuous Flooding

a. Causes of Changes in Extractable Phosphorus

Phosphorus availability in soil increases upon submergence due to the following changes (Goswami and Banerjee, 1978; Ponnamperuma, 1985; Sanyal and De Datta, 1991).

1. Reduction of ferric compounds. The reduction of free hydrous Fe oxides during flooding and the liberation of sorbed, and co-precipitated P as a result, increases the levels of solution or extractable P in flooded acidic soil. The subsequent release of occluded P from within the structure of amorphous Fe oxides has also been proposed.

   The chemical equilibria equations of the following types have been used to describe the activity of Fe^{2+} ions in solution of flooded soils.

   $$Fe(OH)_3 + 3H^+ + e^- = Fe^{2+} + 3H_2O \qquad (6.8)$$

   for reduction in the early stages of flooding, and

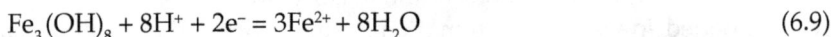

   $$Fe_3(OH)_8 + 8H^+ + 2e^- = 3Fe^{2+} + 8H_2O \qquad (6.9)$$

   in soils after prolonged flooding. There is thus an increase of exchangeable Fe^{2+} ions in soil with a concomitant rise in soil pH and a decline in redox potential (E_h).

   Among the ferric hydrous oxides, ferrihydrite (of standard free energy of formation, $\Delta G°_f = -677$ kJ. mol^{-1}), the least stable oxide, has been postulated to undergo reductive dissolution (*e.g.* reaction; Eq. 6.8) first, releasing the sorbed P, in advance of the more stable oxides such as goethite ($\Delta G°_f = -742$ kJ. mol^{-1}) (Sanyal and De Datta, 1991). Even though the ferric oxide reduction appears to be the dominant source of P released during flooding, however, the amount of P released was strongly inhibited by resorption. It was suggested that direct measurement of the amount of ferric iron reduced during flooding and of P sorption are required to predict the *net* amount of P released during flooding.

   Reduction of $FePO_4.8H_2O$ to more soluble $Fe_3(PO)_2. 3H_2O$ or $Fe_3(PO_4)_2. 8H_2O$ has also been proposed under submerged conditions in soil (Sanyal and De Datta, 1991). Fischer (1983), from a theoretical study of simple chemical systems, suggested that reduction of ferric hydrous oxides in the presence of P in solution takes place more easily than in similar solutions without P, because of the precipitation of vivianite, $Fe_3(PO_4)_2.8H_2O$. However, if P sorbed onto ferric hydrous oxide is a significant source of P in soils, then it is the rise in pH (associated with the reduction of Fe^{3+} compound) rather than the fall in redox potential, (E_h) (favouring the reductive dissolution of ferric hydrous oxides) that is responsible for the relatively high P concentration in waterlogged soil solutions (Sanyal and De Datta, 1991). An increase in pH would, in fact, favour P desorption from clay, aluminum oxides, and excess (not yet reduced) ferric oxide surfaces through a deceased surface positive charge. Such pH changes were, however, found to favour desorption of freshly applied P only, but did *not* affect P release in untreated soils (Willett, 1989).

Reductive dissolution of Mn (III) and Mn (IV) has not been found to affect P release during flooding. Thus, P release during flooding follows the reduction of ferric compounds, which, in turn, occurs after the reduction of manganese oxides (*vide.* Table 6B.1) (Willett, 1986).

2. Higher solubilities of $FePO_4.2H_2O$ and $AlPO_4.2H_2O$ resulted from hydrolysis due to increased soil pH in acid and strongly acid soils.

3. Organic transformations influencing P release. Organic acids released during anaerobic decomposition of organic matter under flooded soil conditions can increase the solubilities of Ca-P compounds by complexing Ca^{2+} ions, thereby disturbing the solubility equilibria of Ca-P. The observed lowering of fixation of applied P in the presence of added organic matter in flooded acidic lowland rice soils was attributed to the complexation of soil Fe and soil Al by the decomposition products of organic matter (Sanyal and De Datta, 1991).

Organic matter in soil may also have an important effect on ferric iron reduction through its promoting influence on the bacterial activity in flooded soil, and the level of organic matter was noted to govern the amount of P released in several soils, due to its effects on ferric reduction (Willett, 1986).

Mineralization of organic P is generally too slow to be significant in plant nutrition, although mineralization rates increase under flooding (Sanyal and De Datta, 1991). Mineralization of organic P has been considered as a minor source of P in flooded soils except in flooded organic soils (Sanyal and De Datta, 1991). On the other hand, an increased mineralization of organic P upon flooding, particularly Fe-phytates in acid soil, was also observed (Goswami and Banerjee, 1978). The contribution of organic matter to P release during flooding appears to be mainly through organic P mineralization rather than by accelerating reduction of ferric compounds (Willett, 1989). Further the biological mineralization of organic N and the transformation of NH_4^+-N to NO_3^--N are likely to play an active role in the seasonal pattern of water-soluble P in soils, which increased during the high rainfall periods compared with that in the dry months.

4. Release of phosphate ions from the exchange between organic anions and phosphate ions in Fe-P and Al-P compounds.

5. Increased solubility of Ca-P in calcareous soils as a result of pH depression due to CO_2 accumulation by organic matter decomposition.

Thus the solubility of several Ca-P compounds, such as octacalcium phosphate, β-tricalcium phosphate, hydroxyapatite and fluorapatite, has been suggested to increase following a fall in pH after flooding a calcareous soil (Ponnamperuma, 1985).

In acidic soils, such an accumulation of CO_2 under anaerobic conditions would tend to bring down the pH, opposing thereby a pH rise due to reaction, represented by Eq. 6.8. This will also cause increase in HCO_3^- concentration in the solution phase through the solvent action of CO_2 on carbonates, and would cause desorption of several exchangeable cations (*e.g.* Fe^{2+}, Ca^{2+}, Mg^{2+} and NH_4^+) to maintain the electroneutrality in the solution. These effects, coupled with that accompanying reduction of soil Fe^{3+} compounds, would increase the specific conductance and the ionic strength of the soil solution (Ponnamperuma, 1985).

6. Increased P diffusion under submerged conditions. The flooding increases the buffer capacities for soil P (Tian-ren *et al.*, 1989). The buffer capacities for P sorption at low solution P concentration (<1 µg P. ml^{-1}) increased up to five-fold (Sanyal and De Datta, 1991). Thus, although the increased soil moisture content tends to bring down the soil impedance factor and increase the P diffusion coefficient, a simultaneous steep rise in the buffer capacity (*see* later) may *more than offset* such increases. As a result, the P supply by diffusion in the soil to the rhizosphere zone may become the controlling factor for P uptake by lowland rice (Tian-ren *et al.*, 1989).

Increased buffer capacity due to flooding has been attributed to P adsorption from soil solution by the re-precipitated poorly crystalline ferrous hydroxides or carbonates from Fe^{2+} ions formed by soil reduction (Patrick *et al.*, 1985; Ponnamperuma, 1985; Tian-ren *et al.*, 1989). An increase in pH opposed further P absorption due to an increase in negative charge of the variable-charge P-adsorbing surfaces in the flooded soil. At the same time, an increase in ionic strength of the solution depresses the activity coefficients of the ionic species in the solution phase (*vide.* Section 3.1.2; Eqs. 3.3, 3.4, 3.5). The latter would tend to raise the concentration of phosphate ions and hence, affect the ionic equilibria between the solid-phase P and the soil-solution P, opposing P desorption due to pH rise. As a result of such moderating influence of ionic strength on reduction of P sorption at a higher pH, increasing the salt concentration increases the P sorption at high pH but decreases it at a lower pH. There is thus a point where the P sorption is independent of salt concentration (the point of *zero-salt effect* on P sorption). The latter decreases with increasing level of P sorption because of the increased negative charge of the soil colloid through increased ligand-exchange of P on the variable charge-surfaces in soil (*vide.* Section 2.5.2) (Sanyal and De Datta, 1991).

7. Phosphorus mobilization resulted from an increased microbial activity in the presence of physiologically active rice roots and from the capacity of rice plants to re-oxidize the rhizosphere during the later phase of the growing period (Sanyal and De Datta, 1991).

8. In soils poor in free iron oxides, under highly reduced conditions, another process, shown in the following conversion, increases the availability of P in flooded soils (Patrick and Mahapatra, 1968; Sanyal and De Datta, 1991).

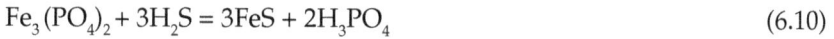

$$Fe_3(PO_4)_2 + 3H_2S = 3FeS + 2H_3PO_4 \tag{6.10}$$

Flooding a soil increases the soluble P concentration in the soil and reaches a maximum before falling. The subsequent fall in P concentration after reaching the peak has been attributed to re-adsorption of P on clays and Al hydroxides, precipitation or microbial degradation of organic anions at the exchange sites, causing P resorption from soil solution (Patrick *et al.*, 1985; Ponnamperuma, 1972, 1985; Tian-ren *et al.*, 1989).

During prolonged flooding, the level of Fe^{2+} iron in solution stabilizes, but the level of acid extractable-Fe^{2+} iron continues to increase (Willett, 1986). Precipitation of ferrosic (or ferroso-ferric) hydroxide, $Fe_3(OH)_8$, on prolonged flooding was proposed (Ponnamperuma, 1972) and this compound was stated to have a large surface area with a high P-sorption capacity. The latter could well contribute to the decline of P concentration in soil solution on continued submergence (Sanyal and De Datta, 1991). However, P thus sorbed may still remain acid-extractable and contribute to the labile pool of P in soil, and thus remain available to the plants.

However ferrosic hydroxide has not been isolated or synthesized so far, and its existence has been speculated on the basis of conformation of the E_h, pH and Fe^{2+} activity in soil solutions to Eq. (6.9) (Sanyal and De Datta, 1991).

The increase in water soluble P concentration after flooding a soil is, however, less than that in stagnant lake waters, and is also strongly dependent on soil properties. Maximum P concentration was found highest in sandy calcareous soils, low in Fe, moderate in acid sandy soils, and small in nearly neutral clay. Values were the lowest in acid ferrallitic clays (Ponnamperuma, 1972).

b. Forms of Phosphorus in Flooded Soils

Extractable P content in flooded soils increases, depending primarily on the distribution of different inorganic P fractions and the intensity of soil reduction. In general, the higher P availability in flooded soils is attributed mainly to Fe- P; the role played by Al- P or Ca-P compounds is usually secondary (Mahapatra and Patrick, 1969; Goswami and Banerjee, 1978; Sanyal and De Datta, 1991). The acetate and oxalate-extractable Fe and P sorptivity of soil largely increased upon flooding (Willett and Higgins, 1978). On prolonged waterlogging, oxalate-Fe and P sorptivity levels reached values, dependent on the free iron oxide content of the soils.

The anaerobic decomposition of the added cellulose in flooded soils decreased Al-P and increased Fe-P and reductant soluble (RS)-P (Sanyal and De Datta, 1991). This was attributed to an increase, mediated by the anaerobic decomposition of the organic matter, of crystalline Fe transformation into amorphous forms which, in turn, increased Fe- P and severe occlusion of P, resulting in the increase of RS-P fraction upon subsequent soil drainage.

The thermodynamic study of P equilibria in flooded acid soil, treated with water-soluble and insoluble P fertilizers, suggested the formation of strengite as a fertilizer reaction product in soil, as well as a general trend of increases in Fe-P, Al-P, and RS-P in several flooded soils of India, with Ca-P registering an increase in black and laterite soils only (Sanyal and De Datta, 1991).

6B.2.2. Phosphorus Transformations under Alternate Wetting and Drying

When flooded rice fields become dry, the reduced soil constituents are re-oxidized with concomitant changes in E_h, pH, and ferrous iron concentrations. Drying a soil subsequent to flooding generally decreases the solubility of both native and applied P. Phosphorus applied before flooding was found to be immobilized to a greater degree than when P was applied after draining a soil, rich in organic carbon and reducible Fe. In soils low in these, however, P applied before flooding was immobilized but P applied after drying was not (Sanyal and De Datta, 1991). Indeed non-flooded crops, grown in rotation with flooded rice, have often been reported to develop P deficiency, but not rice (Sanyal and De Datta, 1991). The P deficiency is more acute than in similar soil that has not been recently flooded. Rotation crops are thus expected to respond to P fertilizer under similar conditions.

However, some investigator also noticed an increased native P availability to rice upon submergence, followed by soil drying (Sanyal and De Datta, 1991). It has been suggested that this may have resulted from organic P mineralization in soil, whereas Fe-P- and Al-P availability may have actually decreased.

Soil P sorption capacity and bonding energy for P were found to increase upon flooding, and then upon drying conditions (Sanyal and De Datta, 1991). Also drying increased the amount of acid ammonium oxalate-extractable Fe and the Fe-bound P at the expense of Al-bound P. Consequently, flooding and drying were suggested to increase the activity of ferric hydrous oxides in sorbing P (by way of decreasing their crystallinity) that resulted in added P immobilization after draining the rice soils (Sanyal and De Datta, 1991). This was associated with the decreased plant-availability of P.

Much earlier, Patrick and Mahapatra, (1968) suggested that the biological reduction of Fe during flooding, followed by re-oxidation during drying, enhanced the reactivity of the sesquioxides fraction of the soil, thereby increasing the P-fixing capacity, and hence decreased P solubility. In agreement with this observation, a finding suggests that the induced P deficiency in soils, subjected to flooded-drained conditions, was due to high P sorptivity and low P desorption as a consequence of Fe transformations in soil (Sanyal and De Datta, 1991). However, the alternate wetting and drying effects on P adsorption and desorption in soils were proposed to be associated with changes in soil structure, caused by the rewetting of dry soil samples (Olsen and Court, 1982).

The P sorption capacity as well as bonding energy for P sorption of soils from the flooded-drained systems increased with temperature and duration of prior flooding. It was further demonstrated (Willett, 1979) that P-availability to maize, grown in rotation with flooded rice, was more closely related to bonding energy between soil and P than to the soil's capacity to sorb P. This suggests that the depressed P supply to maize grown in previously flooded soil was due to stronger P sorption by the drained soils, rather than to P immobilization during flooding.

Phosphorus sorptivity and bonding energy of sorption, which increased under flooded-drained soil conditions, after quite some time declined when the previously flooded soil was drained, but not to the same levels as that prior to flooding (Sanyal and De Datta, 1991). Therefore, effectiveness of P fertilizers to crops after rice should increase with time after draining the rice soil. More efficient use of P fertilizer may be achieved by delaying the sowing of the following crop as far as practicable (Sanyal and De Datta, 1991).

Addition of organic manure and an elevated temperature greatly enhanced P sorption in drained soils from the flooded-drained systems, thereby causing a higher P sorption for a relatively shorter period of previous flooding. The effect of organic matter was attributed to an increase of amorphous Fe in soil during the anaerobic decomposition of the organic matter, as observed for soils under continuous flooding conditions (Sanyal and De Datta, 1991).

6B.2.3. Soil Test for Phosphorus in Flooded Soils

A soil test for formulating recommendations for fertilizer P requirement of lowland rice is beset with the following two problems, specific to wetland soils (Sanyal and De Datta, 1991).

1. A test for available P on an aerobic (air-dried) soil sample may not provide a satisfactory index of P availability after flooding because available P increases significantly due to submergence, the extent of such increase being dependent on the soil characteristics.

2. Wet soils are difficult to sample without altering their physiochemical properties.

Notwithstanding these, several attempts have been made to a test for available P in air-dried soil samples from submerged fields and correlate the findings with crop-response data. The methods used may be classified into four major groups, depending on the nature of the chemical extractants used (Sanyal and De Datta, 1991).

1. Solution of a strong acid such as 0.002 N H_2SO_4 (Truog, Ayres-Hagihara), 0.13 N HCl (Spurway) and 0.05 N HCl + 0.025 N H_2SO_4 (Mehlich).

2. Solution of an organic acid or an acidified organic salt such as sodium acetate + acetic acid (Peech, Morgan), 1 per cent citric acid (Dyer) and CO_2 (McGeorge).

3. Alkaline solutions such as 0.5 M NaHCO$_3$ (pH 8.5) (Olsen) and 0.1 N NaOH (Saunders).

4. Solution of a strong acid containing complexing radicals for Fe and Al, such as 0.03 N NH$_4$F + 0.025 N HCl (Bray No. 1) and 0.03 N NH$_4$F + 0.1 N HCl (Bray No. 2).

There have been several studies that tested the correlation of soil-available P, as determined by the above mentioned methods, with response of rice to P application. Findings established the superiority of some of the chemical extractants to others for an index of available P in flooded soils (Patrick and Mahapatra, 1968; Goswami and Banerjee, 1978; Ponnamperuma, 1985; Sanyal and De Datta, 1991). It was particularly suggested that any method that can extract Fe-P and RS-P in aerated samples should provide a reliable measure of available P in lowland rice soils (Sanyal and De Datta, 1991). The findings of several investigators in this regard are summarized in terms of the following:

1. When a group of soils is dominant in Fe-P (usually with a pH < 5.5), good correlations are often obtained from most of the soil testing methods mentioned.

2. When a group of soils is dominant in Ca-P (usually with a pH > 6.5), methods employing an alkaline extractant (*e.g.* Olsen method) are better.

3. When a group of soils has mixed distribution patterns, having either Fe-P or Ca-P or both as dominant fractions, an alkaline extractant (*e.g.* Olsen method) or a weakly acidic extractant containing a complexing radical for trivalent cautions (*e.g.* Bray No. 1 method) is superior.

Thus, in general Olsen and Bray No. 1 methods, especially the former, have been found to be universally applicable to all soil types.

The suitability of Olsen method to predict the soil P-availability in submerged soils from the values obtained on air-dried soil samples has also been established by correlation studies between aerobic and anaerobic P determinations (by Olsen method) on many soil samples, varying in pH from 4 to 8, and with different soil P distribution patterns. A high degree of correlation was found, which was independent of soil pH, and only slightly affected by a drop of E$_h$. Phosphorus determined by the Bray No.1 and Bray No. 2 methods was also moderately correlated only when the soil pH values were more than 5. However, a decrease in E$_h$, caused upon submergence and addition of starch, greatly increased the extractable-P values, as obtained by these two methods. The variation of Olsen P of rice soils was also found to be closely correlated with the *net* gains or losses of P, that is, the total P applied to soil minus the P removed by crops. Such accumulation and depletion of available P (Olsen P) in rice soils appear to have similar slopes when plotted against the above stated *net* gains or losses of P (Sanyal and De Datta, 1991).

References

Fischer, W. R. (1983). Theoretische Betrachtungen zur reduktiven Ausflosung von Eisen (III)-oxiden. *Z. Pflanzenernaehr. Bodenkd.* **146**, 611-622.

Goswami, N. N. and Banerjee, N. K. (1978). Phosphorus, potassium and other macro-elements. In: *Soils and Rice.* International Rice Research Institute, Los Baños, Laguna, Philippines, pp. 561-580.

Mahapatra, I. C. and Patrick, W. H., Jr. (1969). Inorganic phosphate transformation in waterlogged soils. *Soil Sci.,* **107**, 281-288.

Olsen, R. G. and Court, M. N. (1982). Effect of wetting and drying of soils on phosphate adsorption and resin extraction of soil phosphate. *J. Soil Sci.,* **33**, 709-717.

Patrick, W. H., Jr. and Mahapatra, I. C. (1968). Transformation and availability to rice of nitrogen and phosphorus in waterlogged soils. *Adv. Agron.,* **20**, 323-359.

Patrick, W. H., Jr. and Reddy, C. N. (1978). Chemical changes in rice soils. In: *Soils and Rice.* International Rice Research Institute, Los Baños, Laguna, Philippines, pp. 361-398.

Patrick, W. H., Jr., Mikkelsen, D. S. and Wells, B. R. (1985). Plant nutrient behavior in flooded soil. In: *Fertilizer Use and Technology,* Third Edition, Soil Sci. Soc. Am., Madison, Wisconsin, pp. 197-228.

Ponnamperuma, F. N. (1972). The chemistry of submerged soils. *Adv. Agron.,* 24, 29-97.

Ponnamperuma, F. N. (1985). Chemical kinetics of wetland rice soils relative to soil fertility. In: *Wetland Soils: Characterization, Classification, and Utilization.* International Rice Research Institute, Los Baños, Laguna, Philippines, pp. 71-89.

Sanyal, S. K. and De Datta, S. K. (1991). Chemistry of phosphorus transformations in soil. *Adv. Soil Sci.,* **16**, 1-120.

Tian-ren, Y., Kirk, G. J. D. and Chaudhary, F. A. (1989). Phosphorus chemistry in relation to water regime. Paper presented at the *Symposium on Phosphorus Requirement for Sustainable Agriculture in Asia and Oceania.* March 6-10, 1989. International Rice Research Institute, Los Baños, Laguna, Philippines.

Willett, I. R. (1979). The effects of flooding for rice culture on soil chemical properties and subsequent maize growth. *Plant Soil,* **52**, 373-383.

Willett, I. R. (1986). Phosphorus dynamics in relation to redox processes in flooded soils. *13th Intern. Congress Soil Sci. Trans. (Hamburg),* **6**, 748-755.

Willett, I. R. (1989). Causes and prediction of changes in extractable phosphorus during flooding. *Aust. J. Soil Res.,* **27**, 45-54.

Willett, I. R. and Higgins, M. L. (1978). Phosphate sorption by reduced and reoxidized rice soils. *Aust. J. Soil Res.,* **16**, 319-326.

6C. Chemistry of Micronutrients

6C.1. Introduction

The micronutrients are present in soils, mostly as the corresponding oxides, sulphides and silicates. These are inherited from the rocks and minerals on weathering during the soil formation processes. The micronutrient content of a soil depends primarily on the parent material and the degree of weathering (Deb and Sakal, 2002). Table 6C.1 shows the major sources of these micronutrients and their chemical forms along with the content in surface soils (Deb and Sakal, 2002). In what follows, the various aspects of the micronutrients are discussed in brief.

6C.1.1. Boron

Boron is the only non-metal among the micronutrient elements. Its concentration in the earth's crust is 10 ppm and it ranges from 5 to 15 ppm in igneous rocks. Among the usual sedimentary rocks, shales have the highest boron concentrations of up to 100 ppm, with boron present mainly in the clay minerals (Tisdale et $al.$, 1985).

The total concentration of boron in most soils varies between 2.8 and 630 ppm, while the available boron content (hot water-extractable) in soil ranges from traces to 24 ppm (Table 6C.1). The economic sources are borates, $e.g.$ Borax ($Na_2B_4O_7.10H_2O$), Kernite ($Na_2B_4O_7.4H_2O$), Colemanite ($Ca_2B_6O_{11}.5H_2O$), Tourmaline [Na (Mg, Fe, Mn, Li, Al)$_3$Al$_6$(Si$_6$O$_{18}$) (BO$_3$)$_3$(OH,F)$_4$], Ulexite(NaCaB$_5$O$_9$.8H$_2$O), Kotoite [Mg$_3$(BO$_3$)$_2$] (Table 6C.1). Tourmaline is a highly complex aluminosilicate, as shown above, which is believed to control the B solubility in soils. Being quite insoluble and resistant to weathering, release of boron from it is slow and inadequate to support intensive cropping. Boron solubility in many soils seems to be buffered near $10^{-5.5}$ M, but the specific solid phases responsible for this control are unknown.

Boron exists in four major forms in soil and parent material, namely in rocks and minerals, adsorbed on surfaces of clays and hydrous iron and aluminum oxides, combined with organic matter, and as free non-ionized boric acid (H_3BO_3) and B $(OH)_4^-$ in the soil solution. Undissociated boric acid (H_3BO_3) is the predominant form in soil solution at pH values ranging from less than 5 to 8.5-9 (pKa_1, pKa_2, and pKa_3 values of boric acid are 9.24, 12.4, and 13.3 at 25°C). The other soluble boron species $H_2BO_3^-$ is of less significance at pH values below 8 and only at a pH near 9.2, does its concentration become equal to that of undissociated boric acid ($viz.$ Henderson equation, Eq. 3.72).

Although the process of boron uptake by plants is not fully comprehended, it appears that undissociated boric acid is the most effective form. Boron, apparently taken up by the plants mostly as the undissociated boric acid, is transported from the soil solution to the absorbing plant roots by both mass flow and diffusion. Passage of boron into the plant is probably the result of both passive and active processes.

Adsorbed boron constitutes the reserve that maintains boron concentration in the soil solution, and also reduces the leaching losses. Adsorbed boron is basically constituted of complexes of either molecular boric-acid or the hydrated borate

Table 6C.1: Major Natural Sources of Micronutrients, Chemical Forms and the Content in Surface Soils

Sl.No.	Micronutrient	Minerals	Major Forms in Nature	Total Content in Surface Soil (ppm)	Available Content in Surface Soil* (ppm)
1.	Zinc	Sphalerite (ZnS) Smithsonite ($ZnCO_3$), Hemimorphite [$Zn_4(OH)_2$ $Si_2O_7H_2O$]	Sulphides, Oxides, Carbonates and Silicates	7 to 1000	0.8-20.5
2.	Copper	Chalcocite (Cu_2S) Covellite (CuS) Cuprite (Cu_2O) Malachite [$Cu_2(OH)_2CO_3$] Chrysocolla ($CuSiO_3.2H_2O$) Azurite [$Cu_3(OH)_2(CO_3)_2$]	Sulphides, Oxides Hydroxy carbonates, Silicates	1.8 to 960	Tr.-32.0
3.	Iron	Hematite (Fe_2O_3) Goethite (α-FeOOH) Magnetite (Fe_3O_4) Pyrite (FeS_2) Olivine [$(Mg,Fe)_2SiO_4$]	Oxides, Sulphides and Silicates	4000 to 2,73,000	0.36-174
4.	Manganese	Pyrolusite (MnO_2), Manganite (MnOOH) Rhodochrosite ($MnCO_3$) Rhodonite ($MnSiO_3$)	Oxides, Carbonates and Silicates	37 to 1,15,000	0.60-164
5.	Boron	Borax ($Na_2B_4O_7.10H_2O$) Kernite ($Na_2B_4O_7.4H_2O$) Colemanite ($Ca_2B_6O_{11}.5H_2O$) Tourmaline [$Na(Mg,Fe)_3$ $Al_6(BO_3)_3 Si_6O_{18}(OH)_4$] Ulexite ($NaCaB_5O_9.8H_2O$) Kotoite [$Mg_3(BO_3)_2$]	Borates and Borosilicates	2.8 to 630	Tr.-24**
6.	Molybdenum	Molybdenite (MoS_2) Ilsemannite ($Mo_3O_8.8H_2O$) Wulfenite ($PbMoO_4$) Ferrimolybdite [$Fe_2(MoO_4)_3$. $8H_2O$] Powellite ($CaMoO_4$)	Sulphides, Oxides and Molybdates	Trace to 12	Tr.-2.8***
7.	Chlorine	Muriate of potash (KCl), Sodium chloride (NaCl)	Chlorides	20 to 1000	0.5-10.0**
8.	Nickel	Serpentine, Olivine	Silicates		
9.	Cobalt	Serpentine, Olivine, Sandstone, Quartzite	Silicates	4 to 78	0.2-0.5

*: DTPA-$CaCl_2$-extractable; **: Hot Water-extractable; ***: Ammonium Oxalate (pH 3.3)-extractable.

Source: Deb and Sakal (2002).

ion B $(OH)_4^-$ with various minerals. Four main adsorption sites for boron have been identified: (1) broken Si-O and Al-O bonds at the edges of aluminosilicate minerals, (2) amorphous hydroxide structures (*e.g.* allophane in weathered soils),

(3) magnesium hydroxide clusters (*e.g.* in arid-zone soils), and (4) iron and aluminum oxy and hydroxy compounds. Increasing pH and the presence of aluminum compounds rather than analogous iron substances both favour hydroxyl borate reactions. Other factors enhancing boron adsorption are fineness of soil texture and organic-matter content.

A significant quantity of plant available boron in soils is held in the organic-matter fraction. The diol-type compounds are proposed to form as a result of boron reacting with α-hydroxyaliphatic acids and ortho-dihydroxy derivatives of aromatic compounds as shown below (Tisdale *et al.*, 1985).

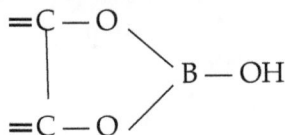

$$
\begin{array}{c}
=\!C\!-\!O \\
\quad\quad\quad\searrow \\
\quad\quad\quad\quad B\!-\!OH \\
\quad\quad\quad\nearrow \\
=\!C\!-\!O
\end{array}
$$

This ring compound can also be transformed into a boron-containing organic acid:

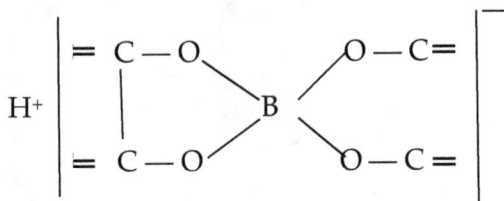

$$
H^+ \left|
\begin{array}{c}
=\!C\!-\!O \quad\quad O\!-\!C\!= \\
\quad\quad\quad\searrow\quad\nearrow \\
\quad\quad\quad\quad B \\
\quad\quad\quad\nearrow\quad\searrow \\
=\!C\!-\!O \quad\quad O\!-\!C\!=
\end{array}
\right|^-
$$

Diol-type structures generated during the microbial breakdown of soil polysaccharides can readily react with boron to form similar organic complexes (Tisdale *et al.*, 1985).

6C.1.1.1. Factors Affecting Availability and Movement of Boron

Soil Texture

The coarse-textured well-drained sandy soils are low in boron while the heavy soils have greater content. As regards the clay type, montmorillonite follows illite in ability to retain boron, while kaolinite has the lowest boron-adsorption capacity.

Soil pH

Boron availability to plants generally becomes less at higher pH, more than 6.3 to 6.5. Liming highly acidic soils frequently induces at least a temporary boron deficiency in susceptible plants. Freshly precipitated $Al(OH)_3$ (*e.g.* on liming an acid soil) facilitates boron adsorption, with such boron adsorbing capacity falling on aging of $Al(OH)_3$. Boron adsorption by iron and aluminum hydroxides is strongly pH-dependent. However, a higher pH on liming acidic soils, rich in organic matter, may encourage organic matter decomposition, leading to release of boron.

Organic Matter

Organic matter application to soils can raise substantially the concentration of boron in plants and even cause phytotoxicity, besides aiding the role of organic fraction of soil in complexing boron and rendering its greater availability in surface soil of relatively higher organic matter content.

Soil Moisture

Boron deficiency is often associated with dry weather and low-soil-moisture conditions, due possibly to restricted boron release from soil organic complexes and the lower ability of plants to extract boron from the relatively dry root zone, supplied with less boron (by diffusion, for instance) from the bulk soil solution to the crop rhizosphere under moisture stress.

6C.1.2. Copper

The copper concentration in the earth's crust ranges from 55 to 70 ppm. Copper levels in igneous rocks vary from 10 to 100 ppm, while they range between 4 and 45 ppm in sedimentary rocks (Tisdale *et al.*, 1985). The plant-available copper concentration in soils ranges from trace to about 32 ppm (Table 6C.1).

Sulphides are the predominant minerals of copper in the earth's crust (Table 6C.1) with strong covalent bonds linking reduced copper (Cu^+) and sulphide (S^{2-}) anions. Other sources include oxides, hydroxy carbonates, silicates, etc.

6C.1.2.1. Forms of Soil Copper

Besides its presence in the lattice structure of primary and secondary minerals, copper also exists in the following forms: in the soil solution (ionic and complexed), at the cation exchange sites of clays and organic matter, occluded and co-precipitated in soil oxide material, at specific adsorption sites, in biological residues and living organisms, and so on (Tisdale *et al.*, 1985).

The copper concentration in soil solutions is usually very low. At pH values below 6.9, Cu^{2+} ions predominate, while above pH 6.9, $Cu(OH)_2^0$ is the principal solution species formed through the $CuOH^+$ species by hydrolysis reactions as shown below (Tisdale *et al.*, 1985).

$$Cu^{2+} + H_2O \rightleftharpoons CuOH^+ + H^+$$

$$CuOH^+ + H_2O \rightleftharpoons Cu(OH)_2^0 + H^+$$

Solubility of Cu^{2+} increases 100-fold for each unit decrease in pH.

Root interception is the principal uptake mechanism of copper by plants, mainly because of low copper solubility in soil. Copper concentration in soil solutions is governed mostly by adsorption of copper on the sesquioxides surfaces, as well as by the soil organic matter. A substantial fraction of soil copper is occluded or buried in various mineral structures, such as layer-lattice aluminosilicate soil clays and iron and manganese oxides, and is thus non-diffusible. Copper undergoes isomorphous substitution in the octahedral layer of crystalline clays, remains associated with magnesium and iron carbonate. Copper may also be trapped within the oxide structures since it readily co-precipitates in aluminum and iron hydroxides (Tisdale *et al.*, 1985). Theoretical considerations show that $Cu_2Fe_2O_4$ (cuprous ferrite) is more stable in reduced soils than cupric ferrite, thereby playing an important role in controlling the copper solubility.

After Pb^{2+}, Cu^{2+} is the most strongly adsorbed of all the divalent transition and heavy metals on iron and aluminum oxides and oxyhydroxide. Adsorption of copper by the sesquioxides increases with pH as a consequence of hydrolysis of ionic species [such as $Cu(H_2O)_6^{2+}$] on the surfaces of clay minerals, and concomitant release of protons, which bring down the exchangeability of the bound Cu^{2+}.

6C.1.2.2. Factors Affecting Availability and Movement of Copper

Copper availability in soil depends on a number of soil properties such as texture, pH, organic matter, hydrous oxides, etc.

Texture

The soil solution concentration of copper is generally lower in excessively leached sands and calcareous sands than in other soil types.

Soil pH

An increase of soil pH depresses the solubility and mobility of copper in soil solutions in addition to causing increased sorption onto mineral colloid surfaces. Further, the degree of copper complexation by the organic binding groups may also be modified with changes of soil pH.

Organic Matter

Copper is more strongly bound to organic matter than any other micronutrient. Such complexation raises the copper concentration in the soil solution. The latter helps plant uptake of copper under conditions of low copper availability, such as at high soil pH. Indeed the humic and fulvic acid fractions of soil organic matter, having several functional groups such as carboxyl, carbonyl and phenol, facilitates binding of Cu^{2+} ions, often by two or more such functional groups. Whereas in most mineral soils, organic matter is intimately associated with clay as a clay-metal-organic complex (Clay-Humic complex). Indeed, such complexation of copper by organic matter will be the highest in soils, having similar clay and organic-matter contents, when predominant clay mineral is kaolinite and lowest with montmorillonite (Tisdale *et al.*, 1985).

6C.1.3. Iron

Iron comprises about 5 per cent of the earth's crust and is the fourth most abundant element in the lithosphere. Common minerals of iron are oxides, sulphides and silicates (*e.g.* ferromagnesian silicates), namely hematite (Fe_2O_3), goethite (α-FeOOH), magnetite (Fe_3O_4) and limonite [$FeO(OH).nH_2O + Fe_2O_3.nH_2O$], pyrite ($FeS_2$), olivine [$(Mg, Fe)_2SiO_4$], siderite ($FeCO_3$), jarosite [$KFe^{III}_3(SO_4)_2(OH)_6$], etc. Hematite and goethite are the most common iron oxides in soils. The iron content in soil varies widely from 0. 4 to 27.3 per cent. The plant available iron content in soil varies from 0.36 to 174 ppm (Table 6C.1).

6C.1.3.1. Soil Solution Iron

Both diffusion and mass flow account for the transfer of iron from soil to root surfaces, where it is taken up as Fe^{2+}. However, diffusion appears to be the predominant mechanism for transfer of iron from bulk soil solution to crop rhizosphere.

The solubility of iron in soils is principally controlled by ferric oxides. Inorganic Fe (III) iron in soil solution is hydrolyzed forming $Fe_2(OH)_2^{4+}$, Fe^{3+}, $Fe(OH)^{2+}$, $Fe(OH)_2^+$, $Fe(OH)_3^0$ and $Fe(OH)_4^-$. The cationic ferric iron species are prevalent under acidic conditions, while the latter two are predominant above pH 7. Plant absorption of any of these ions will cause the others to dissociate, thereby restoring equilibrium relationships among all the species (Tisdale *et al.*, 1985).

Soluble organic complexes, for instance, fulvic acid and low-molecular-weight humic acids complex and thereby transport ferric (and ferrous) ions to plant root surfaces. These natural iron chelates maintain much higher iron concentration in soil solutions than those normally occurring in solutions in equilibrium with only inorganic iron compounds. The insoluble Fe^{3+} form occurs in well-drained soils, while soluble Fe^{2+} increase substantially in presence of excess water, when the redox potentials falls below 200 mV. In soils of pH near 7.0, $FeOH^+$ is the major Fe^{2+} species (Tisdale *et al.*, 1985).

6C.1.3.2. Factors Affecting Availability of Iron

Soil pH

Iron is least soluble in soil solution (of aerobic soil) within the pH range of 7.4 to 8.5, causing iron deficiency in plants. Indeed Fe^{3+} and Fe^{2+} activities in soil solution fall 1000- and 100-fold, respectively, for each unit increase in pH. The pH of most soils, containing calcium carbonate, falls in the range 7.3 to 8.5, which coincides with the greatest incidence of iron deficiency and lowest solubility of iron (Tisdale *et al.*, 1985).

Organic Matter

Additions of organic manure to well-drained soils often leads to chelation which increases the soil solution level of iron. Further, such organic matter addition also improves the soil structure, which in turn, improves soil aeration and hence iron availability, provided that the level of microbial activity is low.

Nitrogen influences the incidence of iron chlorosis. Plants receiving NO_3^- are more likely to develop this condition than those nourished with NH_4^+.

6C.1.4. Manganese

Manganese concentration in the earth's crust averages 1000 ppm. The predominant minerals of manganese are oxides, hydroxides, carbonates and silicates. These cover pyrolusite (MnO_2), manganite (MnOOH), psilomelane [Ba $(Mn^{2+})(Mn^{4+})_8O_{16}(OH)_4$ or as $(Ba,H_2O)_2Mn_5O_{10}$], rhodochrosite ($MnCO_3$), rhodonite ($MnSiO_3$), etc. The oxides, namely pyrolusite and manganite are the most abundant. The surface soils content of total manganese ranges from 0.0037 to 11.5 per cent, while the available manganese in surface soils varies from 0.60-164 ppm. The main ionic species in soil solution is Mn^{2+}. Its concentration decreases 100-fold for each unit increase in pH. Other species of only minor importance are $MnSO_4$ (aq), $MnHCO_3^+$, $MnOH^+$. The soil solution concentration of Mn^{2+} in acid and neutral soils ranges from very low (< 0.01 ppm) to nearly 15 ppm, with an average of 0.01 to 1 ppm. About 84 and 99 per cent of the Mn^{2+} ions are complexed by soil organic matter (Tisdale *et al.*, 1985).

6C.1.4.1. Factors Affecting Availability and Movement of Manganese

Effect of pH and Carbonates

As mentioned earlier, the concentration of Mn^{2+}, the main ionic species in soil solution, decreases 100-fold for each unit increase in pH. On the other hand, low manganese availability in high pH and calcareous soils and in over-limed, poorly buffered coarse-textured soils can be largely overcome by acidification through the use of acid-forming nitrogen or sulphur materials. The microbial activity of soil microorganisms, oxidizing soluble manganese to unavailable forms, also reaches a maximum at about pH 7 (Tisdale *et al.*, 1985).

Soil Submergence

In the waterlogged soils, lower redox potential favours a rise in the amount of soluble Mn^{2+} ions in soil solution.

Organic Matter

The availability of manganese in high pH soils, which are rich in organic matter, is low owing to the formation of unavailable chelated Mn^{2+} compounds. However, additions of compost and wheat straw favour an increase in the concentration of water-soluble and exchangeable Mn^{2+} and easily reducible fractions.

6C.1.5. Molybdenum

The average concentration of molybdenum in the lithosphere is about 2 ppm and in soils from traces to 12 ppm (Deb and Sakal, 2002). The available molybdenum [Ammonium oxalate (pH 3.3)-extractable] content in soils varies from traces to 2.8 ppm (Table 6C.1).

The mineral sources of molybdenum include Molybdenite (MoS_2), Ilsemannite ($Mo_3O_8.8H_2O$), Wulfenite ($PbMoO_4$), Ferrimolybdite [$Fe_2(MoO_4)_3.8H_2O$], Powellite ($CaMoO_4$) (Table 6C.1).

The main forms of molybdenum in soil include: non-exchangeable positions in the crystal lattice of primary and secondary minerals, as an exchangeable anion, bound to iron and aluminum oxides, as water-soluble molybdenum in the soil solution and as organically bound molybdenum. In addition to the exchangeable, non-exchangeable, and bound forms, the main soil solution molybdenum forms include MoO_4^{2-}, $HMoO_4^-$, and $H_2MoO_4^-$. Above pH 4.2, the MoO_4^{2-} ions predominate in soil solution and increases rapidly with rise in pH. Plants absorb molybdenum as MoO_4^{2-} (Tisdale *et al.*, 1985). Molybdenum is transported from bulk soil solution to crop rhizosphere, chiefly, by the diffusion process.

6C.1.5.1. Factors Affecting Availability and Movement of Molybdenum

Soil pH

Molybdenum availability to plants increases with soil pH, especially when the mineral wulfenite ($PbMoO_4$) is present in soil. Indeed plant uptake of molybdenum increases considerably on liming acid soils on application of liming material, such as limestone and basic slag. Such plant availability decreases in presence of acid producing fertilizers, such as ammonium sulphate and nitrate.

Reaction with Iron and Aluminum and Other Nutrients

A portion of the iron and aluminum oxides-adsorbed molybdenum is unavailable to plants and the remaining portion exists in equilibrium with the soil solution molybdenum. Phosphorus enhances the absorption and translocation of molybdenum by plants, due possibly to the release of the adsorbed MnO_4^{2-} ions from the common sorption sites in soil by the applied phosphate of higher binding energy. However, SO_4^{2-} in the rhizosphere soil depresses molybdenum uptake by plants. Both copper and manganese tend to affect the plant uptake of molybdenum adversely, while magnesium helps molybdenum absorption by plants (Tisdale *et al.*, 1985).

Soil Moisture

Molybdenum availability to plants is favoured by increase in soil wetness, probably because such wet soil facilitates the transport of molybdenum from bulk soil to crop rhizosphere by diffusion. A rise in soil temperature also favours the diffusion rate and hence improves the plant availability of molybdenum.

6C.1.6. Zinc

The zinc content of the lithosphere is about 80 ppm. The total zinc content in soils varies between 7 and 1000 ppm, while the plant available zinc content in soils ranges from 0.8 to 20.5 ppm (Table 6C.1). The mineral sources of zinc include Sphalerite (ZnS), Smithsonite ($ZnCO_3$), Hemimorphite [$Zn_4(OH)_2Si_2O_7.H_2O$] (Table 6C.1). The zinc forms in soils cover water-soluble, exchangeable, adsorbed ions (Zn^{2+} ions) on surfaces of clay, organic matter, carbonates and oxide minerals, organically complexed ions, and Zn^{2+} ions substituting Mg^{2+} in the crystal lattices of clay minerals. In addition to the simple Zn^{2+} ions in soil solution, a number of zinc hydrolysis species exist in solution with Zn^{2+} predominating at soil reactions below pH 7.7, while $ZnOH^+$ species above pH 7.7 and below pH 9.1. Above pH 9.1, the neutral hydroxide $Zn(OH)_2^0$ predominates (Tisdale *et al.*, 1985). Zinc solubility is high in acidic soils and rapidly decreases with rise in pH, for instance by about 100-fold for each unit rise in pH, in agreement with the following equilibrium relationship (Lindsay and Norvell, 1969).

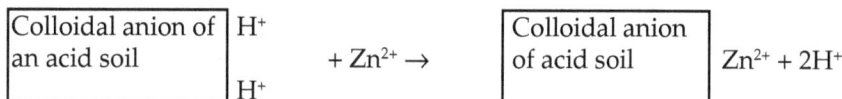

$$\boxed{\begin{array}{l}\text{Colloidal anion of} \\ \text{an acid soil}\end{array}\begin{array}{l}H^+ \\ \\ H^+\end{array}} + Zn^{2+} \rightarrow \boxed{\begin{array}{l}\text{Colloidal anion} \\ \text{of acid soil}\end{array}}\ Zn^{2+} + 2H^+$$

The value of the equilibrium constant (K^0) was found to be as follows:

$$\log K^0 = \log (a_{H^+}^2 / a_{Zn}^{2+}) = -5.8$$

Or, $\log (a_{Zn}^{2+} / a_{H^+}^2) = +5.8$

Where the activity terms refer to the equilibrium soil solution. This expression is obtained by ignoring the solid-state ionic activities, assuming them to be unity for the respective standard states of the adsorbed ions as a first approximation (*vide.* Section 3.2.5). The respective activity terms may be approximated by the corresponding concentration terms, assuming dilute soil solution. Thus, it follows:

$$(Zn^{2+}) = 10^{5.8} (H^+)^2$$

Where the ( ) terms refer to the respective equilibrium ionic concentrations in dilute soil solution.

Hence, $\log (Zn^{2+}) = 5.8 - 2 \, pH$

Zinc is transported from the bulk soil solution to the crop rhizosphere mostly by the diffusion process.

6C.1.6.1. Factors Affecting Availability and Movement of Zinc

Soil pH

As demonstrated above, the plant availability of soil zinc falls with rise in soil pH. Most of the pH-induced zinc deficiencies occur within pH range of 6.0 to 8.0, especially in calcareous soils. However, chelation of soil zinc by the native and the applied organic matter in soil maintains its plant availability, even in soils which are not acidic or neutral. At high pH, the available soil zinc decreases owing to the formation of insoluble compounds such as $Zn(OH)_2$, and $ZnCO_3$ which reduce the zinc availability to plants. Liming acid soils, especially the ones low in zinc, will reduce the zinc availability (Tisdale *et al.*, 1985). Zinc adsorption by clay minerals, sesquioxides and magnesium oxide readily increases with rising soil pH.

Adsorption by Clay Minerals

Adsorption leading to fixation of Zn^{2+} is facilitated by the clay minerals, namely bentonite, illite, and kaolinite, as well as the sesquioxides present in soils. Strong adsorption of zinc by magnesite ($MgCO_3$), less so by dolomite [$CaMg (CO_3)_2$] and calcite ($CaCO_3$) has been reported (Tisdale *et al.*, 1985).

Complexation by Soil Organic Matter

Zinc forms stable complexes with soil organic-matter components. The humic and fulvic acid fractions are prominent in zinc-adsorption. The functional groups in humic substances involved in binding metal ions, including Zn^{2+} ions, are quite similar to those involved in clay-humic acid (HA)/fulvic acid (FA) associations. These include enolate (OH), amino, imino, azo, heterocyclic N (aromatic ring N), carboxyl/carboxylate, phenolic (OH), ether group, sulfhydryl (-SH), carbonyl groups.

Interaction with Other Nutrients

The cations, namely Cu^{2+}, Fe^{2+} and Mn^{2+} inhibit plant uptake of Zn^{2+}, possibly owing to competition for the same carrier site. On the other hand, solubility and mobility of zinc in soils generally increase in the presence of SO_4^{2-} ions. The amount and the nature of the nitrogen fertilizer (*i.e.*, whether acid-producing ones which increases the Zn uptake by the standing crop, or otherwise) and its placement in relation to the zinc fertilizer has a notable effect on zinc availability (Tisdale *et al.*, 1985).

Submergence

The submergence of soil is known to raise the pH of acidic soils, and bring it down for the alkaline soils, close to neutrality in both the cases. Thus in acid soils, the fall of Zn availability could be attributed to the rise in pH of the flooded soils, while the fall in Zn^{2+} concentration in soil solution of calcareous soils (despite having pH more than 7.5, and such pH falling on submergence of soil) may arise from the formation of franklinite ($ZnFe_2O_4$), a lowly soluble mineral.

6C.1.7. Cobalt

The average total concentration of cobalt, which is required for symbiotic nitrogen, fixation, in the earth's crust is 40 ppm. Serpentine, Olivine, Sandstone, Quartzite are the mineral sources providing cobalt. The total cobalt content in soils varies from 4 to 78 ppm, while the available cobalt in soil ranges between 0.2 and 0.5 ppm (Table 6C.1).

6C.1.7.1. Behavior and Availability of Cobalt in Soil

Cobalt is retained in soil mainly in specifically adsorbed, exchangeable forms or as clay-organic matter complexes, and its availability increases with soil acidity and under submergence or reducing conditions. The adsorption of cobalt by the clay (and primary) minerals follows the order, namely muscovite> hematite> bentonite = kaolin.

References

Deb, D. L. and Sakal, R. (2002). Micronutrients. **In:** *Fundamentals of Soil Science* (G. S. Sekhon *et al.*, Eds.), Indian Society of Soil Science, New Delhi, pp. 391-403.

Lindsay, W. L. and Norvell, W. A. (1969). Equilibrium relationships of Zn^{2+}, Fe^{3+}, Ca^{2+}, and H^+ with EDTA and DTPA in soils. *Soil Sci. Soc. Am. Proc.*, **33**, 62-68.

Tisdale, S. L., Nelson, W. L. and Beaton, J. D. (1985). *Soil Fertility and Fertilizers*, Fourth Edition, Macmillan Publishing Company, New York, pp. 350-413.

Transport Processes in Soil

7.1. Introduction

The processes by which nutrients are transported from the bulk soil to crop root-zone are mainly diffusion and mass flow, and the less important root-interception (or contact exchange). While mass flow is relatively important in soils under unlimited (mostly saturated soils) moisture conditions, diffusion accounts for the above mentioned transport of less soluble nutrients under deficient moisture (unsaturated soil) conditions. The nutrients like calcium, magnesium, nitrate and sulphate ions are transported to the crop root-zone mostly by the process of mass flow, while the nutrients such as phosphate, potassium, most of the micronutrient ions are transported by the process of diffusion.

7.2. Diffusion

Diffusion in a mixture implies relative movement of the components in a system, that is, a difference in the mean velocities of each diffusing species, whereas mass (or viscous or convective) flow refers to the movement of the system *en masse* from one point to another. Diffusion, most frequently associated with a non-uniformity of composition (but it can also arise from non-uniformity of other properties as well; in particular, of temperature and pressure; *see* later), is the result of random movement of the molecules or the ions of different components present in the system, governed by their kinetic energy. The latter is determined by the prevailing temperature. Such random movement causes a *net* movement of molecules from the point of high concentration towards the point of low concentration by the process of diffusion. The diffusive flux of a component in a uniform solution under steady-state conditions in one dimension can be expressed by the well-known empirical law, namely the Fick's First Law as in Eq. 7.1

$$(dQ/dt) = -D_0 A (dc/dx) \tag{7.1}$$

$$\text{Or, } (1/A) (dQ/dt) = (dq/dt) = -D_0 (dc/dx) \tag{7.2}$$

Where Q= Quantity of the diffusing ion in homogeneous solution (medium) in time t, A = cross-sectional area of the reference frame of diffusion (RF_D) across which diffusion takes place and is measured, c = concentration of the diffusing species in solution, x = distance in the direction of *net* movement of the diffusing species, D_0 = Diffusion coefficient or diffusivity of the ion, q= Diffusive flux (or diffusive flux-densities) per unit cross-sectional area of the RF_D per unit time.

7.2.1. Reference Frame

For any study of diffusion, it is first of all necessary to define a reference frame relative to which a diffusive flux-density will be measured. Several reference frames may be defined, *e.g.* "volume-fixed", "mass-fixed", mole-fixed", "apparatus-fixed", etc. Evidently, it is convenient to have an "apparatus-fixed" reference frame, which can be physically defined, for practical work. It can be shown that such an "apparatus-fixed" reference frame can be closely approximated by a "volume-fixed" reference frame (Agar, 1960) across which there should not be any *net* transfer of volume in course of the diffusion process. However, it can be also shown that in a multi-component system (such as soil solution having a range of dissolved solutes), these flux-densities will not be linearly independent. Ensuring such linear independence is necessary for overcoming certain thermodynamic restrictions [that is, the applicability of what is known as the *Onsager's Reciprocity Relations, ORR* (Onsager, 1931; Agar, 1963)] in describing the diffusion process in a multi-component system on rigorous principles of irreversible thermodynamics of transport processes (Miller, 1960; Agar, 1963; Sanyal and Adhikari, 1979; Sanyal and Mukherjee, 1988). The latter are quite beyond the scope for discussion in the present chapter. To overcome such difficulty, an *unsymmetrical* reference frame has to be adopted to describe the various diffusive fluxes. In the present case of diffusion, the diffusive flux-densities of various components in the system may be referred to the "solvent", or, what is more commonly known as the "Hittorf frame" of reference, so that the flux-density of the solvent (*e.g.* soil water in soil solution) may be equated to zero, identically (*i.e.*, the flux-densities represent the rate of transfer of matter across unit area of a surface which *moves with the solvent*). This simplifies the algebra considerably; a more *symmetrical* treatment, in which *all the flux-densities* might be non-zero and *all* would, therefore, have to be considered, leads to difficulties (as explained above) because they would not be linearly independent and the corresponding driving forces (*e.g.* the concentration gradients in Eqs. 7.1 and 7.2) would not be linearly independent either. However, in an unsymmetrical reference frame, any other component may be considered stationary, and other flux-densities referred to it, *e.g.* in soil-water system, the *net* flux-density of soil particles may be assumed to be zero which may serve as the reference frame (Sanyal and Adhikari, 1979).

7.3. Diffusion in Soil: Fick's Law for Diffusion through Soil Solution

While extending such concepts to soil system, the description of diffusion of ions or molecules through soil water needs to consider several additional geometric, physical and chemical properties of soil since the latter is a complex, charged and irregularly arranged porous material. Some factors, which account for the differences in the diffusion coefficient of ions or molecules in soil water and in bulk water, are as follows (Ghildyal and Tripathi, 1987).

7.3.1. Reduction in the Cross-Sectional Area

Since diffusion occurs mainly through the liquid phase in soil (*i.e.*, soil water), the cross-sectional area available for diffusive flux will be restricted to the water-filled soil pores, determined by the volumetric soil moisture content (θ). That is the cross-sectional area available for flow of solute by diffusion in soil will be given by $A\theta$.

7.3.2. Tortuosity of Flow Path

The soil pores being arranged in irregular manner, the diffusion of nutrients takes place through a channel which, instead of being a straight capillary system, is quite tortuous in nature. As a consequence, the actual jump of a diffusing species from one equilibrium position to the next in soil may be regarded as equivalent to the passage over a distance (L_e) that is longer than the macroscopic straight line distance (L) between the two positions in question. To account for this factor, the dx term in Eq. 7.1 or 7.2 ought to be multiplied by (L_e/L) so as to represent the actual distance traversed. Obviously, (L_e/L) is more than unity. Furthermore, the actual path will form an angle with the macroscopic pathway, as a result of which the reduced cross-sectional area $A\theta$ (considered earlier) should be multiplied by L/L_e (Ghildyal and Tripathi, 1987).

7.3.3. Relative Fluidity of Water

The clay surfaces being charged with relatively high specific surface charge density (especially the expanding type of 2:1 clay minerals; *see* Section 2.2.2 and Table 2.5), the water in immediate contact with such clay surfaces tends to be highly ordered (polarized) through ionic-dipole interactions. This leads to a higher viscosity, with a reduced fluidity, of the clay-sorbed water in soil pores as compared to the bulk soil solution. As a result, the diffusive movement of water through soil pores in clayey soils will be much more restricted than that in soils of sandy or sandy loam texture. This is accounted for in terms of introduction of a factor f (< 1.0) in the diffusive flux equation, namely Eq. 7.1. Thus, the factor f takes care of the increased viscosity of clay-sorbed water in soil pore *vis-à-vis* that in bulk soil solution. The value of f was estimated from the measured diffusion coefficient of chloride ion in Na-clay and was noted to be 0.5 for $\theta = 0.61$ and 0.65 for $\theta = 0.75$ (Van Schaik and Kemper, 1966). The value of f in soils was found to be about 0.8 at 0.33 bar suction (Porter *et al.*, 1960).

7.3.4. Electrostatic Restriction

The soil pores are irregular in shape and size. There could be a sequence of large-to-small-to-large interconnected voids (Ghildyal and Tripathi, 1987). The flow of ions by diffusion is retarded due to the relatively larger degree of exclusion of anions from the regions of electrostatic restriction in the smaller pores and in thin water films adjacent to the negatively charged surface of clay minerals. Indeed, the heterogeneity of soil pore-size leads to the negative absorption of ions in soil, which, in turn, causes a greater degree of exclusion of anions from smaller pores and narrow films of water at the pore neck connecting large pores. Thus, a factor γ (≤ 1.0) has been included in Eq. 7.2 to account for such electrostatic restriction (electrostriction). The value of γ is unity when the pore width is uniform (Ghildyal and Tripathi, 1987).

The consideration of the above factors leads Eq. 7.1 to be rewritten as

$$(dQ/dt) = -D_0\,[(L/L_e)^2\theta\,f\gamma]\,A\,(dc/dx) = -D_p\,A\,(dc/dx) \tag{7.3}$$

$$\text{Or, } (dq/dt) = -D_p\,(dc/dx)\ (\text{since } q = Q/A) \tag{7.4}$$

Where D_p is the effective diffusion coefficient of an ion in soil solution, and is given by,

$$D_p = D_0\,[(L/L_e)^2\theta\,f\gamma] = D_0.\,\text{IF} \tag{7.5}$$

Where the term, $[(L/L_e)^2\theta\,f\gamma]$, is known as the impedance factor (IF) of the soil concerned (IF < 1.0). Evidently, D_p, being given by (IF. D_0) (*vide.* Eq. 7.5), is always less than the corresponding diffusion coefficient (D_0) in homogeneous and uniform solution. The ratio, (D_p/D_0), in respect of a diffusing ion (*e.g.* Cl$^-$) is used to compute the impedance factor (IF) characteristic of a given soil under the corresponding ambient conditions (*e.g.* the given soil moisture condition) (Van Schaik and Kemper, 1966).

Further, Eq. 7.4 may be modified as follows:

$$(dq/dt) = -D_p\,(dc'/dx)\,(dc/dc')$$

Where c' is the concentration of the diffusing species (ion) per unit volume of the whole soil (that is concentration of the concerned species in soil solid phase + soil solution phase), while c is its concentration per unit volume of soil solution. Thus,

$$(dq/dt) = -[D_p/(dc'/dc)]\,(dc'/dx) \tag{7.6}$$

$$\text{Or, } (dq/dt) = -(D_p/b)\,(dc'/dx) \tag{7.7}$$

Where (dc'/dc) is known as the *soil buffering capacity* (b).

It follows from Eq. (7.7)

$$(dq/dt) = -D_p'\,(dc'/dx) \tag{7.8}$$

Where D_p', given by the ratio (D_p/b), is known as the porous diffusion coefficient of the concerned ion which is inversely proportional to the soil buffering capacity

(b) in respect of the given diffusing ion (Eqs.7.6 and 7.7). Evidently an increase in soil buffering capacity (b) brings down the ionic diffusivity (D_p') in soil.

It is also evident that the soil moisture content (θ) has profound influence on the factors, (L/L_e), f and γ. To emphasize this fact it is customary to rewrite Eqs. 7.4 and 7.8 in an unsaturated soil as follows:

$$(dq/dt) = -D_p (\theta) (dc/dx), \text{ and} \tag{7.9}$$

$$(dq/dt) = -D_p' (\theta) (dc'/dx) \tag{7.10}$$

7.4. Factors Affecting Diffusion Coefficient in Soil

Several factors affect the ionic diffusivity through soil. More important ones are soil moisture content, bulk density, temperature, application of amendments (*e.g.* liming or gypsum application), etc. These are discussed in what follows.

7.4.1. Soil Moisture Content

As discussed above, diffusion takes place primarily through the water-filled porosity in soil. Hence the ionic diffusivity increases with an increase in soil moisture content (θ), that is, as the air-filled porosity in soil decreases. Furthermore, θ has got direct effect on the factors, (L/L_e), f and γ. Thus, a reduced θ leads to (i) a decrease in the cross-sectional area available for diffusion, (ii) an increase in the path length, (iii) higher viscosity, and (iv) an increased electrostriction (negative adsorption). Accordingly, D_p and D_p' are strongly dependent on soil moisture content (θ).

Several investigators studied the effect of soil moisture content on ionic diffusivity in soil, and a number of empirical equations have been proposed from time to time to account for such dependence of D_p on θ. In particular, the following empirical relationship, suggesting that D_p is a positive exponential function of soil moisture content (θ), while being quite independent of the solute content for all practical purposes in the clay-water system, has been found to be quite useful (Ghildyal and Tripathi, 1987).

$$D_p (\theta) = D_0 a e^{b\theta} \tag{7.11}$$

Where a and b are empirical constants. Olsen and Kemper (1968) reported the value of b to be 10, while that of a ranges from 0.0005 to 0.001, depending on the surface area of the soils (ranging from clay to sandy loam). Eq. 7.11 describes satisfactorily the dependence of the ionic diffusivity in soil on the corresponding soil moisture content over a wide range of soil moisture suctions (*e.g.* 0.030 to 1.5.MPa), corresponding to the conditions under which diffusion appears to be the major mode of solute transport in soil. Figure 7.1 illustrates the effect of water content on the self-diffusion coefficient of phosphorus in two soils of widely varying texture (Ghildyal and Tripathi, 1987). It is thus clear from Figure 7.1 that the solute diffusivity in soil (*e.g.* D_p) tends to be higher in a coarser-textured soil than that in a finer-textured one at an identical soil moisture content.

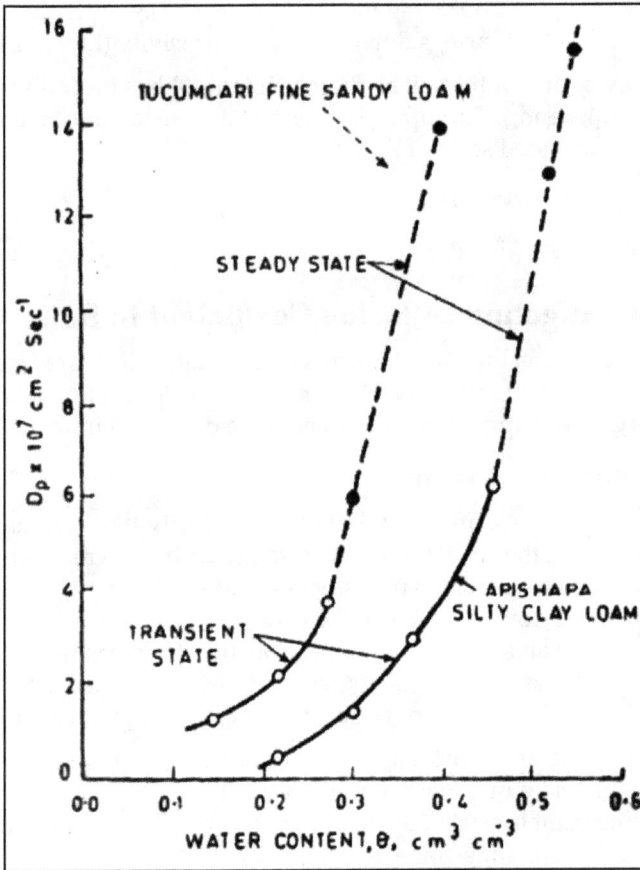

**Figure 7.1: Effect of Water Content on Self-diffusion Coefficient
of Phosphorus in Two Soils.**
Source: Olsen and Kemper (1968); Ghildyal and Tripathi (1987)

7.4.2. Bulk Density of Soil

With increase in bulk density of soil, the soil solid particles are brought closer together with the gradual exclusion of air space. The latter, in turn, increases the continuity of water-filled porosity in soil, thereby facilitating the solute diffusivity in soil. Further, the impedance factor (IF) also tends to fall with initial rise in bulk density due to a corresponding reduction in the tortuosity factor (L_e/L), which causes an increase in the D_p values. However, beyond a certain bulk density in a given soil, the D_p values register a fall, due presumably to nearly complete exclusion of the air space, as well as a reduction of the water-filled porosity, arising from the coming together of the soil solid particles, that leads to a concomitant soil compaction, accompanied with a rise in soil bulk density. The latter causes a fall of D_p values.

7.4.3. Temperature

As mentioned briefly earlier, the relatively independent particles (atoms, molecules, ions) vibrate about their temporary equilibrium positions in a given medium, homogeneous or multiphase (such as soil). In order to facilitate passage of such a species from one equilibrium position to another, it is necessary that a suitable hole or site is available; the production of such a site requires the expenditure of energy, since work must be done in pushing back other molecules or ions. As a consequence, the jump of a diffusing molecule from one equilibrium position to the next may be regarded as equivalent to the passage of the system over a potential energy barrier, known as the activation energy, characterizing the given transport process such as diffusion. The latter happens when the particular molecule, in question, gains, due to fluctuations in the frequency of vibration about the initial equilibrium position, energy, sufficient to overcome the above mentioned energy barrier, separating the two adjacent equilibrium positions (Sanyal, 1980).

With the rise of temperature, the kinetic energy of the diffusing species (such as nutrient ions in soil solution) increases, and as a result, there is a greater probability that the diffusing ions over-role the above mentioned activation energy barrier, so that the D_p increases, leading to a greater diffusive flux. A rise in ambient temperature also brings down the viscosity and increases the fluidity of clay-sorbed water in soil pores through which diffusion occurs. This, in turn, raises the value of D_p of the diffusing ions in soil.

7.4.4. Amendments

As discussed in Sections 6C.1.1 to 6C.1.6, the concentration of the micronutrient cations such as zinc, iron, manganese, copper, etc., decreases with rise in soil pH, for instance, on liming an acid soil, whereas that of the anionic micronutrient, namely molybdate increases with a rise in soil pH. Thus, zinc solubility is high in acidic soils and rapidly decreases with rise in pH, for instance, by about 100-fold for each unit rise in pH, in agreement with the following equilibrium relationship (Lindsay and Norvell, 1969):

$$\log (Zn^{2+}) = 5.8 - 2 \text{ pH} \tag{7.12}$$

Where the ( ) terms refer to the respective equilibrium ionic concentration in dilute soil solution. Most of pH-induced zinc deficiencies occur within range 6.0 to 8.0, especially in calcareous soils. Iron (Fe^{3+} ion) is also least soluble in soil solution (of aerobic soil) within the pH range of 7.4 to 8.5, causing iron deficiency in plants. Indeed Fe^{3+} and Fe^{2+} activities in soil solution fall 1000- and 100-fold, respectively, for each unit increase in pH. The concentration of Mn^{2+}, the main ionic species in soil solution, decreases 100-fold for each unit increase in pH. An increase of soil pH depresses the solubility and mobility of copper (Cu^{2+} ions) in soil solutions in addition to causing increased sorption onto mineral colloid surfaces. Further, boron availability to plants generally becomes less at higher pH, more than 6.3 to 6.5. Liming highly acidic soils frequently induces at least a temporary boron deficiency in susceptible plants. Freshly precipitated $Al(OH)_3$ (*e.g.* generated on liming an acid soil) facilitates boron adsorption, with such boron adsorbing capacity falling on aging of $Al(OH)_3$. On the other hand, molybdenum availability to plants increases with soil pH, especially when the mineral Wulfenite ($PbMoO_4$) is present in soil.

Indeed plant uptake of molybdenum increases considerably on liming the acid soils on application of liming material, such as limestone and basic slag.

It is thus apparent that liming an acid soil decreases the micronutrient cation concentrations ('c' term in Eqs. 7.4 and 7.6) in soil solution, while raising that of the molybdate anion (MoO_4^{2-}). This will cause a rise in the *soil buffering capacity* (b = dc'/ dc) for the cations, which, in turn, will cause a fall in the porous diffusion coefficient (D_p') of the diffusing cations in soil, since the latter is inversely proportional to the soil buffering capacity (b) in respect of the given diffusing cation (*vide*. Eqs. 7.6-7.8). For anions, however, an opposite effect will result on liming an acid soil. A fall in soil pH will have a reverse influence on the cationic and anionic diffusion coefficients in soil. The latter results when a sodic soil is treated with gypsum or other such amendments used (discussed in Section 4.12.1) to bring down the pH of sodic soils.

Several low molecular weight organic acids (such as tartaric, malic, citric gluconic acids, etc.) are formed *in situ* on incorporation of organic manures to soil. Besides, several biochemicals are synthesized by the microorganisms in soil (which proliferate on addition of organic manures), including amino acids. These low-molecular weight aliphatic acids as well as the biochemicals form the soluble chelate complexes with cations in soil. The relatively low-molecular weight humus fraction in soil, as well as that of the incorporated organic manures, such as fulvic acid also participates in soluble chelate formation with the cations in soil (Stevenson, 1994). This enhances the soluble cation concentration (c) in soil solution, which, in turn, brings down the buffering capacity (b = dc'/dc) of soil. The latter facilitates the diffusion of cations in soil by way of increasing the corresponding D_p' values, as explained above.

7.5. Self and Tracer Diffusion

Several studies have been reported on self-diffusion of various nutrient ions through soil to plant roots and the self-diffusion coefficients in soil have been compiled. (Olsen and Kemper, 1968; Nye and Tinker, 1977; Sanyal, 1980) In describing such diffusive fluxes, Fick's laws of diffusion have been widely used for both soil systems and homogeneous media. In what follows, a systematic development of the general thermodynamic equations, describing the diffusive fluxes, are discussed, along with the applications of the same to describe the self- and tracer-diffusion as special cases.

7.5.1. Theoretical Considerations

As stated earlier, in a given medium, homogeneous or multiphase, the relatively independent particles (atoms, molecules, ions) vibrate about their temporary equilibrium positions. In order to facilitate passage of such a particle, from one equilibrium position to another, it is necessary that a suitable hole or site is available; the production of such a site requires the expenditure of energy, since work must be done in pushing back other molecules. As a consequence, the jump of a moving molecule from one equilibrium position to the next may be regarded as equivalent to the passage of the system over a potential energy barrier. The latter happens when the particular molecule, in question, gains, due to fluctuations in frequency of vibration about the initial equilibrium position, energy, sufficient to overcome the energy barrier, above referred to, separating the two adjacent equilibrium positions. Such energy barrier, associated with the process of diffusion, is known as the corresponding *activation energy* hill.

Apart from other factors, which are beyond the scope of this chapter to discuss, the frequency of such jumps depends on the height of the energy barrier, separating the two neighbouring equilibrium positions of the species, which is governed by the general structure of the medium (not necessarily homogeneous). With no external forces acting, the direction of these jumps is random; they have equal probability in all directions. In this way, a given molecule leaves its equilibrium position with a finite probability of arriving at some other position within a given period of time. This motion constitutes true self-diffusion. In this connection, mass-flux and self-diffusion coefficient can be defined as for ordinary (isothermal) diffusion (Sanyal, 1980).

Strictly speaking, self-diffusion cannot be investigated experimentally owing to the indistinguishability of the molecules. Self-diffusion may, however, be approached by replacing a very small portion of atoms or ions in a system by an isotope, which can be detected by its specific property of say, radioactivity or atomic weight or its spectrum which differs from that of bulk atoms or ions (or, molecules). The departure one obtains in this way from true self-diffusion becomes smaller, the less the relative difference in the atomic weights of the various isotopes. In particular, the diffusion of deuterated water (HDO) and tritiated water (HTO) molecules is not entirely identical with self-diffusion of ordinary water for this reason (Sanyal, 1980).

The diffusion of molecules, tagged with isotopes, is in fact the limiting case of tracer-diffusion in a uniform or heterogeneous medium (*e.g.* soil system). Tracer-diffusion means diffusion of a single component (*e.g.* an ion), present in very small concentration, in a large excess of other components; in other words, the "tracer-component" is subjected to a concentration gradient of its own, while the concentration gradient of all other components, present in excess, is zero in comparison with the former. Examples are provided by the diffusion of radioactive sodium ion ($*Na^+$), present in trace quantities, in an otherwise uniform solution of (a) KCl, or (b) NaCl. In case (b), the diffusion coefficient of the tracer ion is assumed to be identical with the true self-diffusion coefficient of sodium ion in the NaCl solution (Sanyal, 1980).

The relationship between diffusion (or diffusive) flux (J) and the thermodynamic driving force (X), responsible for diffusion, can be described by a power series of the force, namely,

$$J = A + LX + NX^2 + ...$$ (7.13)

Where, A, L, N, etc., are constants (*see* later). However, the constant A must be zero since there cannot be any diffusion flow in the absence of the driving force, X, *i.e.*, J = 0, for X = 0 (equilibrium case). Also in a state, sufficiently close to equilibrium (which results for small values of X), it is adequate to retain only the linear term, as confirmed by experiments (Sanyal, 1980), *i.e.*,

$$J = LX \tag{7.14}$$

The driving force (per mole) for diffusion is given by the gradient of chemical potential (μ_i) of the diffusing solute, say 'i' (however, *see* later). If therefore the concentration of the diffusing species, adjacent to the transit plane, across which the flux is reckoned (*i.e.*, the Reference Frame introduced in Section 7.2.1), is C_i mole per unit volume, the driving force at this plane is given by,

$$X_i = -C_i \, grad \, \mu_i \tag{7.15}$$

Where the term 'grad μ_i' refers to the gradient (in three dimensions) of the chemical potential (μ_i) of the given diffusing species, 'i'.

However, it is *not* obvious how the chemical potential, corresponding to equilibrium states (a quantity of free-energy character), can be applied to the description of the rate of irreversible processes and among them, of diffusion (*see* later). It is well known, for example, that the rate of chemical reactions is *not* proportional, in general, to the change of free energy in the process, but depends rather on the kinetic barrier, for instance. However, in a detailed analysis of the problem, it was noted that in diffusion, the change in free energy resulting for mixing of solutions of various concentrations (which is measured by the chemical potential gradient) can be identified with the work of the diffusing species against the friction of the viscous medium (Onsager, 1931; Agar, 1963; de Groot, 1966). Since diffusion is a slow process, deviations from the equilibrium state are much smaller in the course of the process than in most of the chemical reactions. Under such conditions, the total change in free energy is *almost* equal to the energy dissipated by the viscous forces in friction encountered in course of diffusion (Sanyal and Adhikari, 1979).

Hence, one obtains from Eqs. 7.14 and 7.15

$$J_i = -L \, C_i \, grad \, \mu_i \tag{7.16}$$

Remembering $\mu_i = \mu^0_i + RT \ln C_i$ (Section 3.2.1.3; Eq. 3.16) on the assumption of ideal behaviour of the diffusing solute 'i' in a dilute solution (*e.g.* soil solution) with the corresponding activity coefficient, γ_i being equal to 1, one obtains from Eq. 7.16

$$J_i = -L \, C_i \, \frac{RT}{C_1} \, grad \, C_i$$

Or, $J_i = -LRT \, grad \, C_i$ \hfill (7.17)

However, by Fick's First Law (for three dimensional diffusion),

$$J_i = -D_i \, grad \, C_i \tag{7.18}$$

By comparing Eqs. 7.17 and 7.18, one obtains

$D_i = LRT$

Whence $L = (D_i / RT)$ \hfill (7.19)

The diffusion flux, J_i in terms of Fick's First Law diffusion coefficient is thus given (*vide.* Eq. 7.16) by Eq. 7.20.

$$J_i = -(D_i C_i/RT) \, \text{grad} \, \mu_i \tag{7.20}$$

For diffusion of an ionic solute, the chemical potential μ_i in the "driving force" should be replaced by the corresponding electrochemical potential $\overline{\mu}_i$ in order to include the electrical effects; $\overline{\mu}_i$ is defined as,

$$\overline{\mu}_i = \mu_i + z_i \, F \, \phi \tag{7.21}$$

Or, grad $\overline{\mu}_i$ = grad $\mu_i + z_i \, F \, \text{grad} \, \phi$ \hfill (7.22)

Where z_i is the valence (including sign) of the ionic species 'i', and F, the Faraday's constant (96,500 Coulombs per equivalent), and ϕ is the electrostatic potential at the point in the given phase, where $\overline{\mu}_i$ is the electrochemical potential of 'i'. It therefore follows from Eq. 7.20,

$$J_i = -(D_i C_i/RT) \, \text{grad} \, \overline{\mu}_i \tag{7.23}$$

Finally, the flux equation in presence of bulk motion of the solution is given by,

$$J_i = -C_i \left(\frac{D_i}{RT} \, \text{grad} \, \overline{\mu}_i - v \right) \tag{7.24}$$

Where v is the vector of velocity of motion of the solution at a given point.

Using Eq. 7.22 for grad $\overline{\mu}_i$ and assuming ideal behaviour (dilute solution) so that grad $\mu_i = RT$ (grad C_i/C_i), one obtains from Eqs. 7.24,

$$J_i = -D_i \, \text{grad} \, C_i - (D_i z_i \, F \, C_i/RT) \, \text{grad} \, \phi + C_i \, v \tag{7.25}$$

For unidimensional diffusion in x-direction,

$$(J_i)_x = -D_i \, (\partial C_i/\partial x) - (D_i z_i \, F \, C_i/RT) \, (\partial \phi/\partial x) + C_i v_x \tag{7.26}$$

Where, v_x is the component of velocity v in the x-direction.

For a non-ideal solution, it is well-known

$$\mu_i = \mu_i^0 + RT \ln a_i = \mu_i^0 + RT \ln C_i + RT \ln \gamma_i$$

Where a_i is the activity and γ_i, the activity coefficient of 'i'.

$$\text{Hence, grad} \, \mu_i = \frac{RT}{C_i} \left[1 + \left(\frac{\partial \ln \gamma_i}{\partial \ln C_i} \right) \right] \text{grad} \, C_i \tag{7.27}$$

Substitution of Eqs. 7.25 and 7.27 in Eq. 7.24 leads to Eq. 7.28 for the diffusion flux,

$$J_i = -D_i \, \text{grad} \, C_i \, [1 + (\partial \ln \gamma_i/\partial \ln C_i)_T] - \frac{D_i z_i \, F \, C_i}{RT} \, \text{grad} \, \phi + C_i \, v \tag{7.28}$$

In an ionic solution, the flux of all the charged particles $(\sum_i J_i)$ at a given position, and the density of the electrical current at the same point (I) are mutually equivalent on the basis of the Faraday's law, the current, corresponding to the flux of positive particles, being positive by convention, *i.e.*,

$$I = \sum_i z_j F J_i$$

(7.29)

Combining Eqs. 7.28 and 7.29 one obtains Equation 7.30 for the above mentioned current density, I. Thus,

$$I = -\sum_i z_i F D_i \operatorname{grad} C_i [1 + (\partial \ln \gamma_i / \partial \ln C_i)_T] + \sum_i \frac{z_i^2 F^2}{RT} D_i C_i E + \sum_i z_i F C_i v$$

(7.30)

Where E is the electrical field strength, and is given as $E = -\operatorname{grad} \phi$. In the usual cases, the field strength is such that the electroneutrality condition is observed so that $\sum_i C_i z_i = 0$, and the third term on the right-hand side of Eq. 7.30 drops out. The field strength E may be split into two terms, namely

$$E = \frac{I}{H} + \frac{1}{H} \sum_i z_i F D_i \operatorname{grad} C_i [1 + (\partial \ln \gamma_i / \partial \ln C_i)_T]$$

(7.31)

Or, $E = E_{ohm} + E_{diffusion}$

(7.32)

Here, H (Eq. 7.31) is the electrolytic conductivity of the solution and is given by,

$$H = \sum_i \frac{z_i^2 F^2}{RT} D_i C_1$$

(7.33)

The term E_{ohm} in Eq. 7.32 is the ohmic electrical field strength which is due to the flow of electrical current through a medium of a given conductance. This term drops out for diffusion of ionic solutes in the absence of an externally applied electric field. The term $E_{diffusion}$, on the other hand, is the diffusion electrical field strength, and does *not* disappear in the absence of an external electric field, being due to unequal rates of diffusion of charged particles (ions). This inner electrical potential drop is produced by diffusion of an electrolyte. In the absence of an external electric field, and of convection (v = 0), the diffusion flux of an ith component is given (*vide.* Eq. 7.28) by Eq. 7.34,

$$J_i = -D_i \operatorname{grad} C_i [1 + (\partial \ln \gamma_i / \partial \ln C_i)_T] - \frac{D_i Z_i F C_i}{RT} \operatorname{grad} \varphi_{diffusion}$$

(7.34)

Where, $-\operatorname{grad} \phi_{diffusion} = E_{diffusion} = (1/H) \sum_i z_i F D_i \operatorname{grad} C_i [1 + (\partial \ln \gamma_i / \partial \ln C_i)_T]$

(7.35)

In the case of diffusion of ions, present at a low concentration in an excess of an *indifferent electrolyte* (however, *see* later), the diffusion potential drop in the electrolyte is suppressed so that the movement of the tracer ions is *not* held or tied to that of ions of opposite sign. This directly follows from Eq. 7.35, since the quantity, $E_{diffusion}$, depends inversely on the conductivity H of the solution which may arbitrarily be increased by the indifferent electrolyte (Eq. 7.33) so as to make $E_{diffusion}$ negligible. The last term on the right-hand side of Eq. 7.34, therefore, disappears, and also for a dilute solution of the tracer ion in the indifferent electrolyte, the term, $(\partial \ln \gamma_i / \partial \ln C_i)_T$, vanishes. Finally, the simple relation, Fick's First Law, results, namely Eq. 7.36, which is, therefore, adequate to describe the processes of self-and tracer-diffusion.

$$J_i = -D_i \text{ grad } C_i \tag{7.36}$$

For self-diffusion in mixtures of isotopic species, showing practically ideal thermodynamic behaviour, the gradient of chemical potential for each isotope is also ideal, *i.e.*, it is simply the gradient of (RT ln C); the "driving force" for the inter-diffusion of isotopic species is, in fact, the term arising solely from the entropy of mixing in the free energy of mixing; and this obeys the ideal relation (Sanyal, 1980):

$$\Delta S_{mixing} = -R (x_1 \ln x_1 + x_2 \ln x_2) \tag{7.37}$$

x_1 and x_2 being the mole fractions of the two species 1 and 2.

In tracer-diffusion also, since the ionic surrounding of the diffusing tracer ion is *effectively* unchanged, the "driving force" for diffusion is once again the gradient of (RT ln C) of the tracer ion, its activity coefficient virtually remaining unaltered [*i.e.* $(\partial \ln \gamma / \partial \ln C) = 0$].

It may appear interesting to consider the more recent modifications to Fick's First law, proposed for a complete description of solute flows in a three-component system, as is encountered in self- and tracer-diffusion [*e.g.* in case of $*Na^+$ ion diffusing in $Na^+ Cl^-$ (NaCl), or $*Na^+$ ion diffusing in K^+Cl^- (KCl) solution]. These may be written as,

$$J_1 = -D_{11} (\partial C_1/\partial x) - D_{12} (\partial C_2/\partial x) \tag{7.38}$$

$$J_2 = -D_{21} (\partial C_1/\partial x) - D_{22} (\partial C_2/\partial x) \tag{7.39}$$

Where $D_{12} \neq D_{21}$

These relations have their origin in the irreversible thermodynamics for transport processes (Miller, 1960; Agar, 1963; Sanyal, 1980), and they express the fact that flow of either solute component (NaCl being denoted by subscript 1, and KCl by 2) depends, not only on its own (conjugate) concentration gradient, but also on that of the other solute in the solution, *i.e.*, there is an interaction or "coupling" of flows. Here, D_{11} and D_{22} represent the main (or, "straight") diffusion coefficients of solutes 1 and 2, respectively, while D_{12} and D_{21} are the "cross term" (coupling) diffusion coefficients. Equations 7.38 and 7.39 illustrate that, because the flow J_2 of KCl must approach zero as $C_2 \rightarrow 0$, regardless of the concentration gradient of NaCl, it is required that $D_{21} \rightarrow 0$ as $C_2 \rightarrow 0$, and similarly, $D_{12} \rightarrow 0$ as $C_1 \rightarrow 0$.

In soil systems, such concept of coupling is indeed expected to be useful in view of the former's poly-component nature, and thus this approach of irreversible thermodynamics may lead to additional insight into the phenomenon of diffusion as occurring in soil (Sanyal, 1980).

It can be shown (Gosting, 1956) that for a solution in water of two kinds of univalent cations, denoted by subscripts 1 and 2, and one univalent anion, denoted by 3 (*e.g.* *Na$^+$ ion in Na$^+$Cl$^-$ or, *Na$^+$ ion in K$^+$Cl$^-$ solution),

$$D_{11} = \overline{D}_1 \left[1 - \frac{C_1 \left(\overline{D}_1 - \overline{D}_3 \right)}{M} \right]$$

(7.40)

$$D_{12} = -\overline{D}_1 C_1 \left[\frac{\overline{D}_2 - \overline{D}_3}{M} \right]$$

(7.41)

$$D_{21} = -\overline{D}_2 C_2 \frac{\left(\overline{D}_1 - \overline{D}_3 \right)}{M}$$

(7.42)

$$D_{22} = \overline{D}_2 \left[1 - \frac{C_2 \left(\overline{D}_2 - \overline{D}_3 \right)}{M} \right]$$

(7.43)

$$\text{Where, } M = \left(\overline{D}_1 + \overline{D}_3 \right) C_1 + \left(\overline{D}_2 + \overline{D}_3 \right) C_2$$

(7.44)

Here, the limiting ionic diffusion coefficients for the three ionic species are defined by,

$$\overline{D}_i = \frac{RT \lambda_i^o}{F^2} \cdot 10^{-7} \quad (i = 1, 2, 3)$$

(7.45)

Where λ_i^o is the limiting equivalent ionic conductance of the ion 'i'. It should be pointed out here, that it is immaterial whether the subscripts 1 and 2 in the basic flow equations, namely Eqs. 7.38 and 7.39 denote Na$^+$ and K$^+$ ions or *else*, NaCl and KCl.

It is seen from Eqs. 7.41 and 7.42 that $D_{12} \to 0$ as $C_1 \to 0$ for all values of C_2 other than zero, and $D_{21} \to 0$ as $C_2 \to 0$ for all values of C_1 other than zero, as stressed above.

It is important to note from Eq. 7.43 that D_{22} approaches the binary diffusion coefficient of KCl in water at the given composition (given C_2), namely

$$\frac{2RT}{F^2} \left[\lambda_2^o \lambda_3^o / (\lambda_2^o + \lambda_3^o) \right] \cdot 10^{-7}$$

as $C_1 \to 0$. Hence, this is consistent with what one might have expected from the Nernst's limiting expression for diffusion coefficient. At the same time (as $C_1 \to 0$ for a given C_2), D_{11} approaches the tracer-ion diffusion coefficient in a KCl solution of that composition (Eq. 7.40), while $D_{12} \to 0$, as mentioned above. But, interestingly, D_{21} does *not* now approach zero, and approaches, instead, a value, given by,

$$-\overline{D}_2 \frac{\left(\overline{D}_1 - \overline{D}_3 \right)}{\left(\overline{D}_2 + \overline{D}_3 \right)}$$

(*vide* Eq. 7.42)

Thus, when $^*Na^+$ diffuses as a tracer ion in an aqueous KCl solution, the conventional form of the Fick's First Law (Eqs.7.1 and 7.2 as well as Eq.7.36) is adequate to describe the flow of $^*Na^+$ because $D_{12} \to 0$ (Eq. 7.38) as a consequence of C_1 being essentially zero. The classical form of the Fick's First Law is *not* adequate to describe the flow of K^+ or KCl, however, because D_{21} is not zero, and whatever concentration gradient of $^*Na^+$ ion is present as a tracer, will produce some flow of K^+ (or, KCl), even in the absence of its own concentration gradient, $(\partial C_2 / \partial x)$ (Eq. 7.39).

Similar considerations apply to self-diffusion, which is, indeed, a special case of tracer-diffusion, as emphasized before. For example, diffusion of a trace of radioactive $^*Na^+$ ion in an otherwise uniform solution of ordinary NaCl is described adequately by the classical form of the Fick's First Law, as established above (in the earlier part of the present Section 7.5.1) in a different approach, but diffusion of the untagged Na^+ ion (or, ordinary NaCl) is not; a concentration gradient of the radioactive $^*Na^+$ ion should produce a flow of the ordinary Na^+ ion in response to a non-zero value of the corresponding "cross-term" diffusion coefficient (D_{21}). This important point often goes unnoticed in the study of self-and tracer-diffusion.

In particular, such studies in soil systems frequently depend on the measurement of diffusive fluxes of tagged radio-isotopes in "untagged" soil, the latter having the same total background concentration as the tagged soil, with the *tacit* assumption that such "indifferent" electrolyte is stationary, in not being subjected to any concentration gradient of its own. However, as discussed above, a non-zero value of the corresponding "cross-term" diffusion coefficient is likely to cause this *"supposedly stationary"* background electrolyte in soil system to diffuse, and thereby introduce an error in the measured self-(or, tracer) diffusion coefficient of the radioactive nutrient ion, which neglects such migrations (Sanyal, 1980).

7.6. Diffusion of Water in Soil under Temperature Gradient–Thermal Diffusion

As mentioned above, the process of diffusion in a mixture implies a relative motion of the components, that is, a difference in the mean velocities of each species of molecule. Diffusion is most frequently associated with a non-uniformity of composition, as discussed above, but it can also arise from non-uniformity of other properties in particular, of temperature and pressure. Diffusion induced by a temperature gradient is thermal diffusion.

It has long been known that the simple form of Darcy's law, describing the flow of water in soil under a hydraulic gradient, or a soil-water potential gradient, *fails* to take into account the influence of forces other than those arising from a differential moisture content, or potential gradient. In particular, the effect of the thermal gradients has been shown to induce the movement of moisture in soil, both in the liquid and the vapour phases (Cary and Taylor, 1962; Cary, 1963; Taylor and Cary, 1964; Cassel *et al.*, 1969; Sanyal and Adhikari, 1979, 1984; Ghildyal and Tripathi, 1987). As early as 1927, Lebedeff (cited in Cassel *et al.*, 1969) observed in a field experiment during Russian winter that more than 6 cm of soil water moved upward into the soil profile in response to the seasonal thermal gradients.

Whether or not this flow of soil-moisture, associated with day-night and seasonal temperature cycles, is of practical significance, depends upon local conditions. However, there is little doubt that such flow often has considerable practical importance, especially in supplying small amounts of water needed for germination of seeds in dry-land areas.

Surprisingly enough, very little has thus far been done to explore this important phenomenon, in spite of the fact that the diurnal and seasonal temperature fluctuations lead to considerable thermal gradients in soil. The need at this stage of a general theory, encompassing all these various aspects of soil-water movement, seems obvious. Buckingham, even as early as 1907 (cited in Taylor and Cary, 1964), recognized this while stating that *"a rigorous treatment of the subject, with no restrictions imposed on either the water content or the soluble salt content of the soil, would have to use thermodynamic reasoning. But the simple conception of a mechanical potential will suffice for the present purpose, though it is not impossible that with more comprehensive experimental data available, we should have to use thermodynamic potential or the free energy"*. As mentioned earlier in Section 7.2. 1, the thermodynamics of irreversible processes appears to be capable of providing such a general theory and understanding of transport process in soil. Indeed, it has proved to be a powerful tool in systems of physics and chemistry for arriving at certain general and useful working relations between various fluxes, and the corresponding forces, responsible for them. In the case of soil systems also, a generalized theory of simultaneous transport of heat and matter was developed (Cary and Taylor, 1962; Cary, 1963; Taylor and Cary, 1964; Sanyal and Adhikari, 1979; Ghildyal and Tripathi, 1987).

7.6.1. Thermodynamics of Thermal Diffusion

The relationships for transport phenomena cannot be deduced from classical thermodynamics as it deals with equilibria and reversible processes (hence the name *"thermostatics"*; de Groot, 1966), while the transport processes arc irreversible in nature. One of the starting points of extension of thermodynamics to irreversible processes is the fact that, according to the empirical laws of Fourier, Ohm and Fick, regarding conduction of heat, electricity and matter, respectively, the correlation between the fluxes and the forces causing them is linear (stated earlier in Section 7.5.1). Recalling Eq. 7.14, one has,

$$J = LX \tag{7.46}$$

In Eq. 7.46, L is a phenomenological constant, characteristic of the substance through which flux J occurs, and may be termed its generalized (specific) conductance, or "admittance" (= 1/resistance). The forces X are given by the gradients of thermodynamic variables, such as temperature, chemical potential, electrostatic potential, etc., causing macroscopic flows of heat, matter, or electric charge. These forces are thermodynamic in nature which have got nothing to do with the Newtonian forces in that they do not, in general, impart any acceleration (they might impart acceleration in a special case when they can incorporate a Newtonian force, *e.g.* sedimentation under gravity) (Sanyal and Adhikari, 1979).

Furthermore, there may be interaction (coupling) between two or more transport processes, occurring simultaneously in the same system so that a flux density of one kind may result from a force of different kind. Well known examples of such coupled transport processes are thermal diffusion and thermo-osmosis (interaction of mass diffusion and heat conduction), thermoelectricity (Peltier, Seebeck and Thomson effects) electrokinetic phenomena (Electroosmosis, electrophoresis, streaming potential, streaming current), isothermal diffusion in electrolyte solutions, etc.

A brief description of the phenomenon of thermal diffusion will now follow. When a temperature gradient exists in a solution, or mixture, of initial uniform composition, a partial de-mixing of the components usually takes place owing to their relative movement. This phenomenon goes under the name of "thermal diffusion", or the "Soret effect" (after Charles Soret, the discoverer of the phenomenon), the latter term being restricted to condensed phases. This systematic relative movement of the components of a solution leads, in absence of convective re-mixing, to the establishment of concentration gradients, which, in turn, trigger ordinary (isothermal) diffusion in the opposite direction. Eventually, a steady state is reached when the rates of these opposing processes balance each other, and a time-invariant concentration gradient is established in the system. This steady (or, stationary) state is sometimes referred to as "Soret equilibrium", although it is not a true thermodynamic equilibrium state in that heat continues to pass through the system by thermal conduction. Whereas viscosity, ordinary diffusion, and thermal conductivity are first-order effects, depending primarily on the occurrence of molecular collisions, and only secondarily on the *nature* of these collisions, thermal diffusion is a second-order effect, which may be positive, negative, or zero, according to the nature of the molecular interactions.

In the case of thermal diffusion, the flux-densities (J_i) of various components in the system may be defined along with an additional flux density to characterize the flux of "heat". An entropy flux density (J'_s) is defined for this purpose, and all the flux densities are referred to the "solvent", or, what is more commonly known as the "Hittorf frame" of reference, so that J_o (solvent) = 0, identically (*i.e.*, the flux-densities represent the rate of transfer of matter and heat across unit area of a surface which "moves with the solvent", as introduced earlier in Section 7.2.1). The advantages of adopting such an *unsymmetrical* reference frame have already been elaborated earlier in Section 7.2.1.

At this stage, an energy flux-density (J_u), associated with the flow of matter and heat across the boundary of the system, in question, may also be defined by

$$T J'_s = J_u - \sum_i \mu_i J_i$$

$$(7.47)$$

Where μ_i is the chemical potential of the component 'i'. It appears from Eq. 7.47 that J_u may possibly be better regarded as an enthalpy flux-density.

An associated parameter, called the Soret coefficient (σ') of the solute, is defined as the fractional separation at the steady state per unit difference of temperature, and is given as,

$$\sigma' = -\left(\frac{d \ln m}{d T}\right)_{st}$$

(7.48)

Where m is the molality (mean) of the non-isothermal solution (*e.g.* the soil solution subjected to a thermal gradient), and T, the (mean) absolute temperature. Most of the salts in aqueous medium, including soil system, have positive values of σ' in that they migrate towards the colder parts of the system (Agar, 1963; Sanyal and Mukherjee, 1988). Obviously, σ' has the unit of deg^{-1}.

The thermal diffusion factor, σ_T is yet another parameter, and is given by, $\sigma_T = \sigma'. T; \sigma_T$ is a pure number.

Even though the Soret effect has been studied for many years, and various theories, starting with van't Hoff's osmotic pressure theory in the last century, have been put forward from time to time, the situation is still far from being clearly understood. To mention as a matter of historical interest, van't Hoff (1887) suggested (*Z. physic. Chem.*, 1887, **1**, 481; cited in Agar, 1963) that the distribution of solute in such a non-isothermal solution will be such that the osmotic pressure will be uniform throughout. Arrhenius (1894-1898) (*Z. physic. Chem.*, 1898, **26**, 187; cited in Agar, 1963) disproved this hypothesis by experimentally demonstrating that not only the amount, but also the direction of such migrations were highly specific for different electrolytes. Soret effect has since been detected in liquid alloys, mixtures of molten salts, as well as in solutions of non-electrolytes and macromolecules. Furthermore, the discovery of thermoelectric EMF in metallic circuits led to the idea that similar effects might be observed when one member of thermocouple is an electrolytic conductor. Such a thermogalvanic cell covers the **Electrolytic thermocouple** and the **Thermocell** (Agar, 1963).

Thermoelectric EMFs in metals are associated with an inverse effect, namely the Peltier effect – which may be described as the reversible evolution or absorption of heat when current flows across a junction between two dissimilar metals under isothermal conditions. The relation between the two effects was first given by Lord Kelvin as,

$$\frac{dE}{dT} = \frac{\pi_P}{T}$$

(7.49)

Where E is the EMF of a thermocell with junctions at different temperatures, and π_P is the Peltier heat. A similar relation holds in case of thermocells, and a number of workers have attempted to measure the electrolytic Peltier heats by calorimetric techniques.

The most general theory dealing with such non-isothermal transport processes has been the non-equilibrium (irreversible) thermodynamic approach, based on the *Onsager's Reciprocity Relations* (ORR) (Onsager, 1931; Agar, 1963). Such a non-isothermal system, not being in thermodynamic equilibrium, cannot be obviously treated by the classical thermodynamics (referred to as "thermostatics" by de Groot, 1966). Before, however, Onsager's relations were proposed, Eastman (Eastman, E. D. (1926) *J. Am. Chem. Soc.*, **48**, 1482; Eastman, E. D. (1928) *J. Am. Chem. Soc.*, , **50**,

292; cited in Agar, 1963 and Agar *et al.*, 1989) treated the analogous effect of themo-electric potentials in thermocells by a quasi-reversible thermodynamic theory. He introduced the concept of entropy of transfer, and by implication, that of heat of transfer or transport. The latter is defined to be the amount of heat absorbed per mole of a diffusing substance from the region it leaves, and released in the region it moves into under isothermal conditions and with migration of no other component in the system. The physical significance of this quantity can best be explained with the aid of a hypothetical experiment (Agar, 1963; Sanyal and Adhikari, 1979).

Let us consider a column of solution of uniform composition and temperature, enclosed in a cylinder, and divided into two halves L and R by an imaginary reference plane 00_1 (Figure 7.2). Let us now transport one mole of a particular component 'i' only from L to R across 00_1 without affecting the temperature and the pressure at any point. The increase in entropy of R by such migration should then be the partial molal entropy S_i of 'i' and there should be an equal decrease in L. Under reversible conditions, the *net* transfer of heat to the surroundings from the whole system should be zero. However, it does *not* follow that no heat is given out, or taken in, by L and R separately; thus, for instance, if the plane 00_1 is imagined to be a membrane in which the component 'i' is partially soluble, the latter may be supposed to be passing through 00_1 by the process of solution, followed by diffusion. The dissolution of 'i' at one surface of 00_1 will then be accompanied by the absorption or evolution of heat (the heat of solution), and there will be a corresponding reverse effect at the opposite surface of 00_1.

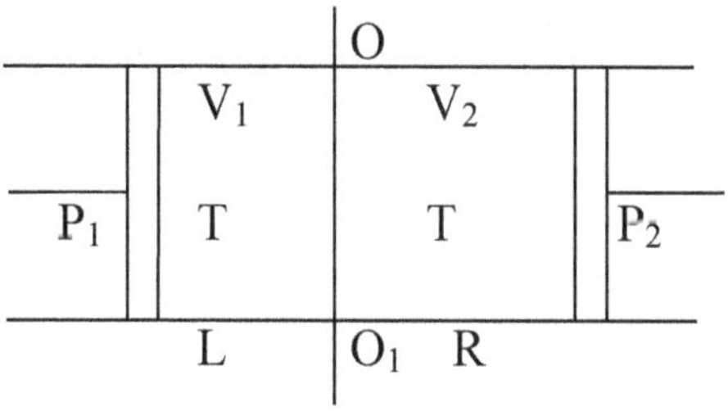

Figure 7.2: Schematic Diagram Showing the Significance of the Heat of Transport.
***Source*: Agar (1963).**

It is, therefore, possible that migration of one mole of 'i' into R increases the entropy of R by S_i^{**} which is thus the entropy transported with one mole of 'i' under isothermal conditions. Let this be called the "transported entropy (S_i^{**})" of 'i'. For the entropy increment in R to be S_i, which it must be from the definition of partial molal quantities, an amount of heat, $Q_i^* = T(S_i^{**} - S_i)$ must be transferred from R to a heat reservoir, and an equal amount of heat be added to L from another heat reservoir at the same temperature (T), in order to maintain the constancy of the temperature. This amount of heat, Q_i^*, is, from definition, the heat of transport of 'i'.

The entropy of transport, S_i^*, also called the Eastman entropy, is then given as,

$$S_i^* = \frac{Q_i^*}{T} = S_i^{**} - S_i$$

$$(7.50)$$

Sanyal and Adhikari (1979) extended these arguments, leading to a physical insight into Q_i^*, to a non-isothermal system with L and R of Figure 7.2 maintained at different temperatures. In the latter case, heat will be transferred between heat reservoirs at different temperatures, and this is significant thermodynamically.

If such transport processes elaborated above (in isothermal system), were carried out in an adiabatic system, i.e., if the solution column (Figure 7.2) were enclosed by the adiabatic walls, the temperature of one section (R) would have risen, while that of the other (L) fallen. Such setting up of a thermal gradient by diffusion of matter under isothermal condition would have been exactly the reverse situation of what is encountered in thermal diffusion. This effect, called the Dufour effect, is observed when gases diffuse into one another isothermally; in liquids, the thermal diffusivity being much higher than the isothermal diffusion coefficients, as compared to the corresponding situation in gases, no such effect has been observed so far (although in similar condensed systems, heats of transport do make measurable contributions to the electrolytic Peltier heats).

One important point to note in this connection is that a moving solute does *not* necessarily carry all its partial molal entropy (S_i) with it, since some of it resides in the surrounding region, and is left behind when the central solute moves away. In other words, the entropy transported by the solute, which has been termed its transported entropy (S_i^{**}), is not equal to its partial molal entropy (S_i), and as a result, the entropy of transport (S_i^*), which is given by the difference, $(S_i^{**} - S_i)$, is not zero. Thermal diffusion experiments, which determine Q^* or S^* through the measurement of the Soret coefficient (*see* later), therefore, provide a sensitive index of solute–solvent interactions. In terms of a Frank and Wen type of A, B and C-zone model (Sanyal, 1984) for the state of solvent (water) surrounding an ionic solute, one would expect the strongly bound A-zone (*e.g.* the electrostricted hydration zone) to migrate along with the moving solute, and it is the thermal effects associated with the changing orientations of solvent molecules in the outer hydration zones, namely B and C zones, which are left behind, that determine Q^* and S^* of the solute. Thus, the so-called structure making solutes, which lower the "local" entropy-density of the solvent, but do *not* drag along the entire ordered region with it, are expected to have high Q^*. The reverse is likely for the structure–breakers.

The various relations (*viz.* Eqs. 7.49, 7.50), as being discussed here on a qualitative basis, can be derived by the rigorous methods of irreversible thermodynamics, based on ORR (Agar, 1963; Sanyal and Mukherjee, 1988). Thus, the molar heat of transport of an electrolyte (Q^*) is related to its Soret coefficient (σ') by the relation,

$$Q^* = v_1 Q_1^* + v_2 Q_2^* = v\, BRT^2 \sigma' \qquad (7.51)$$

Where v_1, and v_2 are the number of cations and anions obtained on complete ionizations of one mole of the electrolyte $(v = v_1 + v_2)$, while Q_1^* and Q_2^* are the corresponding ionic heats of transport; B is a thermodynamic factor, given by

$$B = \left(1 + \frac{\delta \ln \gamma_\pm}{\delta \ln m}\right)_T$$

T being the mean (absolute) temperature and $\gamma_{\pm}$, the mean ionic activity coefficient of the electrolyte.

The temperature-dependence of Q^* is expressed in terms of what is defined as the heat capacity of transport, C^*, given by,

$$C^* = T\left(\frac{\delta S^*}{\delta T}\right)_P = T\left(\frac{\delta (Q^*/T)}{\delta T}\right)_P$$

Or, $C^* = (\delta Q^*/\delta T)_P - Q^*/T$ \hfill (7.52)

These concepts have been extended to soil system in fairly recent times (Sanyal and Adhikari, 1984; Ghildyal and Tripathi, 1987; Saha and Sanyal, 2003). The significance of heat of transport of soil water (Q_w^*) to characterize the moisture retention properties of soil has been discussed, while the prediction from the irreversible thermodynamic theories (as discussed above) for correlating the heat of transport values with other soil moisture characteristics have been verified (Sanyal and Adhikari, 1984) by using the findings reported in the literature (Cary and Taylor, 1962; Tripathi and Ghildyal, 1978), with encouraging trends. Indeed it was shown that the given Q_w^* values may function as additional soil-moisture retention characteristic, along with the already existing (largely empirical) constants such as field capacity, permanent wilting point, hygroscopic coefficient, etc., being at the same time more amenable to theoretical analysis and understanding than what has been so far possible with the existing soil-moisture constants (Sanyal and Adhikari, 1984).

Furthermore, the thermal diffusion studies of aqueous solutions of electrolytes (*e.g.* KCl, $CuSO_4$, $ZnSO_4$) were conducted in selected Entisols and Inceptisols of West Bengal and Sikkim (Bhattacharya and Sanyal, 1987; Saha and Sanyal, 2003). The heat of transport (Q^*) values were determined at several mean temperatures, and all were found to be positive, exhibiting increase with rise in temperatures. The Q^* values have been interpreted in terms of changes in the 'local' degree of ordering of water molecules in the hydration hull of the dissolved ions in the soil solution as well as that in immediate contact with the reactive soil components *vis-à-vis* that in the bulk phase (*i.e.*, bulk soil solution). Attempts have been made to correlate the heat of transport and heat capacity of transport (C^*) values of the above mentioned solutes to the characteristic ion-solvent (water) and soil-water interactions in the given soils with encouraging results.

7.7. Isothermal Coupled Transport of Aqueous Solutes in Soils and Clays

During seepage in soil, the retention–release of aqueous electrolytes by the soil matrix was inferred by treating the process in terms of the thermodynamic theories of isothermal coupled transport of aqueous electrolytes. The parameters used for seepage studies by several workers covered the reflection coefficient (σ), mechanical filtration capacity (L_p) of soil, solute permeability (ω) and related parameters. Thus, Malusis *et al.* (2003) used the σ term for studying the flow and transport of electrolytes through a clay membrane that was used as a semi-permeable barrier.

According to those workers, σ indicates the degree of semi-permeability that a membrane exhibits for a particular solute. It also provides a measure of the extent of 'demixing' of solute and solvent in the course of passage of a solution through the membrane, and is thus expected to provide important information pertaining to the interaction of the solute and solvent in the membrane phase (Majumdar and Sanyal 1989). Indeed, the extent to which a clayey soil would act as a semi-permeable barrier has been traditionally quantified in terms of a reflection (σ) or osmotic efficiency coefficient (r) (Katchalsky and Curran, 1965; Kemper and Rollins, 1966; Olsen *et al.*, 1990; Nanda and Sanyal, 1995). The equation relating the volume flux (J_v, cm. sec^{-1}) across the unit area of a suitably chosen reference frame in the experimental soil per unit time under a hydrostatic pressure gradient [$\nabla P = (\Delta P/\Delta x)$] at a constant osmotic gradient [$\nabla\pi = (\Delta\pi/\Delta x)$] across the given soil volume element is derived from the irreversible thermodynamic theories of coupled transport processes under isothermal conditions (Katchalsky and Curran, 1965), and is given below.

$$J_v = L_p (\nabla P - \sigma \nabla\pi) \tag{7.53}$$

On plotting the experimental J_v values under varying ∇P values at a constant $\nabla\pi$, against the corresponding ∇P values, a linear plot is expected from Eq. 7.53, the slope and the intercept of which lead to the mechanical filtration capacity (L_p, cm.sec^{-1}) of the soil volume and the corresponding reflection coefficient (σ). It may be noted here that the value of σ (a dimensionless quantity) ranges between 0 and 1. The higher the value of σ, the more is the solute accumulation in a given soil.

The solute permeability (ω), characterizing a given soil, is obtained from Eq. 7.54 (Katchalsky and Curran, 1965). Thus,

$$\omega = [J_s/\nabla\pi]_{J_v = 0} \tag{7.54}$$

Where J_s is the solute flux, defined as the moles of the given solute passing per unit time through unit area of the given reference frame in the soil (mol. cm^{-2}. sec^{-1}), and at $J_v = 0$.

The data for σ, L_p, ω and the solute distribution coefficient (K_s) between the soil solid phase and the equilibrium soil solution phase (determined by independent experiments using the given soils), as well as the volume fraction of water in soil (ϕ_w, a dimensionless term, determined by independent experiments) were further used to assess the efficiency of the given soil for solute *rejection* during the passage of an electrolyte solution through the soil. Indeed these parameters and the partial molal volume of water (V_w, cm^3. mol^{-1}) were used to compute the hydrodynamic frictional coefficients between soil, water and solute during the passage of an electrolyte solution (including the native soil solution, especially in salt-affected soils) through a soil. The following criteria, developed by Spiegler and Kedem (1966), involving the various hydrodynamic frictional coefficients, were used in this context of ascertaining the extent of efficient solute rejection by the soil, namely: $f_{sm} \gg f_{wm}$ and $f_{sm} \gg f_{sw}$ where f represents the friction between the species indicated by the subscripts (s, solute; w, water; m, soil).

The above noted formulations have also been used to infer the relative solute retention (accumulation) capacity of soils. The latter was noted to be governed primarily by, among others, the clay content and the dominant clay mineral, organic matter, specific surface area, and amorphous iron and aluminium content in soil. The salt accumulation in soil profile has also been discussed from these studies in terms of the decoupling of solute and solvent during the osmotic flows, as characterized by the corresponding hydrodynamic frictional coefficients introduced above, and related to the soil properties, especially the content and the nature of the clay fraction present (Nanda and Sanyal, 1995). It is thus evident that such treatments, based on irreversible thermodynamics, provides a potential approach to characterize the salt accumulation dynamics in salt-affected soils, and the reclamation of the latter in terms of the relevant soil properties and the underlying processes. Furthermore, such treatment has also been recently used to characterize the arsenic dynamics in several arsenic contaminated soils in the country (Das *et al.*, 2014).

References

Agar, J. N. (1963). Thermogalvanic cells. **In:** *Advances in Electrochemistry and Electrochemical Engineering* (P. Delahey and C. W. Tobias, Eds.), Interscience, New York, Vol. **3**, pp. 31-121.

Agar, J. N. (1960). The rate of attainment of Soret equilibrium. *Trans. Faraday Soc.,* **56**, 776-787. DOI: 10.1039/TF9605600776

Agar, J. N., Mou, C. Y. and Lin, Jeong Long (1989). Single-ion heat of transport in electrolyte solutions: a hydrodynamic theory. *J. Phys. Chem.,* **93**, 2079–2082.

Bhattacharya, A. and Sanyal, S. K. (1987). Heat of transport of an aqueous chloride in an acid soil. *J. Indian Soc. Soil Sci.,* **35**, 177-180.

Cary, J. W. (1963). Onsager's relations and the non-isothermal diffusion of water vapor. *J. Phys. Chem.,* **67**, 126–129.

Cary, J. W. and Taylor, S. A. (1962). Thermally driven liquid and vapor phase transfer of water and energy in soil. *Soil Sci. Soc. Am. Proc.,* **26**, 417–420.

Cassel, D. K., Nielsen, D. R. and Biggar, J. W. (1969). Soil-water movement in response to imposed temperature gradients. *Soil Sci. Soc. Am. Proc.,* **33**, 493-500.

Das, Indranil, Ghosh, Koushik, Das, D. K. and Sanyal, S. K. (2014). Transport of arsenic in some affected soils of Indian subtropics. *Soil Res.,* CSIRO Pub., **52**, 822–832.

de Groot, S. R. (1966). *The Thermodynamics of Irreversible Processes.* North Holland Publishers, Amsterdam, The Netherlands.

Ghildyal, B. P. and Tripathi, R. P. (1987). *Soil Physics,* Wiley Eastern Limited, New Delhi, pp. 614-621.

Gosting, L J. (1956). Measurement and interpretation of diffusion coefficients of proteins. *Adv. Protein Chem.,* **11**, 429-554.

Katchalsky, A., Curran, P. F. (1965). *Non-equilibrium Thermodynamics in Biophysics.* Harvard University Press, Cambridge, MA, USA.

Kemper, W. D. and Rollins, J. B. (1966). Osmotic efficiency coefficients across compacted clays. *Soil Sci. Soc. Am. Proc.*, **30**, 529–534. DOI:10.2136/sssaj1966.03615995003000050005x

Lindsay, W. L. and Norvell, W. A. (1969). Equilibrium relationships of Zn^{2+}, Fe^{3+}, Ca^{2+}, and H^+ with EDTA and DTPA in soils. *Soil Sci. Soc. Am. Proc.*, **33**, 62-68.

Majumdar, K., Sanyal, S. K. (1989). Reflection coefficient for transport of aqueous solutions across bentonite and kaolinite clay membranes in hydrogen form. *J. Indian Soc. Soil Sci.*, **37**, 645–649.

Malusis, M. A., Shackelford, C. D. and Olsen, H. W. (2003). Flow and transport through clay membrane barriers. *Engng. Geol.*, **70**, 235–248. DOI: 10.1016/S0013-7952(03)00092-9

Miller, D. G. (1960). Thermodynamics of Irreversible Processes. The experimental verification of the Onsager Reciprocal Relation. *Chem Rev.*, **60**, 15-37.

Nanda, M. and Sanyal, S. K. (1995). Isothermal coupled transport processes in soils and clays. *J. Indian Soc. Soil Sci.*, **43**, 160-166.

Nye, P. H. and Tinker, P. B. (1977). Solute Movement in the Soil Root System. Blackwell Science Ltd., Oxford, United Kingdom.

Olsen, H. W., Yearsley, E. N. and Nelson, K. R. (1990). *Chemico-osmosis versus diffusion–osmosis*. Transportation Research Record, vol. 1288. pp. 15–22. (Transportation Research Board: Washington, DC).

Olsen, S.R. and Kemper, W.D. (1968). Movement of nutrients to plant roots. *Adv. Agron.*, **20**, 91-151.

Onsager, L. (1931). Reciprocal relations in irreversible processes. *Phys. Rev.*, **37**, 405 (Published February 15, 1931).

Porter, L.K., Kemper, W.D., Jackson, R.D. and Stewart, B. C. (1960). Chloride diffusion in soils as influenced by moisture content. *Soil Sci. Soc. Am. Proc.*, **24**, 460-463.

Saha, P. B. and Sanyal, S. K. (2003). Thermal diffusion of water and aqueous salt solutions in soils. *J. Indian Soc. Soil Sci.*, **51**, 1-5.

Sanyal, S. K. (1980). Validity of the simple form of Flick's First Law in self and tracer diffusion. *J. Indian Soc. Soil Sci.*, **28**, 417-422.

Sanyal, S. K. (1984). Structure of water in solution of organics-Hydrophobic hydration. *Chem. Edn.* (UGC), **1** (No.2), 14-18.

Sanyal, S. K. and Adhikari, M. (1979). Diffusion of water in soil under a temperature gradient. *J. Indian Chem. Soc.*, **56**, 1071-80.

Sanyal, S. K. and Adhikari, M. (1984). Heat of transport of soil water. *J. Indian Soc. Soil Sci.*, **32**, 9-13.

Sanyal, S. K. and Mukherjee, A. K. (1988). Heat of transport and heat capacity of transport of some aqueous electrolytes. *Can. J. Chem.*, **66**, 435-438.

Spiegler, K. S. and Kedem, O. (1966). Thermodynamics of hyperfiltration (reverse osmosis) criteria for efficient membrane. *Desalinization*, 1, 311–326. DOI:10.1016/S0011-9164(00)80018-1

Stevenson, F. J. (1994). *Humus Chemistry: Genesis, Composition, Reactions*. Second Edition, John Wiley and Sons, New York.

Taylor, S. A. and Cary, J. W. (1964). Linear equations for the simultaneous flow of matter and energy in a continuous soil system. *Soil Sci. Soc. Am. Proc.*, **28**, 167-172.

Tripathi, R. P. and Ghildyal, B. P. (1978). Transport and evaporation of water from soil columns under conditions of high water-table. *J. Indian Soil Sci.*, **26**, 313-319.

Van Schaik, J. C. and Kemper, W. D. (1966). Chloride diffusion in clay-water systems, *Soil Sci. Soc. Am. Proc.*, **30**, 22-25.

Environmental Soil Chemistry: Soil Pollution–Environmental Impact and Amelioration

8.1. Soil Pollution: Types of Soil Pollution

Soil pollution is defined as the build-up of pollutants in soils like persistent toxic compounds, chemicals, salts, radioactive materials, or disease causing agents, etc., with adverse effects on plant growth and human/animal health. The pollutants are released into the environment, including soil, from both natural (geogenic) and anthropogenic sources. The nature of the parent material is often responsible for different levels of geogenic heavy metal accumulation in soils. Indeed the natural contamination of soil and groundwater with arsenic, selenium and fluoride in many parts of the world, particularly in the Australasia– Pacific region, is a cause of great concern to human health. Further, the intense industrial activities during the last century led to serious soil and environmental pollution, resulting in a large number and variety of contaminated sites, thereby posing threat to the local eco-systems. Such anthropogenic sources of trace metals include metal smelting industries, industrial effluents, sewage-sludge, municipal solid wastes, burning of fossil fuels, use of leaded gasoline, spraying of arsenic pesticides, etc. (Rattan *et al.*, 2009; Sanyal *et al.*, 2015a).

8.1a. Geogenic Sources of Pollution

8.1a.1. Arsenic

8.1a.1.1 Arsenic Toxicity in Soil-Plant-Animal-Human System

Arsenic (As), a toxic trace element, is of great environmental concern due to its presence in soil, water, plant, animal and human continuum. Its high toxicity and increased appearance in the biosphere has triggered public and political concern. Out of the 20 countries (covering Argentina, Chile, Finland, Hungary, Mexico, Nepal, Taiwan, Bangladesh, India and others) in different parts of the world, where groundwater arsenic contamination and human suffering therefrom have been reported so far, the magnitude is considered to be the highest in Bangladesh, followed by West Bengal, India (Sanyal *et al.*, 2015a). The scale of the problem is grave and unprecedented, exposing millions of people in the Bengal delta basin to risk. The widespread arsenic contamination in groundwater in different parts of West Bengal, located primarily in five districts adjoining the river Bhagirathi, as well as the contiguous districts in Bangladesh, is of great concern. Even beyond the Bengal delta basin, the widespread arsenic contamination in groundwater above the permissible limit (50 µg As. L^{-1}; WHO, 2001; *see* below) has also been detected in several places in the country (Table 8.1, Figure 8.1), for instance at Chandigarh (1976), Nepal (2001), Bihar (2002), Uttar Pradesh (2003), Jharkhand (2003-2004) (Sanyal and Dhillon, 2005), Chhattisgarh and Punjab (2006-2007).

8.1a.1.2. Guideline Value of Maximum Arsenic Concentration

As mentioned briefly earlier, the World Health Organization (WHO)-recommended provisional guideline value of *total* arsenic (As) concentration in drinking water is 10 µg As. L^{-1} since 1993 (WHO, 1993), mainly because lower levels preferred for protection of human health are not reliably measurable on a large scale. However, the National Standard for maximum acceptable concentration (MAC) of arsenic in drinking water is 50 µg As. L^{-1} in several countries including India and Bangladesh, based on an earlier WHO (1971) advice. The proposed new standard value of 5 µg As. L^{-1} is under consideration (WHO, 2001). This is due mainly to the fact that inorganic arsenic compounds are classified in Group 1 (carcinogenic to humans) on the basis of adequate evidence for carcinogenicity in humans and limited evidence for carcinogenicity in animals (IARC, 1987). Adequate data on the carcinogenicity of organic arsenic have not been generated. The joint FAO/WHO Expert Committee on Food Additives (JECFA) set a provisional maximum tolerable daily intake (PMTDI) of inorganic arsenic by humans as 2.1 µg As. kg body weight^{-1}. day^{-1} in 1983 and confirmed a provisional tolerable weekly intake (PTWI) as 15 µg As. kg body weight^{-1}.day^{-1} in 1988 (FAO/WHO, 1989). Such guideline values for soil, plant and animal systems are not available.

Two major hypotheses, both of geogenic origin, have been proposed to account for such widespread arsenic contamination in the groundwater in parts of West Bengal and Bangladesh, confined within the delta bound by the rivers Bhagirathi and Ganga-Padma. Of these two hypotheses, namely the arsenopyrite oxidation hypothesis and the ferric oxyhydroxide reduction hypothesis, the latter is more

Table 8.1: Groundwater Arsenic Contamination in the Indian Subcontinent

State	Coverage	Level of Contamination in Groundwater (μg As. L^{-1})	Citation
West Bengal	12 Districts (Malda, Murshidabad, Nadia, North 24-Parganas, South 24-Parganas, Kolkata, Howrah, Hooghly, Bardhaman, North Dinajpur, South Dinajpur, Coochbehar), 111 blocks	50-3700	http://www.soesju.org./arsenic/wb.htm
Assam	18 Districts	>50	Singh, A. K. (2007). *Curr. Sci.* 92 (11):1506-1515.
	5 Districts (Barpeta, Dhemaji, Dhubari, Darrang and Golaghat)	100-200	
	4 Districts (Jorhat, Lakhimpur, Nalbari and Nagaon), 72 blocks	228-657	
Bihar	12 Districts (Bhagalpur, Khagaria, Munger, Begusarai, Lakhisarai, Samastipur, Patna, Baishali, Saran, Bhojpur, Buxar and Katihar), 32 blocks	> 50	Acharya, S. K. and Shah, B. A. (2004). *Environ. Health. Pers.* 112 (1): 19-20.
Jharkhand	1 District (Sahibgunj)	> 50	http://www.soesju.org./arsenic/jharkhand.htm
Uttar Pradesh	21 Districts (Ballia, Lakhimpur, Kheri, Baharaich, Chandauli, Gazipur, Gorakhpur, Basti, Siddharthnagar, Balarampur, Sant Kabir Nagar, Unnao, Bareilly, Moradabad, Rae Bareli, Mirzapur, Bijnore, Meerut, Sant Ravidas Nagar, Shahjahanpur and Gonda)	> 50	http://www.nerve.in/news:253500133730
Madhya Pradesh	1 District (Rajnandgaon)	52-88	Press Trust of India, September 4, 1999.
Manipur	1 District (Thoubal)	798-986	Singh, A. K. (2007). *Curr. Sci.* 92 (11): 1506-1515.
Tripura	3 Districts (North Tripura, Dhalai and West Tripura)	65-444	Singh, A. K. (2007). *Curr. Sci.* 92 (11): 1506-1515.
Nagaland	2 Districts (Mokokchung and Mon)	> 50	Singh, A. K. (2007). *Curr. Sci.* 92 (11): 1506-1515.

Source: Sanyal et al. (2015a); Sanyal et al. (2012).

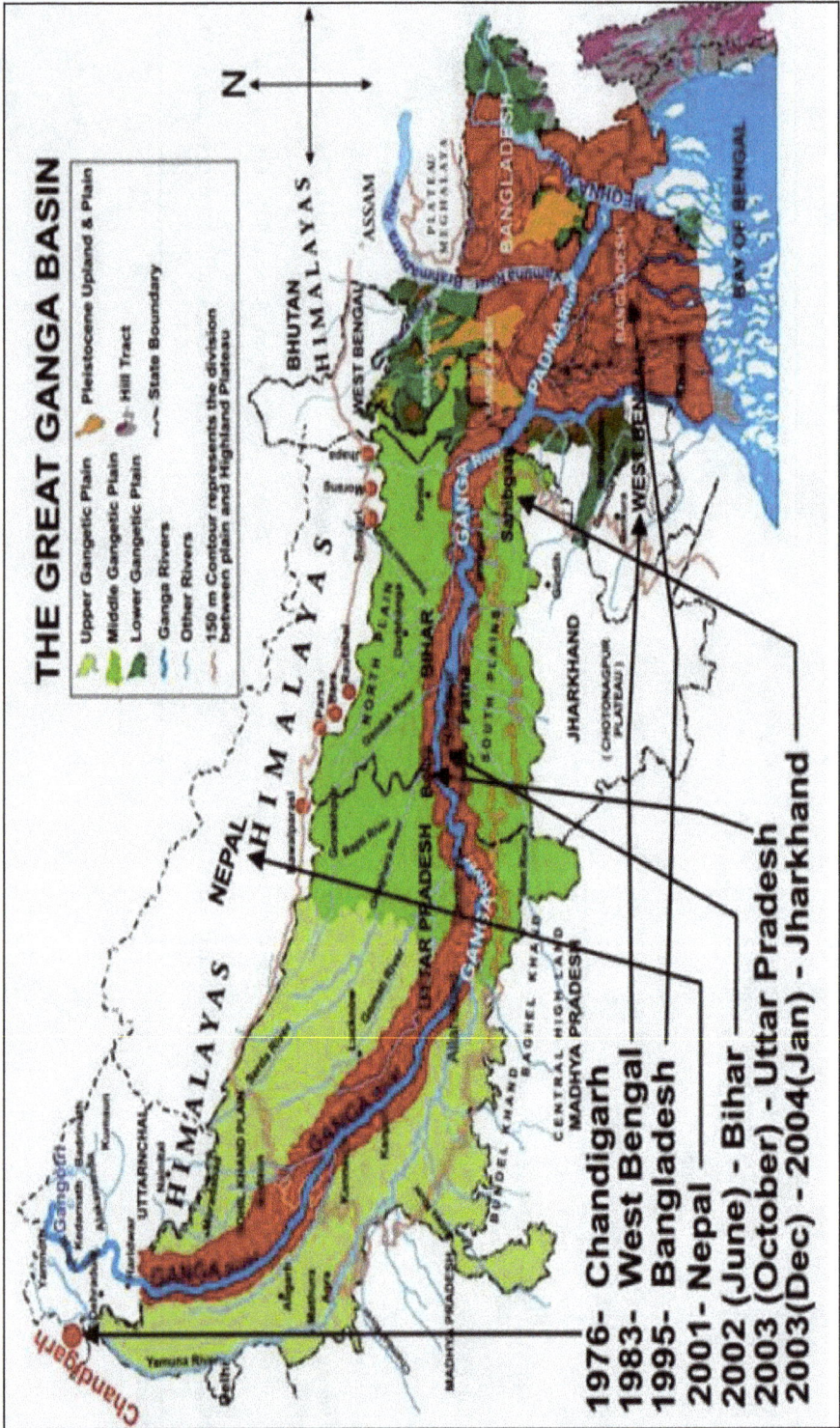

Figure 8.1: Groundwater Arsenic Contamination in the Indian Subcontinent. *Source:* Sanyal (2005).

consistent with the experimental observations reported for the aquifer sediments and the groundwater of the Bengal delta basin (Sanyal *et al.*, 2012; Das *et al.*, 2014). According to this hypothesis, an anoxic condition of the aquifer causes arsenic mobilization from the arsenic-bearing sediments into the groundwater aquifer. The maintenance of such anoxic condition is proposed to be facilitated by the widespread practice of wetland paddy cultivation in the affected belt.

Table 8.2: Arsenic Concentration in Rocks and some other Materials

Sl.No.	Types of Rocks/Materials	Arsenic Content (mg As. kg^{-1})
1.	**Rocks**	
	Igneous rocks	
	Ultrabasic:	
	Peridotite, Dunite, Serpentine	0.3-15.8
	Basic:	
	Basalt (extrusive)	0.18-113
	Gabbro (intrusive)	0.06-28
	Intermediate:	
	Latite, Andesite, Trachyte (extrusive)	0.5-5.8
	Diorite, Granodiorite, Syenite (intrusive)	0.09-13.4
	Acidic:	
	Rhyolite (extrusive)	3.2-5.4
	Granite (intrusive)	0.18-15
	Metamorphic rocks	
	Quartzite	2.2-7.6
	Slate/Phyllite	0.5-143
	Schist/Gneiss	0.0-185
	Sedimentary rocks	
	Marine:	
	Shale/Claystone(near-shore)	4.0-25
	Shale/Claystone(off-shore)	3.0-490
	Carbonates	0.1-20.1
	Phosphorites	0.4-188
	Sandstone	0.6-9.0
	Nomarine:	
	Shales	3.0-12
	Claystone	3.0-10
2.	**Coal**	Up to 2000
3.	**Crustal Average**	2

Source: Sanyal (2005).

8.1a.1.3. Natural Abundance

Dissolved arsenic concentrations in natural waters (except groundwater) are generally low, except in areas characterized by geothermal water and/or mining activities. The sedimentary rocks generally have higher arsenic content (Table 8.2) than do igneous and metamorphic rocks, while suspended and bottom sediments in most aquatic systems contain more arsenic (Table 8.2) than most natural waters do (Table 8.3). The capacity to retain arsenic is primarily governed by the sediment grain-size and the presence of surface coating composed of clays, clay-sized iron and manganese oxides and organic matter. Arsenic held by solid phases within the sediments, especially iron oxides, organic matter and sulphides may constitute the primary arsenic sources in groundwater under conditions conducive to arsenic release from these solid phases. These include abiotic reactions (oxidation/reduction, ion exchange, chemical transformations) and biotic reactions (microbial methylation) (Sanyal et al., 2012).

Table 8.3: Arsenic Concentrations in Water other than Groundwater

Source	Arsenic Concentration (μg As. L^{-1})
Rainwater and snow	< 0.002-0.59
Rivers	0.20-264
Lakes	0.38-1.00
Sea water	0.15-6.00
Ponds (West Bengal, India)	4-70
Canals (West Bengal, India)	40-150

Source: Sanyal et al. (2012).

8.1a.1.4. Arsenic Contamination in Groundwater in the Bengal Delta Basin

The groundwater arsenic (As) concentration (50 – 1600 μg As. L^{-1}), reported from the affected areas of West Bengal, are several orders of magnitude higher than the stipulated Indian standard for the permissible limit in drinking water (50 μg As. L^{-1}, which is also the maximum acceptable concentration, MAC, for drinking water in Bangladesh, India and several other countries), as well as the WHO guideline value (10 μg As. L^{-1}). Further, the arsenic concentration in alluvial aquifers of Punjab varied from 4 to 688 μg. L^{-1} (Sanyal et al., 2012). In West Bengal, the presence of arsenic in groundwater in concentrations exceeding maximum acceptable concentration was first detected in 1978, while the first case of arsenic poisoning in humans was diagnosed at the School of Tropical Medicine in Calcutta in 1983 (Acharya, 1997). The effect of ingestion of inorganic arsenic in drinking water and the associated health effects in adults has also been well established (Guha Mazumdar et al., 1998). The main focus of attention, until recently, has been exclusively on arsenic contamination in groundwater-derived drinking water. However, since groundwater is also used extensively (more than 90 per cent) for crop irrigation in the arsenic belt of West Bengal, the possibility of a build-up of arsenic concentration in agricultural soils and agronomic produce was anticipated. Indeed, arsenic uptake by crop plants grown in soils contaminated with high concentration of arsenic, and irrigated with such arsenic contaminated groundwater has been reported by several workers (reviewed by Sanyal *et al.*, 2015a). Such findings call for an immediate attention since what remains essentially a *point* and fixed source of arsenic contamination as for the drinking water (*e.g.* a tube well discharging contaminated water), may well become a *diffuse* and uncertain source of contamination when arsenic finds its way into the food-web, accompanied with possible bio-magnification up in the food-chain. This assumes added significance in view of the reported finding of higher (than permissible) level of arsenic in the urine samples (an early biomarker of arsenic poisoning in humans) of some people, having no history of consuming arsenic contaminated drinking water (Dr. D.N. Guha Mazumdar, by private communication; *see* later). Interestingly, the surface water bodies, located in the affected belt, have remained largely free of arsenic. This tends to suggest that the soil, which receives arsenic-contaminated water, acts as an effective sink to contain the toxin, thereby preventing the surface run-off to carry it to the adjoining water systems (Sanyal, 2005).

8.1a.1.5. Health Implications of Arsenic Poisoning

Arsenic is a widely occurring toxic metal in natural ecosystems. As small as 0.1 g of arsenic trioxide can prove lethal to humans (Jarup, 1992). Early symptoms of arsenic poisoning include skin disorders, weakness, languor, anorexia, nausea and vomiting with diarrhoea or constipation. With the progress of poisoning, the symptoms attain more characteristic features, which include acute diarrhoea, edema (especially of the eyelids and ankles), skin pigmentation, arsenical melanosis and hyperkeratosis, enlargement of liver, respiratory diseases and skin cancer. In severe cases, gangrene in the limbs and malignant neoplasm are also observed (Sanyal *et al.*, 2012). The "Bell Ville Disease" (typical arsenic-induced cutaneous manifestations among the people of Bell Ville) in Argentina, "Black Foot Disease" in Taiwan and "Kai Dam" disease in Thailand are well established as health disorders due to arsenic poisoning (Sanyal *et al.*, 2012). As a matter of fact, the hair, nail, skin-scale and urine samples of a large number of people, residing in the affected belt of West Bengal (India) and Bangladesh, have been analyzed by several workers. Many of these samples had arsenic loading more than the corresponding permissible levels.

8.1a.1.6. Chemistry of Arsenic in Groundwater-Soil Environment

Arsenic (As) in groundwater and soil is present as dissolved oxyanions, namely arsenites ($As^{III}O_3^{3-}$; $H_nAs^{III}O_3^{(3-n)-}$, with n = 1, 2) or arsenate ($As^VO_4^{3-}$, $H_nAs^VO_4^{(3-n)-}$, with n = 1, 2), or both, besides the organic forms. The solubility, mobility, bioavailability and hence toxicity of arsenic in soil-crop system primarily depends on its chemical form, mainly the oxidation state of arsenic. Thus, the toxicity of arsenic compounds in groundwater/soil environment depends largely on its oxidation state, and hence on the redox status and pH, as well as whether arsenic is present in organic combinations. The toxicity of arsenic compounds in groundwater/soil environment follows the order:

Arsine [AsH_3; valence state of arsenic (As): 3] > organo-arsine compounds > arsenites (As^{3+} form) and oxides (As^{3+} form) > arsenates (As^{5+} form) > arsonium metals (+1) > native arsenic metal (0).

The arsenites are much more soluble, mobile, and toxic than arsenates in aquatic and soil environments. The organic forms, namely dimethyl arsinic acid (DMA) or cacodylic acid, which on reduction (*e.g.* in anoxic soil conditions) forms di- and trimethyl arsines, are also present in soil. Another organic form present in groundwater and soil is monomethyl arsonic acid (MMA). At pH 6-8, and in an aerobic oxidized environment (redox potential, $E_h = 0.2$-$0.5V$), arsenic acid species and arsenate oxyanions, that is, $H_n As^V O_4^{(3-n)-}$ ions, with n = 1, 2 (pentavalent arsenic forms), occur in considerable proportions in most aquatic systems, whereas under mildly reducing conditions (such as one encounters in flooded paddy soils with $E_h = 0$-$0.1V$), the arsenous acid, $H_3As^{III}O_3$, and arsenite oxyanion species ($As^{III}O_3^{3-}$; $H_nAs^{III}O_3^{(3-n)-}$, with n = 1, 2) (arsenic in trivalent form) are the predominant species. Furthermore, As (III) is more prevalent in soils of neutral pH range (and in most groundwater), as in the soils of the affected belt of West Bengal, India and Bangladesh, than otherwise thought, and hence is of concern. This is primarily because As (III) exists largely as a neutral, uncharged molecule, namely arsenous acid, $H_3As^{III}O_3^0$ (pKa= 9.2), at the pH of the neutral soils and most natural groundwater, as one would expect based on the Henderson's equation (*vide.* Eq. 3.72 in Section 3.9.1), and is thus less amenable to retention by the charged mineral surfaces in soils and sediments.

There have been both direct and indirect evidence to suggest that arsenic (and selenium) is held in soils and sediments by oxides (*e.g.* of Fe, Al, Mn) through the formation of inner-sphere complexes *via* ligand-exchange mechanism. This is illustrated below by the following scheme of reactions.

$$[M–OH] + H_2O \rightleftharpoons [M–OH_2^+] + OH^- \qquad (8.1)$$

$$[M–OH_2^+] + [As^VO_4^{3-}] \rightleftharpoons [M\text{-}OH_2^{+\cdots\cdots}As^VO_4^{3-}] \rightleftharpoons [M–As^VO_4^{2-}] + H_2O \qquad (8.2)$$

However, the non-specific adsorption (through electrostatic mechanism) of arsenic also occurs at pH values below the point of zero charge (PZC) for a given adsorbent (Sanyal *et al.*, 2012). As shown above, the said ligand exchange tends to increase the negative charge of the soil colloidal fraction, for instance, of iron oxides, and thus push the PZC of the arsenic-laden soil to lower pH. Indeed, this was shown to be the case with the concomitant increase in the negative magnitude of the variable-surface charge and the surface potential of the corresponding soil colloidal fraction (Sanyal *et al.*, 2012). However, the non-specific adsorption (through electrostatic mechanism) of arsenic also occurs at pH values below the point of zero charge (PZC) for a given adsorbent (*see* Section 3.9).

It ought to be emphasized that groundwater or soil solution, which is subjected to affluxes and influxes, as well as circulation and also to man-made perturbations of groundwater due to its withdrawal, cannot be expected to remain in thermodynamic equilibrium, it being very much of an open system (thermodynamically speaking). Thus, more often than not, the ratio of concentrations of arsenic species, namely the ratio, $[(As^{III})/(As^V)]$, in field soils does *not* quite agree with the ones computed from the observed redox potential (E_h) and the application of the Nernst's equation (at 25°C) to the equilibrium redox reaction, namely

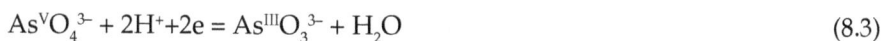

$$As^V O_4^{3-} + 2H^+ + 2e = As^{III} O_3^{3-} + H_2O \qquad (8.3)$$

$$E_h \cong E^0_h - 0.0295 \log [(As^{III}O_3^{3-})]/(As^VO_4^{3-})] - 0.059 \text{ pH} \qquad (8.4)$$

Where the ( ) terms refer to the equilibrium concentrations of the respective ionic species in dilute soil solution, and E^0_h is the standard redox (reduction) potential of the $(As^VO_4^{3-}/As^{III}O_3^{3-})$ redox couple at 25°C. It is evident from Eq. 8.4 that the proportion of As^{III}, and hence soluble arsenic level in soil, should increase substantially with diminishing E_h and increasing pH. Furthermore, at a high pH, the OH⁻ion concentration would increase, causing displacement of As ^{III} and As^V species from their binding sites through competitive ligand exchange reactions. The dependence of arsenic sorption on pH of the sorption medium is governed largely by the nature of the soil colloidal fraction. A fall of arsenate adsorption was noted with increasing pH, but only at lower arsenate concentrations, which got reversed at a higher arsenate equilibrium concentration. This trend was explained in terms of the varying electrostatic potential of the variable-charge soil colloidal surfaces with pH, solubility product principles, and buffering action of the arsenic salt used (Majumder and Sanyal, 2003).

8.1a.1.7. Selected Findings in Soil-Crop System

8.1a.1.7.1. Arsenic Loading in Soils of West Bengal

Some of the research studies, conducted at the selected affected areas, revealed that the total and Olsen-extractable arsenic (*i.e.*, 0.5M NaHCO₃, pH 8.5–extractable arsenic which constitutes the soil arsenic pool amenable to plant uptake) varied from 8.4 to 24.3 mg As. kg⁻¹ and from 2.90 to 15.8 mg As. kg⁻¹, respectively (Sanyal *et al.*, 2015a), in the given affected soils of West Bengal. The soil arsenic contents of these areas were generally higher than those reported for the soils of several other countries like Argentina, China, Italy, Mexico, France, Australia, etc. (Sanyal *et al.*, 2012).

Inorganic soil arsenic fractions from the affected soils were also fractionated into different soil arsenic pools, namely water soluble As (Ws–As), arsenic associated with Al compounds in soil, the so-called aluminium-bound As (Al–As), iron-bound As (Fe–As) and calcium-bound As (Ca–As), by following the sequential extraction methodology. The findings suggested that these inorganic soil arsenic (As) pools fell in the order: Ws–As < Al–As < Ca–As < Fe–As. In particular, the Fe–As fraction contributed 45 per cent to 74.7 per cent towards the total soil arsenic sequential sum (Sanyal *et al.*, 2012).

8.1a.1.7.2. Interaction of Arsenic with Organics in Soil System

As mentioned earlier, soil acts as an effective sink of arsenic present in the contaminated groundwater used for irrigating the crops. The soil organic fractions including humic acid (HA) and fulvic acid (FA) behave as effective accumulators of toxic heavy metals by forming the metal-humate complexes (chelates) with different degrees of stability (Datta *et al.*, 2001; Mukhopadhyay and Sanyal, 2004; Sinha and Bhattacharyya, 2011; Ghosh *et al.*, 2012; Sanyal *et al.*, 2012; Sanyal *et al.*, 2015a). Besides, soil clays, Al oxides, Fe oxides, especially the amorphous Fe and Al oxides in soil also influence the As retention by soils, soil minerals and sediments. The above mentioned organo-arsenic complexes were quite stable, even in the presence of competing oxyanions such as phosphate, sulphate and nitrate (Sanyal *et al.*, 2015a). Further, the moderating influence of the organic fractions from FYM, vermicompost, municipal sludge, mustard cake, and surface soil of West Bengal was assessed in terms of the stability of the corresponding arseno-humic/fulvic complexes, formed in the organic manure-treated contaminated soils (Sinha and Bhattacharyya, 2011; Ghosh *et al.*, 2012). It was found that the organic manures added as soil amendment significantly reduced the accumulation (concentration) of arsenic in sesame seed to the maximum extent of 65.5 per cent (vermicompost), 50 per cent (phosphocompost), 42 per cent (mustard cake) and 40 per cent (FYM), compared with the control counterpart (Sinha *et al.*, 2011). The risk associated with dietary exposure to arsenic-contaminated sesame oil reached a value of 15.6 per cent of provisional tolerable weekly intake for arsenic at the maximum accumulation of arsenic in sesame oil. Thus, improving the soil organic matter stock in the tropical soils of the arsenic-affected belt, relatively poor in native organic matter, by adopting the appropriate management practices (such as recycling of crop residues, incorporation of the appropriate organic manure, etc.) will facilitate arsenic retention in the affected soils.

8.1a.1.7.3. Interaction of Arsenic with Phosphorus and Micronutrients

Phosphorus (P) is one of the essential major plant nutrients for plant growth. Because As and P are both placed in Group Vb of the Periodic Table, the interaction of As and P in soil-plant system is an important issue in respect of arsenic mobilization. Indeed the indications are that these oxyanions would not be adsorbed independently in mixtures, but rather would tend to compete for some portion of the same type of adsorption sites (Sanyal *et al.*, 2012; Das *et al.*, 2014). Several workers showed that the presence of phosphate caused a reduction in arsenate adsorption, and that the reduction was much greater for the competitive effects of arsenate on phosphate adsorption by soil minerals, although a large variation in the degree of competition between these two oxyanions has also been reported (Mukhopadhyay *et al.*, 2002; Sanyal *et al.*, 2015a).

On the other hand, certain micronutrient applications (such as application of zinc/iron salts) to the contaminated soil (where these micronutrients are deficient) help mitigate the arsenic toxicity in soil-crop system. As stated above (*see* Section 8.1a.1.6), arsenic is released to a greater extent in soil solutions upon submergence, which can be countered in presence of graded doses of applied Zn, possibly through the formation of the relatively less soluble zinc arsenate. Arsenic accumulation by rice crops, grown under submerged condition, was also reduced by the application of Zn to soil, while the latter led to the accompanying yield increment (Das *et al.*, 2016). Zinc application in *boro* (summer) rice was especially helpful in reducing the plant accumulation of arsenic and its translocation to the plant biomass, whereas its residual impact influenced positively the arrest of soil arsenic, thereby bringing down its build-up in the succeeding crops. Similar effects were observed when such experiments were repeated with iron in place of zinc (Das *et al.*, 2016).

8.1a.1.7.4. Arsenic in Soil-Plant System

Several workers have reported accumulation and transformation of arsenic by a number of plant species grown in the arsenic affected areas. These crops (such as rice, elephant-foot-yam, green gram, cowpea, sesame, groundnut, etc.) tended to show a build-up of arsenic in substantial quantities in different plant parts. Indeed, pointed gourd, a vegetable creeper plant, has shown considerable arsenic loading when cultivated in the contaminated soils of West Bengal. A number of other vegetables, namely cauliflower, tomato, bitter gourd were also noted to accumulate arsenic in their economic produce. The distribution of arsenic content in plant parts generally followed the order:

root > stem > leaf > economic produce.

As mentioned earlier, reduction of arsenate to more toxic arsenite is facilitated by lowering of the redox potential (E_h) which is encountered under anoxic soil conditions, with arsenite being more soluble and mobile than arsenate. Rice plant is thus rather susceptible to arsenic toxicity since it is grown under submerged soil conditions (low E_h). Further, the processing of rice (*i.e.*, parboiling and milling, etc.) was found to increase the arsenic loading in rice for both the traditional and the high yielding cultivars (Sanyal *et al.*, 2012). The toxicity of arsenic species in plant body is reported to follow the order: Arsine (AsH_3) > As^{3+} > As^{5+} > MMA (Monomethyl arsonic acid) > DMMA (Dimethyl arsinic acid) (NRCC, 1978).

Furthermore, screening of 200 rice genotypes showed a large variation of arsenic accumulation in grain. No significant correlation was found among the pattern of arsenic uptake by root, shoot and grain of the 200 rice lines examined. Initial analysis revealed that arsenic content in grain is controlled by more than one gene (BCKV, Unpublished work). It has also been found that crops like potato, pumpkin and sesame accumulate less arsenic than do others (Sanyal *et al.*, 2012).

On storing the arsenic contaminated groundwater (from a shallow tube well, STW) in a pond, there was a gradual lowering (on standing) of arsenic loading of the stored pond water, while its progressive build-up in the corresponding pond sediment samples. Such decrease of arsenic content in the stored water might have arisen from the sedimentation of arsenic from the water to the pond sediment which obviously increases the arsenic loading in the latter. The dilution of the stored water by rainfall during the wet season (July to September) further decreased the contamination in the water. This opens up the possibility of conjunctive use in agriculture of surface and groundwater during the *lean period* (January to May of no or little rainfall) as a potential remedial option (Sanyal *et al.*, 2012).

8.1a.1.7.5. Hyperaccumulation vis-à-vis Detoxification of Arsenic by Plants/ Microbial Species

The reported hyperaccumulation of arsenic (As) from the contaminated soils by the brake-fern, *Pteris vittata*, and its subsequent translocation into the above-ground biomass suggests that the plant-accumulated arsenic was present almost entirely in the toxic inorganic forms with the proportion of highly toxic As (III) being in fact much greater in the plant body than that of the less toxic As (V) form, as compared to the distribution of these two forms in the contaminated soil in which the fern grows (Ma *et al.*, 2001). Thus, it is worth noting in this context that such accumulation of arsenic does *not* necessarily lead to its detoxification *per se* (unless the plant-accumulated toxin is effectively detoxified, or *else* converted to less toxic forms) within the plant body by its metabolic processes. For this, a systematic search for phytoaccumulating or phytoexcluding plant species is necessary.

A scan of literature reveals a number of plant/microbial species, known for arsenic accumulation/or as bio-indicator, which can effectively remove arsenic (and other heavy metals) from the aquatic system, for instance, to the tune of 170 and 340 µg As. g dry weight^{-1} of water hyacinth in its stem and leaves, respectively (Chigbo *et al.*, 1982), when grown in a pond containing 10 mg As. dm^{-3}. However, such accumulated arsenic in water hyacinth (*Eichornia crassipes*) is also liable to leaching out in the water body, particularly so on decomposition of such aquatic weed. Consequently, appropriate precaution has to be exercised while interpreting the arsenic status of aquatic environment by water hyacinth accumulation.

A number of microbial species (*e.g.* the bacterial species, namely *Proteus* sp., *Escherichia coli, Flavobacterium* sp.; *Corynebacterium* sp. and *Pseudomonas* sp.; the fungus, namely *Candida humicola*; the freshwater algae, namely *Chlorella ovalis, Phaepdactuylum tricornutum, Oscillatoria rubescens*) have been reported to possess varying degrees of arsenic accumulating abilities. Several weed species, normally found along with crops like rice, potato, jute, mustard, etc., growing on arsenic contaminated soils (2-14 mg As. kg soil^{-1}) and subjected to irrigation (given to the desired crops) with arsenic contaminated groundwater, were noted to accumulate considerable amounts of arsenic in their biomass (Das *et al.*, 2005).

8.1a.1.7.6. Bio-remediation of Arsenic in Soil-Plant System

In a study conducted with selected arsenic (As)-contaminated soils of West Bengal, the As-volatilizing indigenous soil bacteria, isolated from these soils, were tested for their ability to turn the toxic indigenous inorganic As to less toxic volatile arsenicals. Approximately 37 per cent of As (III) (under anaerobic condition) and 30 per cent As (V) (under aerobic condition) were volatilized by these bacterial isolates in 3 days. In contrast to the genetically modified organism, the indigenous soil bacteria were capable of removing 16 per cent of arsenic from the contaminated soil during the 60 days-incubation period in presence of FYM (Majumdar *et al.*, 2013a). Further, arsenic-oxidizing bacteria were isolated from the selected arsenic-contaminated soils of West Bengal (Majumdar *et al.*, 2013b), which were closely related to various species of *Bacillus* and *Geobacillus*, based on their 16S rRNA gene sequences. They were found to be hyper-resistant to both As (V) (167–400 mM) and As (III) (16–47 mM). Elevated rates of As (III) oxidation (278–1250 µM. h^{-1}) and arsenite oxidase activity (2.1–12.5 nM. min^{-1} mg^{-1} protein) were observed in these isolates. Furthermore, among the superior arsenic-oxidizers, the AMO-10 completely (100 per cent) oxidized 30 mM of As (III) within 24 h. The presence of the *aoxB* gene was confirmed in the screened isolates. Phylogenetic tree construction, based on the *aoxB* sequence, revealed that the two strains, namely AGO-S5 and AGH-02, were clustered with *Achromobacter* and *Variovorax*, whereas the other two (AMO-10 and

ADP-25) remained un-clustered. The increased rate of the toxic As (III) oxidation to the less toxic As (V) forms by these native strains might be exploited for the remediation of arsenic in contaminated environments (Majumdar *et al.*, 2013b).

8.1a.1.7.7. Arsenic in Soil-Plant System and its Influence on Food-Chain

It was noted that As (III) accounted for the major arsenic species recovered from grains of the transplanted autumn paddy, while As (V) predominates in that from rice straw (Sanyal *et al.*, 2015a). Soil amendment through organic intervention reduced arsenic accumulation in rice grain and straw of autumn rice as manifested through the reduction of inorganic arsenic (Sinha and Bhattacharyya, 2014a). Sinha and Bhattacharyya (2014b) also studied the arsenic toxicity profile in rice, grown in contaminated area of rural Bengal, and the possible risk of its dietary exposure. The unique character of the anaerobic rice ecosystem results in a significant build-up of inorganic arsenic (i-As) in soil and its concomitant accumulation in rice. The recoveries of i-As was dominated by As (III) in rice grain and As (V) in rice straw, thereby emphasizing the presence of higher levels of the more toxic As (III) in the edible portion. Recoveries of organic arsenic species in rice grain and straw further suggested the possibilities of methylation of i-As in the plant system. The risk of dietary exposure to i-As through rice, the staple food in the rural Bengal, poses an almost equal threat to human health as that posed by the contaminated drinking water. Organic amendments and augmented P fertilization showed considerable promise in reducing total and inorganic arsenic accumulation in rice and the consequent dietary risk (Sanyal *et al.*, 2015a).

Few reports are available that characterize daily arsenic (As) exposure through water and diet among the people living in groundwater contaminated regions and correlate the former with arsenic biomarkers. Demographic characteristics and the total daily As intake through water and diet were determined in 167 participants (Group-1 participants, selected from As-endemic region) and 69 participants (Group-2 participants, selected from As-non-endemic region) in a study conducted in West Bengal. The findings showed significantly high dietary arsenic intake in people living in Nadia district of West Bengal, where contaminated groundwater was used for irrigation purpose, but significantly low in the region of Hoogly district, where groundwater was uncontaminated. Even after lowering the As level in drinking water to < 50 µg As. L^{-1} (the permissible limit in India), significant As exposure occurred through water and diet, reflected by the elevated level of As in the arsenic biomarker, namely urine, in people living in the As-endemic region studied. Those with skin lesions were found to have a higher level of arsenic in urine and hair, compared to those without skin lesion (Guha Mazumder *et al.*, 2013, 2014). In yet another study, the dose of daily As intake from both water and diet was found to be significantly and positively associated with urinary As levels in an As-endemic region of West Bengal, even when people were using As-safe water (< 50 µg As. L^{-1}) for drinking and cooking purposes (Figure 8.2). When arsenic levels in drinking water were further reduced to < 10 µg As. L^{-1} (WHO safe limit), the dose from the diet was still found to be significantly associated with urinary arsenic excretion (Figure 8.3). But no significant association was found with arsenic dose from drinking water in this group (Halder *et al.*, 2012; Guha Mazumder *et al.*, 2013;

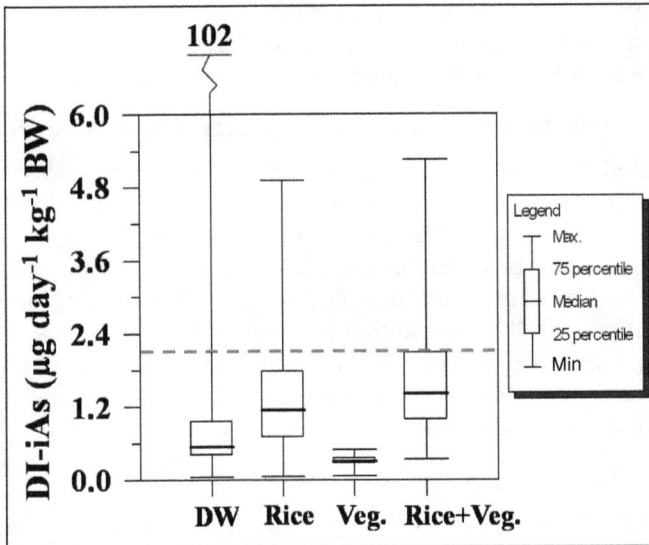

Figure 8.2: Comparison among Daily Intake of Inorganic As (DI-iAs) Due to Consumption of Drinking Water (DW), Rice and Vegetables (Veg.).
The (red) dotted line indicates WHO recommended provisional maximum tolerable daily intake (PMTDI) (PMTDI$_{As}$) value of 2.1µg As. kg body weight^{-1}. day^{-1}.
Source: Halder *et al.* (2012).

Sanyal *et al.*, 2015a). Further, when exposed to arsenic through only diet, urinary arsenic concentration was found to correlate positively with dietary arsenic intake in the participants, showing skin lesions (Figure 8.3), while this correlation was insignificant in participants without skin lesion. Thus Figures 8.2 and 8.3, taken together, amply demonstrate that supply of arsenic-safe drinking water (< 10 µg As. L^{-1}) to the population in rural Bengal *alone is not enough to reduce the risk of arsenic poisoning*, consumption of *rice provides yet another potential pathway of inorganic arsenic (i-As) exposure*, that must also be considered for the effective remedial options. Indeed, any mitigation intervention of chronic arsenic toxicity in rural Bengal needs integrated approaches of attempting to reduce arsenic entry into the food-chain, on one hand, while reduction of arsenic in the drinking water below the safe limits, on the other (Halder *et al.*, 2012; Guha Mazumder *et al.*, 2014; Sanyal *et al.*, 2015a).

8.1a.1.8. Remedial Options at a Glance

☆ Optimum conjunctive use of ground and surface water (*e.g.* harvested rainwater) and recharge of groundwater resource with harvested rainwater, free of arsenic.

☆ Development/identification of suitable low arsenic-accumulating high yielding crops/varieties, and preferring low-water requiring, farmer-attractive cropping sequences (especially for the lean period of January to May), suitable for the arsenic-contaminated areas.

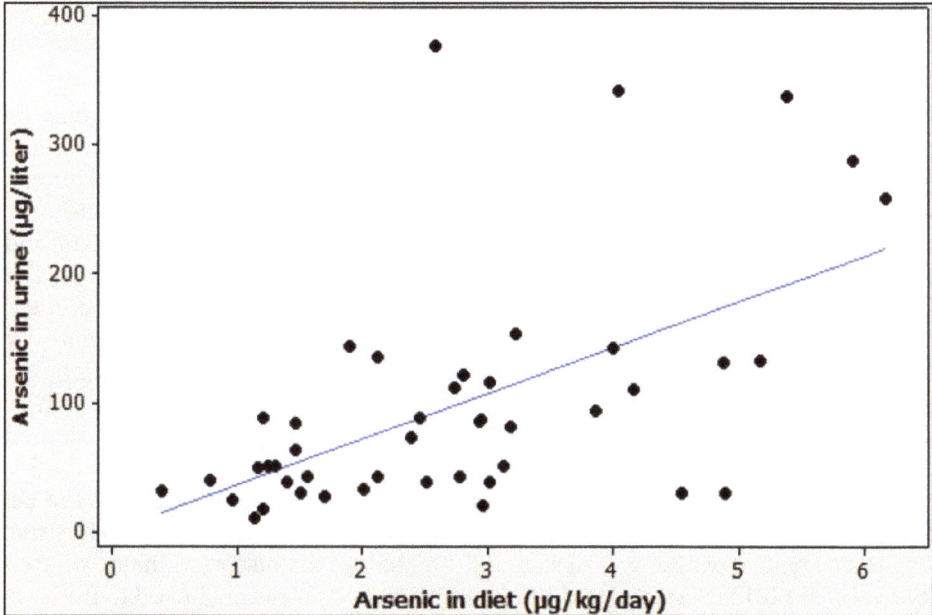

**Figure 8.3: Correlation of Urinary Arsenic Concentration (µg As. L⁻¹)
with Daily Dietary Arsenic Intake (µg As. kg⁻¹. day⁻¹) for Participants,
Drinking Water with Arsenic Level < 0.3 µg As. L⁻¹, having Skin Lesion (*n* = 45),
West Bengal, India (r = 0.573) .**
Source: Guha Mazumdar *et al.* (2014).

☆ Irrigation with pond-stored groundwater in which partial arsenic-decontamination is facilitated by sedimentation-cum-dilution with rain water.

☆ Enhancing the water use efficiency (through optimum water management) for groundwater irrigation, especially for summer (*boro*) paddy.

☆ Increased use of FYM and other manures + green manure crops, as well as application of appropriate inorganic amendments (zinc/iron salts as and wherever applicable).

☆ Identification/development of varieties/crops which accumulate less arsenic in the consumable parts and where ratio of inorganic to organic forms of As is low.

☆ Developing cost-effective phyto- and bio-remediation options.

☆ Creation of general awareness: Mass campaigning, holding of farmers' day, field demonstrations, taking due cognizance of the socioeconomic factors (Sanyal *et al.*, 2012, 2015a).

8.1a.2. Selenium

8.1a.2.1. Selenium Toxicity in Soil-Plant-Animal-Human System

Selenium (Se), a component of the enzyme glutathione peroxidase, may be beneficial or toxic to plants, animals and humans, depending upon its concentration. It is also an inorganic carcinogenic agent at elevated level. The recommended human intake of Se is between 50 and 200 μg Se. day^{-1}(WHO, 1996). The "Kashin-Beck" (chronic arthritis and deformities in the affected joints in children and teenagers) and "Keshan" (cardiomyopathy) are dreaded diseases related to deficiency of Se. The Food and Nutrition Board of the National Academy of Sciences, USA has accepted 5 mg Se. kg diet^{-1} (dry weight basis) as the critical level between toxic and non-toxic animal feeds. Soils containing more than 0.5 mg Se. kg^{-1} produce vegetation (*e.g.* fodder) containing > 5 mg Se. kg^{-1}, and are designated as seleniferous soils (Sanyal and Dhillon, 2005). Selenium toxicity problem is noted in sporadically distributed Se-toxic soils throughout the Great Plains and the Rocky Mountains regions of USA, Prairie regions of Canada, Queensland in Australia, Sangliao, Weihe and Hua Bei plains of China and the Punjab state in India. The Indian states, namely Haryana, sub-Himalayan West Bengal, Assam and Meghalaya are also marginally affected (Sanyal *et al.*, 2015a). Animal and human productivity is closely linked to the level of Se in plants and grains.

8.1a.2.2. Location of Seleniferous Areas in Punjab

Geogenic-driven selenium toxicity has been reported from some pockets of Punjab where as high as 2.41 mg Se. kg^{-1} of soil has been reported. More than 1000 ha of such seleniferous soils exist in the North Eastern Punjab, where toxic Se sites are located at the terminal ends of the seasonal rivulets. The soils are alkaline in reaction, calcareous, silty loam to silty clay loam in texture and are well drained (Dhillon *et al.*, 1992). Parent material of soils in the seleniferous region is derived from the Upper Shiwalik rocks that are mainly composed of polymictic conglomerates of variable composition, containing many unstable materials (granite, basalt, limestone, etc.). These rocks, in turn, are derived from the metamorphic terrain of the Himalayas (Sanyal and Dhillon, 2005). Toxicity of Se causing endemic disease was reported from the Huabi province of China (Neal, 1995), where weathering of coal released Se in the soil. Other areas of Se toxicity are reported from Ireland and Turkestan. Irrigation with groundwater in seleniferous region has led to the accumulation and toxicity of Se in the soils in Punjab (Sanyal and Dhillon, 2005) and California (Mikkelsen *et al.*, 1989). Selenium is present throughout the soil profile (down to 180 cm and even more) in the seleniferous region, but its concentration in different layers does not follow a specific pattern. During wheat season (upland conditions), 14 per cent of the total Se in surface soil was present as selenate-Se (Se^{VI} form). In contrast, under submerged rice conditions, only 2.5 per cent of the total Se was present as selenate-Se, while the remaining amount was predominantly present as selenite-Se (Se^{IV} form) (Sanyal *et al.*, 2015a).

8.1a.2.3. Selenium in Groundwater

Level of Se in water is an important criterion for determining its suitability for different purposes. According to the water quality guidelines for Se of the USEPA (Sanyal *et al.*, 2012), the maximum concentration level (MCL) of Se in water for drinking purposes is 10 µg Se. L^{-1} and the maximum permissible level (MPL) for water used for irrigation is 20 µg Se. L^{-1}. The current ambient threshold water quality criterion for wildlife is 5 µg Se. L^{-1}. However, keeping in view the significance of bio-accumulation with respect to health hazards for wildlife, Peterson and Nebekar (1992) have proposed 1 µg Se. L^{-1} water-borne Se as the toxicity threshold for sensitive birds and mammals. In general, Se content of groundwater and surface water is mainly determined by the composition of certain types of geological formations such as calcareous shale. As a result of rainfall interactions with surface rocks, the dissolved constituents are carried away through surface run-off into nearby groundwater recharge areas. Percolating water through soil also conveys soluble constituents into the local groundwater (Sanyal and Dhillon, 2005). The solubility and mobility of Se increases with pH. The negative relationship between Se and total dissolved salts suggests that Se and other salts in groundwater do *not* originate from the same source (Sanyal *et al.*, 2015a).

8.1a.2.4. Chemistry of Selenium in Groundwater-Soil Environment

In neutral or acid soils, selenite ($Se^{IV}O_3^{2-}$) is dominant, which easily binds to Fe to form the less soluble Fe selenite. In well aerated alkaline soils, more soluble and much more toxic selenate ($Se^{VI}O_4^{2-}$) is predominant, and is easily taken up by plants (Sanyal and Dhillon, 2005). An increase in phosphate concentration in soil solution increased the Se-uptake by plants (Neal, 1995), whereas increase in sulphate decreased the Se uptake due to competitive ligand-exchange preferences of the adsorbing soil colloidal surfaces (Mikkelsen *et al.*, 1989). Addition of organic matter seems to decrease Se availability due presumably to the well-known ability of the humic and fulvic colloids to chelate the metallic cations as well as the polyvalent oxyanions (Sinha and Bhattacharyya, 2011, 2014a; Ghosh *et al.*, 2012). Selenium content varies in soils of different textural classes, suggesting relatively seleniferous nature of the heavier-textured soils (Sanyal and Dhillon, 2005). Another point worth noting here is that while arsenites [As (III) forms] are much more toxic and soluble than the arsenates [As (V) forms], as mentioned earlier (*see* Section 8.1a.1.6), selenates [Se (VI) forms] are more toxic and mobile than the reduced selenites [Se (IV) forms] in soil and water (Sanyal *et al.*, 2015a). As a consequence, the phytomobilisation and phytotoxicity of arsenic (As) and Se forms under flooded and aerobic soil conditions are very different. This is also reflected in the build-up of these toxicants in soils and crops, requiring different cultural practices, and hence the appropriate management options (Sanyal and Dhillon, 2005).

8.1a.2.5. Selenium Content of Fodders

Selenium content of fodders ranged from 0.16 to 37.3 mg Se. kg^{-1} with an average value of 3.21±4.66 mg Se. kg^{-1} (Sanyal and Dhillon, 2005). Fodders containing ≤ 2 mg Se. kg^{-1} are considered safe; 2 to 5 mg Se. kg^{-1} potentially toxic, 5 to 10 mg Se. kg^{-1} toxic and >10 mg Se. kg^{-1} highly toxic for animal consumption. Typical Se toxicity symptoms in crops, *i.e.*, snow-white or papery-white chlorosis with pink coloration at the lower side of leaves and sheath in 45–60 day-old wheat shoots growing in the Se affected region were reported (Sanyal and Dhillon, 2005; Sanyal et al., 2015a).

8.1a.2.6. Selenium in Animal Hair

Selenium content in animal hair ranged from 0.016 to 37.10 mg Se. kg^{-1} with an average value of 0.905± 3.203 mg Se. kg^{-1} (Sanyal and Dhillon, 2005). Calves are considered normal only if the animal hair of the mother contains at least 0.25 mg Se. kg^{-1}, while with the hairs containing more than 1 mg Se. kg^{-1}, the animals may suffer from Se poisoning. The level of Se in animal hair seems to be dependent upon the level of Se in forages consumed by the animals. Animals affected with chronic selenosis had 35 per cent less hemoglobin; suffered from macrocytic and hypo-chromic anemia. The most consistent clinical manifestations of Se toxicity in animals include impairment of hepatic and renal functions, loss of hair, horn and hoof overgrowth with abnormalities, leading to cracks and detachment, tail necrosis, and disturbed reproductive cycle.

8.1a.2.7. Selenium Toxicity Symptoms in Humans

Humans of all age groups are affected by the Se poisoning in the endemic region. Loss of hair from the body, particularly head, malformation of finger as well as toe nails, and progressive deterioration in general health are the typical symptoms of Se poisoning in humans. In the endemic region, urinary excretion of Se was found to be 17 to 22 times higher than that from non-endemic region (Sanyal and Dhillon, 2005). Farmers in the affected region observed that even leaving the region temporarily for only 3-4 months resulted in a remarkable recovery from Se poisoning. Obviously when the source of food is changed from endemic to non-endemic region, the intake of Se is reduced.

8.1a.2.8. Contribution of Groundwater to Selenium Enrichment of Soil-Plant System

On an overall basis, the total amount of Se removed from the soil was found to be highest in the case of mustard (732 g. ha^{-1}) and lowest for the maize crop (92 g. ha^{-1}) (Sanyal and Dhillon, 2005). Depending upon the amount of irrigation water required, the addition of Se through the water was least in the case of pearl millet (55 g. ha^{-1}), followed by mustard (74 g. ha^{-1}). Obviously, addition of Se through the irrigation water was highest in the case of rice (406 g. ha^{-1}), which far outweighed Se removal by the crop. On the other hand, mustard, Egyptian clover, pearl millet and wheat recorded substantial Se removals from the soils under study. The additions and removal of Se were found to be equal in the case of sugarcane and maize (Sanyal and Dhillon, 2005). This implies that cultivation of sugarcane and maize crops can help in maintaining the stability of the soil-plant system with respect to Se content in the present situation. Balance of Se also turned out to be positive for different rice-based cropping sequences. The results suggest that cultivation of rice in the seleniferous region is aggravating the Se toxicity problem as it needs to be irrigated very extensively. On the other hand, hyper-accumulative nature of mustard suggests that more area should be brought under this crop to reclaim soil of Se. Wild mustard has been found to be the most suitable crop for phytoremediation of seleniferous soils (*see* later).

8.1a.2.9. Management of Seleniferous Soils

For amelioration of Se toxic soils, the following measures have been recommended to the farmers (Sanyal *et al.*, 2015a):

    i) Gypsum amendment @ 0.8 to 1 t/ha, to be applied every alternate year, before sowing the fodder crops, which reduces the Se absorption by crops by 60 to 70 per cent.

    ii) Growing oat, having the lowest Se absorption capacity, as a fodder on seleniferous soils.

    iii) Avoiding feeding the first cut to animals in case of multi-cut fodders, as it contains 2 to 3 times more Se than do the subsequent cuts.

    iv) Minimized use of Se-laden groundwater for irrigation.

Therefore, farmers are advised to follow the maize-wheat rotation instead of rice-wheat. In order to restrict the dietary intake of Se, it has been recommended that neither the fodders produced on the seleniferous soils be used for feeding the animals, nor the grains and vegetables for human consumption. For countering the effect of Se toxicity in water and soils, phytoremediation and bioremediation are often considered more potent than the ones mentioned above in view of the fact that the former offer a more permanent removal of Se from the soil matrix. Thus, *Brassica* sp. is an excellent phytoremediation crop. However, when grown for grain production, it sheds at harvest in the field almost all the leaves which may contain up to 50 per cent of the total Se extracted from the soil (Dhillon and Dhillon, 2003). Indeed, the disposal of such Se-enriched vegetation poses a major problem. In this regard, further refined information on microbial transformation of Se into its less (or non-) toxic forms is necessary.

8.1a.3. Fluoride

8.1a.3.1. Health Implications of Excess Fluoride and its Spread

The presence of fluoride (F^-) in water (with a minimal level of 800 µg F^-. L^{-1}) is essential for protection against the dental caries and weakening of the bones, but excess F^- may lead to dental or skeletal fluorosis. The latter is a crippling disease that affects a number of areas including the Rift valley of East Africa, parts of Mexico, erstwhile Soviet Union, and India (Sanyal *et al.*, 2012). Toxic concentrations of fluoride interfere with Ca metabolism (due to precipitation of CaF_2), resulting in simultaneous *osteosclerosis* of the spine and *osteoporosis* of the limb bones. The susceptibility of individuals to fluorosis may also be ascertained by renal impairment (Sanyal *et al.*, 2015a).

In India, this problem has assumed alarming proportion in at least 17 states of the country, mostly from the geogenic-driven causes, rather than from industrial emissions. Indeed the groundwater of around 50-100 per cent districts are thus affected by fluoride toxicity in erstwhile Andhra Pradesh, Tamil Nadu, Uttar Pradesh, Gujarat, and Rajasthan; 30-50 per cent districts are fluoride-toxic in Bihar, Haryana, Karnataka, Maharashtra, Madhya Pradesh, Punjab, Orissa and West Bengal; < 30 per cent districts in Jammu and Kashmir, Delhi and Kerala are fluoride-toxic (Rattan *et al.*, 2009). Indeed, fluorotic ores occupy large areas of the Eastern and the South-Eastern parts of Rajasthan, in constricted synclinal bands in the central region of Aravali synchronium. Secondly, around the mica mines, groundwater is rich in fluorides, and Rajasthan is a rich source of mica. Apart from India, some areas in Pakistan, Bangladesh, Argentina, United States of America, Morocco, Middle East countries, Japan, South African countries, New Zealand, Thailand, etc., are also affected by fluoride-toxicity (Rattan *et. al.*, 2009).

8.1a.3.2. Sources of Fluoride

The three major sources of origin of fluoride in groundwater in India–all geochemical–are traced to fluorspars, phosphate rock, and phosphorites. The fluoride content of aquifers in India varies from 1000 to as high as 25000 $\mu g\ F^- L^{-1}$, depending on the contact time between groundwater and the fluoride-bearing rocks (Rao and Mamatha, 2004). Further, Rajasthan being a rich source of fluoride–rich mica, as stated above, causes fluoride enrichment of groundwater around the mica mines. The over-exploitation of groundwater for agricultural irrigation further adds to the problem. The acceptable concentration of fluoride in water is partly related to the climate, as in warmer climates, the quantity of water intake is higher, thereby leading to a greater risk.

The most important source of fluoride in human diet is sea fish and drinking water. Fluoride in drinking water comes from the soil, as well as the underlying lithology. In the vicinity of aluminium smelters, contamination of pastures with fluoride from industrial emissions has posed severe problems in many countries (Sanyal *et al.*, 2012).

8.1a.3.3. Occurrence in Soil

Fluoride occurs exclusively as the fluoride ion (F^-) in the soils, where it complexes strongly with metal ions such as Al^{3+} and Fe^{3+} ions. It may be present as the structural component of hydrous minerals, isomorphously substituting for structural OH. Chemisorption of fluoride on clays and oxides by ligand exchange of surface hydroxyls, particularly at low pH, leads to reduction in its availability in the soils. In acid soils, solubility and mobility of fluoride could also be enhanced due to formation of soluble Al-F (*e.g.* AlF_6^{3-}) cationic and anionic complexes; in calcareous soils, its solubility and mobility is limited by its incorporation into insoluble Ca-minerals. In sodic soils, its mobility increases due to the formation of highly soluble NaF (Rattan *et al.*, 2009).

8.1a.3.4. Remedial Measures

The remedial measures for correcting the fluoride toxicity in the contaminated groundwater, being used for drinking purpose, include blending of fluoride-rich water with water of low fluoride content, fluoride removal by precipitation using suitable coagulants (generally an alum-lime mixture as in the *Nalgonda* technique), adsorption on activated carbon or activated alumina, reverse osmosis or ion exchange. The *Nalgonda* technique, developed in this country, is a low-cost technique, and is capable of operating at the water-supply level, covering piped water supplies as well as hand pump units (Nawlakhe *et al.*, 1975). However, eco-friendly disposal of the sludge from the *Nalgonda* process, or those using activated alumina, etc., often poses a serious challenge. The latter has been successfully addressed in a fairly recent technique, developed at the Indian Institute of Science, Bangalore, which uses magnesium oxide for the removal of fluoride from water (Rao and Mamatha, 2004).

8.1b. Anthropogenic Sources of Pollution

8.1b.1. Soil Pollution from Agricultural Activities

A wide range of contaminants can reach the soil-crop system through chemical fertilizer residues, insecticides, herbicides, pesticides, farmyard wastes, sewage-sludge, composts and waste waters, all of which are potentially harmful (Sanyal *et al.*, 2015a).

8.1b.1.1. Fertilizers

Presently the intensification of farming practices exposed more than 10 per cent of the population in some countries to nitrate levels in groundwater-derived drinking water above the 10 mg NO_3^--N L^{-1} (or 45 mg N as nitrate. L^{-1}) guideline value for NO_3^--loading in drinking water, thereby running the risk of *methaemoglobinaemia* to which infants are particularly susceptible (WHO, 1993). Indeed, highly intensive agricultural practices, supported by heavy doses of N fertilizers (mostly urea which on nitrification forms soluble nitrates), under extensive irrigation regime facilitate nitrate leaching to groundwater, especially in the coarse-textured soils of high permeability, and hence increase the risk of nitrate pollution. Indeed 20 per cent of all the sampled wells, located in three districts of Punjab, showed nitrate levels exceeding the safe limit of 45 mg of N as nitrate per liter. Such nitrate pollution in drinking water can have serious health impact on humans, especially for babies. The most significant potential health effects of drinking water contaminated with nitrate are the blue-baby syndrome (*methaemoglobinaemia*) and cancer (Sanyal *et al.*, 2015a).

Such excessive nutrients including NO_3^- from manure and fertilizers may also cause eutrophication in water bodies, causing excessive algal growth and cyanobacteria, and loss of biodiversity. Such eutrophication leads to an undesirable rise in the biological oxygen demand (BOD) and the chemical oxygen demand (COD) of the water bodies, thereby affecting several forms of aquatic life, while causing the undesirable changes in the aquatic ecosystems, often with serious economic consequences (*see* Section 6A.11) (Sanyal *et al.*, 2015b).

The main sources of phosphate in aquatic environment is through household sewage water, containing detergents and cleaning preparations, agricultural run-off, containing fertilizers, as well as industrial effluents from fertilizer, detergent and soap industries. The consumption of synthetic detergents is on the rise due to increasing urbanization and most of the former contain phosphate as a 'builder', which increases phosphate loading rates in surface water bodies. The estimated annual consumption of phosphate-containing laundry detergents for the current population in India is about 2.88 million tonnes and the total outflow of P is estimated to be 146 thousand tonnes per year. Therefore, a major point of concern for checking the eutrophication of water bodies, particularly in sensitive areas, is how to reduce P inputs to surface waters (Kundu *et al.*, 2015).

As stated earlier, phosphorus is also lost from crop lands *via* erosion or run-off (*see* Section 6A.11). As a result, non-point sources now account for a larger share of the nation's water quality problems than ever before. The main factors influencing P movement can be divided into transport and P source factors. Transport factors include the mechanisms by which P moves within a landscape. These are rainfall- and irrigation-induced erosion and run-off. Factors which influence the source and amount of P available to be transported are soil P content and rate and method of P applied in either mineral fertilizer or organic forms (Sanyal *et al.*, 2015b).

As mentioned earlier, losses of P from the cultivated lands could be minimized by adopting judicious P management strategies. Continuous P fertilization over the years at rates exceeding those of crop removal results in P build-up, often above the levels required for crop production (*see* Section 6A.11.1). Accumulation of soil test P near the soil surface due to previous P application influences the concentration and loss of P in run-off. Highly significant linear relationships are frequently noted between the soil test P in the surface soil and dissolved P concentration in surface run-off. Adoption of soil test-based P fertilization would, therefore, not only be economically viable, but would also avoid its excessive accumulation in soil (Sanyal *et al.*, 2015b). As run-off enters a stream channel and, ultimately, a water body, there is generally a progressive dilution of P load through water dilution and sediment deposition. Sources of particulate P in streams include eroding surface soil, plant material, stream banks, etc. As the finer-sized fractions of source material are preferentially eroded, the P content and reactivity of the eroded particulate material is usually greater than that of the source soil.

8.1b.1.2. Pesticides

Spray drift from pesticide application may enter water bodies if the treated crops and orchards are located too close to the river and/or streams. The main threat results from the poor storage and accidental spillage, and pesticides may thus enter and contaminate the groundwater. The non-degradable and *bio-accumulating* pesticides are persistent in the food-chain. Organochlorides cause changes in the sexual and reproductive characteristics of wildlife. Top carnivores (in the river, fish) are especially affected. Since pesticides exist in very low concentrations in water, their detection and measurement is complex and expensive (Sanyal *et al.*, 2015a).

The organochlorine (OClP) and organophosphate (OPP) pesticides in surface and groundwater in Maharashtra with intensive agriculture activity indicated that higher concentrations of OClPs and OPPs were noted in surface water compared to groundwater (Lari *et al.*, 2014). Throughout the monitoring study, α-HCH (0.39 µg L^{-1} in Amravati region), α-endosulphan (0.78 µg L^{-1} in Yavatmal region), chlorpyrifos (0.25 µg L^{-1} in Bhandara region) and parathion-methyl (0.09 µg L^{-1} in Amravati region) are frequently found pesticides in groundwater, whereas α, β, γ-HCH (0.39 µg L⁻¹ in Amravati region), α, β-endosulphan (0.42 µg L⁻¹ in Amravati region), dichlorovos (0.25 µg L⁻¹ in Yavatmal region), parathion-methyl (0.42 µg L⁻¹ in Bhandara region), and phorate (0.33 µg L⁻¹ in Yavatmal region) were found in surface water (Lari *et al.*, 2014). Surface water was noted to be more contaminated than groundwater with more number of and more concentrated pesticides. Among the pesticides, water samples had higher levels of organophosphate contamination than that with organochlorine compounds (Sanyal *et al.*, 2015a). Pesticides in the surface water samples from Bhandara and Yavatmal region exceeded the EU (European Union) permissible limit of 1.0 µg L⁻¹ (sum of pesticide levels in surface water), but were within the WHO guidelines for individual pesticides (Lari *et al.*, 2014).

8.1b.2. Soil Pollution from Industries and Heavy Metals

The industrial effluents and water drainage from spoil and rubbish heaps either washes directly to the nearby fields or else enters the local streams, rivers, and ultimately the soil. The latter leads to toxicity to all forms of life (Sanyal *et al.*, 2015a). The heavy metals, which are cytotoxic, mutagenic as well as carcinogenic, are added to soil through (besides the geogenic sources) mining and smelting, disposal of municipal and industrial wastes, use of fertilizers, pesticides and automobiles. The growth media (soil, air, nutrient solutions) of vegetable crops contribute to the uptake of these metals by the crop roots or foliage as the main sources of heavy metals to vegetable crops (Lokeshwari and Chandrappa, 2006). Indeed, the studies on Bellandur Lake water-irrigated crops in Bangalore showed the presence of some of the heavy metals in rice and vegetables, beyond the limits of the Indian standards. Metal transfer factors from soil to vegetation are found to be significant for Zn, Cu, Pb and Cd (Sanyal *et al.*, 2015a). Comparing the results of heavy metals in water, soil and vegetation with their respective natural levels, it was observed that the impact of lake water on vegetation was more than that on the soil. The results revealed the presence of Cd in spinach (4 µg g⁻¹) and radish (2.5 µg g⁻¹) to be beyond the safe limits. The reason for the accumulation is that Cd is relatively easily taken up by food crops, and especially by leafy vegetables. It is found that the average total concentration of metals in plant food-stuff, *i.e.*, Fe, Zn, Cu, Cr, Pb and Cd was 3.5-, 2-, 2.5-, 15-, 2.5- and 21-fold higher than the natural concentration (Lokeshwari and Chandrappa, 2006).

The heavy metal, namely chromium (Cr) is mostly added to the environment by the activities of the tanneries, for instance, the tanneries at Kanpur, near Palar River in Tamil Nadu, and other places. Thus, 30-40 L of tannery effluent is discharged for processing just one kg of skin/hide, the effluent having a load of 97-5125 mg Cr. L⁻¹, which leads to soil, surface water and groundwater contamination. Usually high Cd, Zn and Pb deposition in soils results from mining, Zn- and Pb-smelters; Pb is also added from the vehicular traffic. Thus, zinc smelter effluent-irrigated soils at Debari, Udaipur (Rajasthan) contained extraordinarily high levels of metals (mg kg⁻¹) including Cd (0.32-214), Zn (40.7 -21700) and Pb (3.50-3312). Mean intake of Cd by people in the Jintsu valley, Japan was reported to be of the order of 600 µg day⁻¹ which causes the dreaded disease, namely *itai-itai* disease (Sanyal *et al.*, 2012). Burning of fossil fuels and disposal of wastes (plastic, Cd-containing batteries, sewage sludge, etc.) also contribute to the environmental burden of heavy metals.

8.1b.3. Soil Pollution from the Use of Solid Wastes in Agriculture

The main forms of organic waste are household food, agricultural, human and animal wastes. Indeed, organic recycling is vital for supplementing plant nutrients and maintenance of soil fertility, a key factor in crop yields (Jeevan Rao, 2005). Agenda 21, adopted in Rio in 1992, states that environmentally sound waste management should include safer disposal or recovery of waste and changes to a more sustainable pattern of introducing integrated life cycle management concepts (Sanyal et al., 2015a). Table 8.4 gives the nutrient potential and economic value of biological and industrial wastes in India.

Table 8.4: Nutrient Potential and Economic Value of Biological and Industrial Wastes in India

Type of Wastes	Total Quantity Available (mt)	Total Nutrients (000 t. yr⁻¹)				Economic Value (million Rs**)
		N	P	K	Total	
Cattle manure	280	2813	2000	2069	6882	30970
Crop residue	273	1283	1966	3904	7153	32188
Forest litter	19	100	37	100	237	1066
Rural compost	285	1431	862	1423	3715	16719
City refuse	14	98	84	112	294	1323
Sewage sludge	1	5	3	3	11	49
Press mud	3	33	79	55	168	756
Domestic waste water	6351*	318	140	191	648	2915
Industrial waste	66*	3	1	1	5	22
Total		6084	5172	7858	19113	86008

*: Million cubic meter per annum $(Mm^{-3}.Yr^{-1})$; **: Economic value compounded @ Rs. 4500/t NPK.

Source: Juwarkar et al. (1992)

The trappable availability of major plant nutrients from the organic sources has also been estimated from time to time, so also the projections for future, as shown in Table 8.5.

Table 8.5: Trappable Availability of Major Plant Nutrients (N + P_2O_5 + K_2O) from Organic Sources in India (mt. yr⁻¹)

Source	2000	2010	2025
Human excreta	1.60	1.80	2.10
Livestock dung	2.00	2.10	2.26
Crop residues	2.05	2.34	3.39
Total	5.65	6.24	7.75

Source: Tandon (1997).

8.1b.4. Soil Pollution from the Use of Waste Water in Agriculture

The use of urban wastewater in agriculture is a centuries-old practice that is receiving renewed attention with the increasing scarcity of fresh water resources in many arid and semi-arid regions. Driven by rapid urbanization and growing wastewater volumes, wastewater is widely used as a low-cost alternative to conventional irrigation water; it supports livelihoods and generates considerable value in urban and Peri-urban agriculture despite the health and environmental risks associated with this practice. In India, wastewater irrigation is increasingly used for such crops as vegetables, fruits, cereals, flowers and fodder. The city of Kolkata (formerly Calcutta), for instance, has a long history of using wastewater stabilization tanks for aquaculture. An estimated 2.4 t. ha^{-1} (Sanyal *et al.*, 2012) of fish is produced annually in Kolkata from about 3200 ha of ponds with inflow of about 3 m. sec^{-1}. Throughout India, industries recycle wastewater to reduce the requirements of freshwater. This trend is led by industries in Saurashtra and Chennai. Though pervasive, this practice is largely unregulated in low-income countries, and the costs and benefits are poorly understood.

Properties of water are important because the water supply can be the principal source of dissolved constituents which will appear in raw waste water, collected in the sewage system. Per capita water consumption is one of the principal factors used to estimate the flow of waste water to be discharged into the sewage system. Industrial flows can cause unusual variability in flow rate as well as contribute substantial quantities of specific constituents. The metal content of municipal waste water is attracting much attention and is an important factor to consider in the planning of a land-based waste management system (Sanyal *et al.*, 2012). Table 8.6 gives the selected characteristics of wastewater from domestic and industrial locations in Ludhiana – an industrial city in Punjab.

Table 8.6: Selected Characteristics of Wastewater from Domestic and Industrial Locations in Ludhiana, Punjab

Location	pH	Biological Oxygen Demand (mg. L^{-1})	Chromium Hexavalent (mg. L^{-1})	Nickel (mg. L^{-1})	Cyanides (mg. L^{-1})
Electroplating industry	6.2-7.2	60-380	0.2-2.5	1.0-3.0	0.42-0.97
Sugar industry	7.1-7.9	1058-1640	–	–	–
Paper industry	7.0-10.1	560-1113	–	–	–
Household	6.7-7.8	80-460	0.1-0.2	0.2-2.0	0.05-0.07
Maximum limits for disposal on agricultural lands	5.5-9.0	100	0.1	0.005	0.2

Source: Tiwana *et al.* (1987).

In a study on the impact of wastewater irrigation since 1979 on the dynamics of metal concentrations in the vadose zone in agricultural lands in the Peri-urban area of New Delhi [namely the Western part of the National Capital Territory of New Delhi under the Keshopur Effluent Irrigation Scheme (KEIS)] from the Keshopur Effluent Irrigation Scheme (KEIS), it was noted that there was a significant decline in soil pH and electrical conductivity in wastewater-irrigated soils over their respective tube well-irrigated control soils (Deshmukh *et al.*, 2015a). There was a considerable reduction in sodicity of soil due to sewage irrigation. The use of sewage irrigation also proved to be effective in increasing the soil organic carbon, to the obvious benefits. The order of accumulation of extractable heavy metals in the vadose zone was in the sequence of Fe > Cu > Zn > Ni> Mn. By and large, build-up of none of the metals in extractable pools exceeded the phytotoxicity limits in soil. However, periodical monitoring is required to assess the metal build-up in such sewage-irrigated soils to protect the human and animal health (Deshmukh *et al.*, 2015a).

Furthermore, understanding and quantification of geochemical processes in the vadose zone of the sewage-effluent-irrigated soils are helpful in predicting the transfer of metals and other ions to food-chain and groundwater. Hence, an attempt was made to simulate the various geochemical processes occurring in the flow path of infiltrating sewage water down the vadose zone with the help of Net Geochemical Reaction Along the Flow Path (NETPATH) in the same Peri-urban area of New Delhi from the Keshopur Effluent Irrigation Scheme (KEIS). The results indicated that groundwater of 20- and 10-year old sewage-irrigated lands were slightly oversaturated in respect of calcite and dolomite, and under-saturated in respect of gypsum. Indeed, calcite dissolution was a very common feature in all the sewage-irrigated soils. But, the dissolution of dolomite decreases with time. The shallow groundwater of 5-year sewage-irrigated field was found to be under-saturated in case of calcite, dolomite, and gypsum. There was reduction in concentrations of Fe and Mn in groundwater samples of 20-year old sewage-irrigated field as compared to that in the sewage effluent, possibly due to the formation of goethite and manganite in the vadose zone, respectively, as revealed by simulation with NETPATH. Similarly, increase in Fe and Mn concentrations in groundwater in the case of 10- and 5-year old sewage- irrigated fields arose from the dissolution of siderite and pyrolusite, respectively. It was also noted that the capacity of the vadose zone for purification of sewage effluents by and large decreased with the increase in the duration of irrigation. Consequently, quality of groundwater will be deteriorated due to long-term use of sewage water (Deshmukh *et al.*, 2015b).

Impact of wastewater irrigation on some biological properties of soil was studied in an area where treated sewage water is being supplied to the farmers since 1979 under the Keshopur Effluent Irrigation Scheme (KEIS). Three fields were selected which had been receiving irrigation through wastewater for the last 20, 10 and 5 years (Deshmukh *et al.*, 2011). Sewage irrigation was noted to exert a positive impact on the fungal and bacterial population in soils of the vadose zone. Furthermore, the increase in the duration of sewage water application also increased the bacterial and fungal population in the soil at a greater depth which is the precursor of migrating microorganisms, such as *Escherichia coli, C. freundii* and *S. typhi*. The seasonal changes in the groundwater pollution was noted to arise from the rise and fall of water table, which was proportional to the depth of the vadose zone available for filtration with soil as a media for the applied sewage water until it reaches the groundwater. The organic matter present in the surface layer has a positive aspect for microbial growth, but at the same time, the process of microbe-adsorption on to the soil surface cannot be ruled out in the flow path when the sewage water infiltrates down the vadose zone. Rapid detection of fecal contamination suggested that the *C. freundii* and *Salmonella* were dominant in the shallow groundwater, while *E. coli* was dominant in the deep groundwater, collected from the sewage-irrigated field. This implied that migration of *E. coli* under sewage-irrigated field through the vadose zone was more prominent than other species. Groundwater under sewage-irrigated fields of the study area was found to be unfit for consumption for human and animals due to the contamination with the dreaded microorganisms (Deshmukh *et al.*, 2011).

8.1b.5. Environmental Impacts from Solid Wastes

Presently, most of the solid refuse is disposed of in low-lying areas, leading to breeding of rodents, flies and other domestic pests and also pollution of the surface and groundwater (Jeevan Rao and Shantaram, 1995). Continuous use of fresh wastes on agricultural land for cultivation of crops may result in the pollution of soil, which, if properly addressed, untreated solid waste (USW) can be considered as a valuable resource for use as manure on agricultural land (Jeevan Rao and Shantaram, 1995).

8.1b.6. Sewage-Sludge Utilization in Agriculture

The addition of sewage-sludge to soil provides a useful source of plant nutrients, especially N and P, and also organic matter, which improves the soil physical properties. However, sewage-sludge is rich in heavy metals as well, such as Zn, Mn, Cd, Pb and Cr which can be toxic to animals and humans when applied above certain limits. The nutrient potential of sewage in India is estimated to be more than 3,50,000 t N, 1,50,000 t P and 2,00,000 t K per year (Venkata Sridhar *et al.*, 2006). The risk assessment, based on hundreds of studies conducted on the issue of metals in sludge, considered every possible way in which these metals could conceivably affect the health of humans, animals, plants and the environment (air, water and soil). The possible pathway of transfer is: Sludge → Soil → Plant → Human/Animal→Human (Sanyal *et al.*, 2015a).

8.1b.7. Risk Assessment of Metal-Contaminated Soil

Risk assessment of metal-contaminated soil depends on how precisely one can predict the solubility of metals in soils. Such risk of metal contaminated soils can also be assessed by predicting metal uptake by crops grown on contaminated soils on routine basis. To ascertain this, simpler approaches like integrated solubility and free ion activity model may be successfully used to predict the metal uptake through the edible portion of crops grown on contaminated soils (Datta and Young, 2005). The guiding principle depends on the premises that the response of plants and soil organisms to metal toxicity is determined primarily by the variation in free metal ion activity in soil pore water. Thus, Datta and Young (2005) developed the protocol for prescribing toxic limit of metals, based on extractable metals and soil characteristics, using the solubility and free-ion activity models. Solubility of metals in soil was predicted using the following pH-dependent Freundlich equation (Jopony and Young, 1994) as per the free ion activity of metal and metalloid (FIAM):

$$(M^{2+}) = M_C / [k_M (H^+)^{-n_M}]$$

Where (M^{2+}) is the free metal ion activity in soil solution in soil pore, M_C is the labile pool of soil metal, assumed to be exclusively adsorbed on humus (mol. kg carbon^{-1}), and k_M and n_M are empirical constants which express the pH dependence of the metal distribution coefficient. It follows from detailed theoretical considerations (Datta and Young, 2005; Meena *et al.*, 2016):

$$p(M^{2+}) = [p(M_c) + k_1 + k_2 pH)]/n_F \tag{8.5}$$

Where k_1 and k_2 are empirical, metal-specific constants, expressing the pH-dependence of metal distribution coefficient and n_F is the power term from the Freundlich equation. The metal transfer factor from soil solution to plant biomass is given as

$$\text{Transfer factor} = \log [M_{plant}/(M^{2+})] \tag{8.6}$$

Where M_{plant} is the metal loading of plant biomass.

Eqs. 8.5 and 8.6 can be combined to lead to Eq. 8.7 as follows:

$$p(M_{plant}) = C + \beta_1 pH + \beta_2 p(M_c) \tag{8.7}$$

Where C, β_1 and β_2 are empirical metal- and plant-specific constants. This model (FIAM) predicts the free ion activity of trace metal and metalloid in soil solution as a function of labile soil extractable metal and pH with the simplifying assumption that the whole amount of metal (M_C) is adsorbed on soil humus.

Rang Zan *et al.* (2013), based on this approach, made an attempt to predict the free ion activity of Zn, Cu, Ni, Cd, and Pb in metal-contaminated soil as a function of pH, soil organic carbon, and extractable metal content. Thus, 0.005 M DTPA-, 0.05 M EDTA-, and 0.01 M $CaCl_2$-extractable Zn, Cu, Ni, Cd, and Pb were used as estimates of the labile metal pool, whereas free metal ion activity, as determined by the Baker soil test, was used as an estimate of the free metal ion activity in soil solution. In the case of soil organic carbon, both Walkley–Black organic carbon (WBC) and $KMnO_4$-oxidizable organic carbon were used. The solubility model was appropriately parameterized (Rang Zan *et al.*, 2013). These authors (2013) noted that soil pH and organic carbon are among the important soil properties which control the solubility of metals in contaminated soils. The EDTA-extractable metals were found to be more useful as an input of solubility model as compared to DTPA-extractable and $CaCl_2$-extractable metals. The Walkley–Black carbon was more useful as a predictor variable of the solubility model than the labile pool of organic carbon. It was found that the solubility model as a function of pH, organic carbon, and EDTA-extractable metals was reasonably effective in predicting the metal ion activity in contaminated soils (Rang Zan *et al.*, 2013).

Closely related to what is stated above, the risk to human health for intake of metal through consumption of green leafy vegetables has been computed in terms of what is known as the Hazard Quotient, HQ_{gv}. The latter is given as

$$HQ_{gv} = (ADD/R_f D) \tag{8.8}$$

Where ADD = average total daily dose of metal intake through diet and drinking water (mg metal.kg body weight^{-1}. day^{-1}) and R_fD = the corresponding reference dose which is defined as the maximum tolerable daily intake of the specific metal that does not lead to any deleterious health effects. Obviously, $HQ_{gv} > 1.0$ suggests hazard to human health. The values of R_fD used for Zn, Mn, Ni, Cd, Pb, and Cr (Meena *et al.*, 2016) were 0.3, 0.14, 0.02, 0.001, 0.0035 and 0.003 mg kg body weight^{-1}. day^{-1}, respectively (IRIS, 2015). For Cu, provisional maximum tolerable daily intake (PMTDI) of 0.5 mg kg body weight^{-1}. day^{-1} was used in place of R_fD (Meena *et al.*, 2016). For As, 2.1 µg kg body weight^{-1}. day^{-1} was used as the corresponding PMTDI. As stated earlier (*see* Section 8.1a.1.2), the joint FAO/WHO Expert Committee on Food Additives (JECFA) set a provisional maximum tolerable daily intake (PMTDI) of inorganic arsenic as 2.1 µg As. kg body weight^{-1}.day^{-1} for humans in 1983 and confirmed a provisional tolerable weekly intake (PTWI) as 15 µg As. kg body weight^{-1}.day^{-1} in 1988 (FAO/WHO, 1989). Daily intake of wheat and rice grain by a human of 70 kg body weight was assumed to be 0.2 kg day^{-1}.

It follows that the Hazard Quotient on gravimetric basis (HQ_{gv}) is given by

$$HQ_{gv} = (M_{plant}.W. F)/(R_fD.70) \tag{8.9}$$

for a human of 70 kg body weight.

Here, M_{plant} = Metal content in plant (mg. kg^{-1}), W = Daily intake of green vegetables (mg metal. day^{-1}); F is the factor used to convert fresh to dry weight of plants.

The Hazard quotient (HQ_{gv}) for intake of Zn, Cu and Ni through consumption of spinach (*Brassica oleracea*), gobhi sarson (*Brassica napus*) and Indian rape (*Brassica campestris*) as green leafy vegetables, grown on metal-contaminated soils, receiving sewage irrigation for 20 years, was assessed by Rattan *et al.* (2009). Although Ni exhibited relatively higher HQ_{gv} for all the crops compared to other two metals (Table 8.7), most of the values were far less than 1, which means that these green vegetables are *not* likely to induce any health hazard to humans as far as their metal contents are concerned (Sanyal *et al.*, 2015a). Such type of risk assessment is, however, *not* complete, as green vegetables constitute only a small portion of human diet and other routes of metal entry such as direct ingestion of soil, inhalation and metal intake through rice and wheat as well as other cereals and drinking water were not combined.

Table 8.7: Risk Assessment to Human beings from Consuming Metal Contaminated Green Leafy Vegetables grown on Sewage-Irrigated Soil in Terms of Corresponding Metal Hazard Quotients (HQ_{gv})

Crop	HQ_{gv}		
	Zn	Cu	Ni
Brassica napus	0.04-0.068	0.004-0.021	0.027-0.442
Spinach	0.035-0.152	0.008-0.015	0.046-0.502
Indian rape	0.027-0.053	0.004-0.014	0.016-0.429

Source: Rattan *et al.* (2009)

However, the point worth noting here is that ADD refers to the daily intake of a given metal *from all the food items and drinking water*. It is thus evident that for any one food item (*e.g.* a vegetable or rice), the limiting value of HQ_{gv} will be less than 1.0. In this context, Meena *et al.* (2016) argued that the permissible limits of metal and metalloid in soil was established based on (i) solubility of metal and metalloid in soil, (ii) metal and metalloid content in rice and wheat grain, and (iii) human health hazard, associated with intake of metal and metalloid through consumption of crops raised on metal-contaminated soils. For fixing the toxic limit of the extractable metal and metalloid in soils at a particular pH and organic carbon, the critical value of HQ_{gv} used by these authors was 0.5 for *any one given food item*. These authors (2016) developed a ready reckoner to compute the permissible limit of the extractable metal and metalloid in soils, based on pH and organic carbon content. These permissible limits were based on the predicted HQ_{gv} by the aforesaid solubility-FIAM. Thus, Meena *et al.* (2016), while raising rice crop in Cd-contaminated soil, showed that the permissible limit of the DTPA-extractable Cd in soil varies with soil pH and organic carbon content, corresponding to the respective HQ_{gv} values (associated with metal intake by human through rice grain) remaining below the critical value of 0.5, as mentioned above. The findings are shown in Figure 8.4. Thus, while at a pH of 6.0 and organic carbon content of 0.25 per cent of the contaminated soil, the permissible limit of DTPA-extractable Cd in soil can go up to 0.15 mg.kg^{-1} with the respective HQ_{gv} not crossing the critical value of 0.5, the same may be as high as up to 6.90 mg.kg^{-1} when the soil pH is 8.0 and the organic carbon content 0.5 per cent (Figure 8.4). Hence appropriate interventions may be designed to render a metal-contaminated soil remain within a risk-free domain for raising the crops for human consumption, *without* posing the health hazards.

8.1b.8. Persistent Organic Pollutants

Persistent organic pollutants (POPs) are a group of chemicals which are intentionally or inadvertently produced and introduced into the environment. Due to their stability and transport properties, they are now widely distributed around the world, and found even in the most unlikely places such as the arctic regions (Sanyal *et al.*, 2012).

8.1b.9. Symptoms of Contamination in Humans and Livestock, Fisheries and Birds

Metal-toxicity could lead to either biochemical (genetic) or physiological (environmental) abnormalities. The former occurs due to (i) excessive dietary intake, (ii) excessive absorption, (iii) decreased loss from the body, and (iv) decreased metabolism of the nutrients due to antagonism or metabolic block (Sanyal *et al.*, 2015a).

Excessive intake of Cr is associated with renal dysfunctions, lung cancer, eczematous dermatitis (Deckers and Steinnes, 2004). The hexavalent chromium (Cr^{6+}) is more toxic than the trivalent Cr (Cr^{3+}) because of the high rate of absorption of the former by the living surface. Lead poisoning causes encephalopathy (damage to brain), failure in reproduction, metabolic disorder, neuro-physical deficit in children, affecting the haematologic and renal system, anaemia, and hypertension.

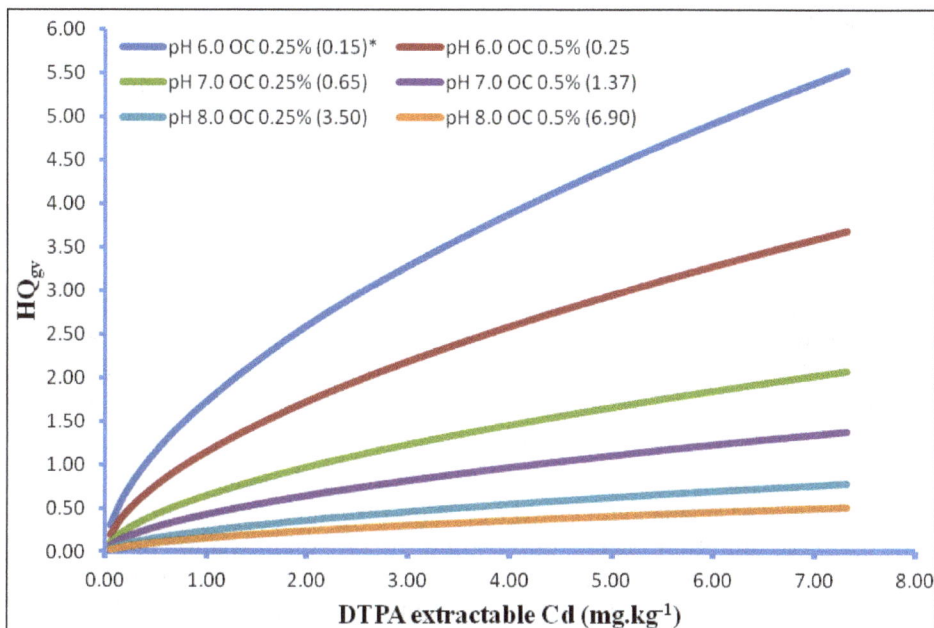

Figure 8.4: Permissible Limit of DTPA-Extractable Cd in Soils in Relation to Soil pH and Organic Carbon for Intake of Cd through Rice Grain by Human.

* Values in parentheses indicate the toxic limit of the DTPA-extractable Cd in soil.

Source: Meena *et al.* (2016).

Children are much more susceptible to lead (Pb) poisoning than the adults. Cereals and water account for 35 per cent of the daily Pb intake. Lead was the first metal to be linked with failure in reproduction (Sanyal *et al.*, 2012). Elemental mercury (Hg) is poorly absorbed after oral ingestion and therefore causes little toxicity. Methyl-mercury is almost completely absorbed through the gastrointestinal tract and readily appears in the blood. The corresponding manifestation is the Minamata disease, with the consumption of methyl mercury-contaminated fish and other foodstuffs, causing deaths (Sanyal *et al.*, 2015a). Excessive cadmium (Cd) causes renal tubular dysfunction, proteinuria, glucosuria, aminoaciduria, and the dreaded *itai-itai* disease. As regards cadmium, cereals alone are responsible for about 50 per cent intake. Air-borne Cd from cigarette smoke, industrialization and population growth is known to be an infrequent cause of hypertension, cancer and immune disorders (Sanyal *et al.*, 2012). Excess zinc interferes with reproduction; impairs the growth of embryo, and causes various types of anaemia.

8.1b.10. Amelioration/Management Options in Agriculture

8.1b.10.1. Nanotechnology for Pollution Control

The water and groundwater purification process may use Fe nano-particles, nano-fibers, Nano-enzymes and nano-filtration techniques. Nanotechnology can also be applied to clean the air from toxic CO and VOCs (volatile organic compounds), using the CNTs (carbon nano-tubes), and nano-particles as adsorbents (Sanyal *et al.*, 2015a). Nano-particles and nano-tubes act as sensors for detection of very small-size and concentration of toxic substances, difficult to detect by using the conventional technologies (Pandya, 2015).

8.1b.10.2. Phytoremediation

The current status of heavy metal pollution of soil and water bodies and the role of plant species in alleviating the contamination as a part of remediation strategy in the Indian context has been recently reviewed by Mandal *et al.* (2014). Indeed, phytoremediation successfully curbs the ill-effects of toxic heavy metals, but is yet to become a commercially viable technology in India. Being a slow process, biotechnological and classical hybridization with efforts from interdisciplinary researchers can develop more efficient hyper-accumulator species of desirable traits, thereby revealing new potential mitigation option (Sanyal *et al.*, 2015a).

The *in situ* remediation of mixed organic and inorganic pollutants – degradation by root-associated rhizosphere microorganisms (*i.e.*, rhizo-degradation) – is likely to be the most important mechanism during the phytoremediation of petroleum hydrocarbons (Siciliano and Germida, 1998). Zayed *et al.* (1998) noted the root accumulation to be 100-fold higher than that in shoot, regardless of the Cr species supplied, in different vegetable plants, including a variety of *Brassicaceae* (cauliflower, cabbage, broccoli, etc.). Several investigations have assessed the potential of maize as a contaminant accumulator in polluted sites for Cd and As (Sanyal *et al.*, 2012). Adhikari *et al.* (2010) observed that *Typha angustifolia* L. and *Ipomoea carnea* L. plants show promise for removal of Pb from contaminated wastewater because they can accumulate high concentrations of Pb in roots (1200 and 1500 mg Pb kg^{-1}, respectively) and shoots (275 and 425 mg Pb kg^{-1}, respectively) (Sanyal *et al.*, 2015a).

8.1b.10.3. Bio-remediation

Microbes can reduce the activity of different types of metals or convert the toxic metal forms to relatively less toxic forms by different processes. The choice of microorganisms for bioremediation depends on the availability of energy and carbon source, environmental conditions and the presence of hazardous contaminants (Sanyal *et al.*, 2015a). The recognized aerobic bacteria include the *Pseudomonas, Alcaligenes, Sphingomonas, Rhodococcus* and *Mycobacterium* species. The initial enzyme in the pathway for such aerobic degradation by the Methylotrophs, namely methane monooxygenase, has a broad substrate range and is active against a wide range of compounds, including the chlorinated aliphatic trichloroethylene and 1,2-dichloroethane (Sanyal *et al.*, 2012). Further, as stated earlier (*see* Section 8.1a.1.7.6), As-oxidizing bacteria were isolated from the selected As-contaminated soils of West Bengal (Majumdar *et al.*, 2013b), which were closely related to various species of *Bacillus* and *Geobacillus*. Indeed, the increased rate of oxidation of highly toxic As (III) to less toxic As (V) form by these native strains provides a promise for effective bio-remediation of As-contaminated environments.

8.1b.10.4. Other Management Options

One of the remediation technologies for metal-contaminated soil includes excavation of soil, followed by washing and subsequent disposal of the treated soils. However, this option is prohibitively expensive. A common method for immobilization of metals in soil is to apply amendments like lime, phosphates or organic manure (Ghosh *et al.*, 2012). Addition of other natural or synthetic chemicals like zeolites, beringite, palygorskite (Paulose *et al.*, 2007), hydrous oxides of Al, Fe and Mn are also known to facilitate the metal immobility in soils. Sadana and Bijay-Singh (1989) reported that applying increasing levels of Zn to Cd-contaminated soils can effectively reduce Cd toxicity and ameliorate Zn deficiency in maize, and thus increase the crop yields. Chemicals like EDTA have been used for rendering the toxic elements in soluble form for their catalyzed removal by the hyper-accumulating plants (Sanyal *et al.*, 2015a).

It is high time that that all the stakeholders join hands to frame strategies for ameliorating the already polluted soils, as well as preventing further pollution (Sanyal *et al.*, 2015a). Indeed, to minimize soil pollution or to manage polluted soils, there is an urgent need to develop the standards in terms of safe, marginal, lethal levels of toxic elements in the target human organs; conduct multi-disciplinary researches to develop simulation models or transfer functions for quantifying or predicting the contributions of different toxic sources to the human body. There is a need to improve the field and laboratory protocols for large-scale measurement of geogenic contaminants, with their speciation in groundwater and soil for ascertaining the *net* toxicity in soil-crop-animal-human continuum. The reactions of the organic wastes or their microbial breakdown products with metals need further investigation. Evidently, all such issues can be best addressed through a multi-level stakeholder approach, involving researchers, technologists, planners with a focus on the real beneficiaries, who ought to be empowered through education and training, thereby enabling them to understand and participate actively in soil pollution mitigation programmes (Sanyal *et al.*, 2015a). These efforts may require a major shift from the purely technical to a holistic approach, needed for ensuring a *technically feasible, socially acceptable, economically viable, and environmentally sound vibrant action plan.*

References

Acharya, S. K. (1997). Arsenic in groundwater-Geological overview. *Consultation on Arsenic in Drinking water and Resulting Arsenic Toxicity in India and Bangladesh.* World Health Organization, New Delhi, India, 29 April to 01 May, 1997.

Adhikari, T., Kumar, Ajay, Singh, M. V., Suba Rao, A. (2010). Phytoaccumulation of lead by selected wetland plant species. *Comm. Soil Sci. Plant Anal.*, **41,** 2623-2632. DOI:10.1080/00103624.2010.517879

Chigbo, F.E., Smith, R.W. and Shore, F.I. (1982). Uptake of arsenic, cadmium, lead and mercury from polluted water by the water hyacinth (*Eichornia crassipes*). *Environ. Pollut. Ser.*, A **27**, 31-36.

Das, I., Ghosh, K. and Sanyal, S. K. (2005). Phytoremediation: A potential option to mitigate arsenic contamination in soil-water- plant system, *Everyman's Sci.*, **40** (No.2), 115-123.

Das, Indranil, Ghosh, Koushik, Das, D. K. and Sanyal, S. K. (2014). Transport of arsenic in some affected soils of Indian subtropics. *Soil Res.*, CSIRO Pub., **52**, 822–832.

Das, Indranil, Sanyal, S. K., Ghosh, Koushik and Das, D. K. (2016). Arsenic mitigation in soil-plant system through zinc application in West Bengal soils. *Bioremed. J.*, **20**, 24-37. Doi.org/10.1080/10889868.2015.1124062

Datta, A., Sanyal, S. K. and Saha, S. (2001). A study on natural and synthetic humic acids and their complexing ability towards cadmium. *Pl. Soil*, **225**, 115-125.

Datta, S. P. and Young, S. D (2005). Predicting metal uptake and risk to the human food chain from leaf vegetables grown on soils amended by long-term application of sewage-sludge. *Water, Air and Soil Pollut.*, **163**, 119-136.

Deckers, J. and Steinnes, E. (2004). State of the art on soil-related geo-medical issues in the world. *Adv. Agron.*, **84**, 1-35.

Deshmukh, S. K., Singh, A. K. and Datta S. P. (2015a). Impact of wastewater irrigation on the dynamics of metal concentrations in the vadose zone: monitoring: part I. *Environ Monit. Assess.*, **187**, 695. DOI 10.1007/s10661-015-4898-3

Deshmukh, S. K., Singh, A. K. and Datta S. P. (2015b). Impact of wastewater irrigation on the dynamics of metal concentration in the vadose zone: simulation with NETPATH–part II. *Environ Monit. Assess.*, **187**, 764. DOI 10.1007/s10661-015-4962-z

Deshmukh, S. K., Singh, A. K., Datta, S. P. and Annapurna, K. (2011). Impact of long-term wastewater application on microbiological properties of vadose zone. *Environ Monit. Assess.*, **175**, 601–612. DOI 10.1007/s10661-010-1554-9

Dhillon, K. S. and Dhillon, S. K. (2003). Distribution and management of seleniferous soils. *Adv. Agron.*, **79**, 119-184.

Dhillon, K. S., Bawa, S. S. and Dhillon, S. K. (1992). Selenium toxicity in some plants and soils of Punjab. *J. Indian Soc. Soil Sci.*, **40**, 132-136.

FAO/WHO. (1989). Joint FAO/WHO Expert Committee on Food Additives. Toxicological evaluation of certain food additives and contaminants. Cambridge Univ. Press.

Ghosh, K., Das, I., Das, D. K. and Sanyal, S. K. (2012). Evaluation of humic and fulvic acid extracts of compost, oil cake, and soils on complex formation with arsenic. *Soil Res.*, CSIRO Pub., **50**, 239–248.

Guha Mazumdar, D.N., Haque, R., Ghosh, N., De, B. K., Santra, A., Chakraborti, D. and Smith, A. H. (1998). Arsenic levels in drinking water and the prevalence of skin lesions in West Bengal, India. *Interl. J. Epidemiology*, **27**, 871-877.

Guha Mazumder, D. N., Deb, D., Biswas, A., Saha, C., Nandy, A., Das, A., Ghose, A., Bhattacharyya, K. and Mazumdar, K. K. (2014). Dietary arsenic exposure with low level of arsenic in drinking water and biomarker: A study in West Bengal. *J. Environ. Sci. Health*, Part A, **49**, 555–564.

Guha Mazumder, D. N., Deb, D., Biswas, A., Saha, C., Nandy, A., Ganguly, B., Ghose, A., Bhattacharyya, K. and Mazumdar, K. K. (2013). Evaluation of dietary arsenic exposure and its biomarkers: A case study of West Bengal, India. *J. Environ. Sci. Health*, Part A, **48**, 896-904.

Halder, D., Bhowmick, S., Biswas, A., Mandal, U., Nriagu, J., Mazumdar, D. N., Chatterjee, D. and Bhattacharya, P. (2012, Apr 3) Consumption of brown rice: a potential pathway for arsenic exposure in rural Bengal. *Environ. Sci. Technol.*, **46** (7), 4142-4148. doi:10.1021/es204298a.

IARC (1987). International Agency for Research on Cancer. Overall evaluations of carcinogenicity; an updating of IARC Monographs vols. 1-42.

IRIS (2015). Integrated risk information system-database, US Environmental Protection Agency

Jarup, L. (1992). Dose-response relation for occupational exposure to arsenic and cadmium. National institute for occupational health, Sweden.

Jeevan Rao, K. (2005). Composting of organic wastes. *Kisan World*, **32**, 29-30.

Jeevan Rao, K. and Shantaram, M.V. (1995). Ground water pollution from open refuse dumps at Hyderabad. *Indian J. Environ. Health*, **37**, 197-204.

Jopony, M. and Young, S. D. (1994). The solid solution equilibria of lead and cadmium in polluted soils. *Eur. J. Soil Sci.*, **45**, 59–70.

Juwarkar, A.S., Shende.A., Thawala, P.R., Satyanrayan., Deshbratar, P.B., Bal, A.S.and Juwarkar, A.(1992). *Fertilizers, Organic Manures, Recyclable wastes and Biofertilizers*. FDCO. New Delhi.72-90.

Kundu, S., Coumar, M. Vassanda, Rajendiran, S., Ajay and Subba Rao, A. (2015). Phosphates from detergents and eutrophication of surface water ecosystem in India. *Curr. Sci.*, **108** (No. 7, April 10, 2015), 1320-1325.

Lari, S. Z., Khan, N. A., Gandhi, K. N., Meshram, T. S. , Thacker, N. P. (2014). Comparison of pesticide residues in surface water and ground water of agriculture intensive areas. *J. Environ. Health Sci. Eng.*, **12** (1), 11. doi: 10.1186/2052-336X-12-11.

Lokeshwari, H. and Chandrappa, G. T. (2006). Impact of heavy metal contamination of Bellandurlake on soil and cultivated vegetation. *Curr. Sci.*, **91**, 622-627.

Ma, L. Q., Komar, K. M., Tu, C., Zhang, W., Cai, Y. and Kennelley, E. D. (2001). A fern that hyperaccumulates arsenic. *Nature*, **409**, 579.

Majumdar, A., Bhattacharyya, K., Kole, S. C. and Ghosh, S. (2013a). Efficiency of indigenous soil microbes in arsenic mitigation from contaminated alluvial soil of India. *Environ. Sci. Pollut. Res. Int.*, **20** (8), 5645-53. doi: 10.1007/s11356-013-1560-x.

Majumdar, K. and Sanyal, S.K. (2003). pH-Dependent arsenic sorption in an Alfisol and an Entisols of West Bengal. *Agropedology*, **13**, 25-29.

Majumder, A., Bhattacharyya, K., Bhattacharyya, S. and Kole, S. C. (2013b). Arsenic-tolerant, arsenite-oxidizing bacterial strains in the contaminated soils of West Bengal, India. *Sci. Total Environ.*, **463–464,** 1006–1014.

Mandal, A., Purakayastha, T. J., Ramana, S., Neenu, S., Bhaduri Debarati, Chakraborty, K., Manna, M. C. and Subba Rao, A. (2014). Status on phytoremediation of heavy metals in India- A review. *Intern. J. Bioresour. Stress Mangmnt.*, **5** (4), 553-560. DOI:10.5958/0976-4038.2014.00609.5

Meena, R., Datta, S. P., Golui, D., Dwivedi, B. S. and Meena, M.C. (2016). Long-term impact of sewage irrigation on soil properties and assessing risk in relation to transfer of metals to human food chain. *Environ. Sci. Pollut. Res.*, **23,** 14269-14283. DOI 10.1007/s11356-016-6556-x)

Mikkelsen, R. L., Page, A. L. and Bingham, F. T. (1989). Factors affecting selenium accumulation by agricultural crops. **In:** *Selenium in Agriculture and Environment* (L.W. Jackobs, Ed.), Special Publication No. 23, Soil Science Society of America, Madison, Wisconsin, pp. 65-94.

Mukhopadhyay, D. and Sanyal, S. K. (2004). Complexation and release isotherm of arsenic in arsenic-humic/fulvic equilibrium study. *Aust. J. Soil Res.*, **42,** 815–824.

Mukhopadhyay, D., Mani, P.K. and Sanyal, S.K. (2002). Effect of phosphorus, arsenic and farmyard manure on arsenic availability in some soils of West Bengal. *J. Indian Soc. Soil Sci.*, **50,** 56-61.

Nawlakhe, W. G., Kulkarni, D. N., Pathak, B. N. and Bulusu, K. R. (1975). Defluoridation of water by Nalgonda technique. *J. Environ. Health,* **17,** 26-65.

Neal, R.H. (1995). Selenium. **In:** *Heavy Metals in Soils* (B.J. Alloway, Ed.), Blackie Academy and Professional, Glasgow. pp. 260-283.

NRCC (1978). National Research Council of Canada, Ottawa, Canada, No. 15391.

Pandya, N. C. (2015). Nanotechnology: Future of Environmental Pollution Control. *Intern. J. Recent and Innovation Trends in Computing and Commun.*, **3** (2), 164-166.

Paulose, B., Datta, S. P., Rattan, R. K. and Chhonkar, P. K. (2007). Effect of amendments on the extractability, retention and plant uptake of metals on a sewage-irrigated soil. *Environ. Pollut.*, **146,** 19-24.

Peterson, J.A. and Nebekar, A.V. (1992). Estimation of water-borne Se concentration that are toxicity thresholds for wildlife. *Arch. Environ. Contam. Toxicol.*, **23,** 440-452.

Rang Zan, N., Datta, S. P., Rattan, R. K., Dwivedi, B. S. and Meena, M. C. (2013). Prediction of the solubility of zinc, copper, nickel, cadmium, and lead in metal-contaminated soils. *Environ Monit. Assess.*, **185,** 10015–10025. DOI 10.1007/s10661-013-3309-x

Rao, M. S. and Mamatha, P. (2004). Water quality in sustainable water management. *Curr. Sci.*, **87,** 942-947.

Rattan, R. K., Datta, S. P. and Sanyal, S. K. (2009). Pollutant elements and human health. *Bull. Indian Soc. Soil Sci.*, No. **27,**103-123.

Sadana, U. S. and Bijay-Singh (1989). Effect of cadmium-zinc interaction on yield and cadmium and zinc content of maize (Zea mays L.). *Curr. Sci.*, **58**, 194-196.

Sanyal, S. K. and Dhillon, K. S. (2005). Arsenic and selenium dynamics in water-soil plant system: a threat to environmental quality. **In**: *Proc. Intern. Conf. Soil, Water and Environmental Quality-Issues and Strategies,* Indian Soc. Soil Sci., New Delhi. pp. 239-263.

Sanyal, S. K., Gupta, S. K., Kukal, S. S. and Jeevan Rao, K. (2015a). Soil degradation, pollution and amelioration. **In**: *State of Indian Agriculture-Soil* (H. Pathak, S. K. Sanyal and P. N. Takkar, Eds.), National Academy of Agricultural Sciences, New Delhi, pp. 234-266.

Sanyal, S. K., Jeevan Rao, K. and Sadana, Upkar S. (2012). Toxic elements and other pollutants- A threat to nutritional quality. **In**: *Soil Science in the Service of Nation* (Goswami, N.N. *et al.,* Eds.), Indian Soc. Soil Sci., New Delhi: 266-291.

Sanyal, S. K.,. Dwivedi, B. S., Singh, V. K., Majumdar, K., Datta, S. C., Pattanayak, S. K. and Annapurna, K. (2015b). Phosphorus in relation to dominant cropping sequences in India: Chemistry, fertility relations and management options. *Curr. Sci.*, **108** (No. 7, April 10, 2015), 1263-1270.

Sanyal, S.K. (2005). Arsenic contamination in agriculture: A threat to water-soil-crop-animal-human continuum. *Presidential Address, Section of Agriculture and Forestry Sciences, 92nd Session of the Indian Science Congress Association (ISCA),* Ahmedabad, January 3-7, 2005; Indian Science Congress Association Kolkata.

Siciliano, S. D. and Germida, J. J. (1998). Mechanisms of phytoremediation: Biochemical and ecological interactions between plants and bacteria. *Environ. Rev.*, **6**, 65-79.

Sinha, B. and Bhattacharyya, K. (2011). Retention and release isotherm of arsenic in arsenic–humic/fulvic equilibrium study. *Biol. Fertil. Soils*, **47**, 815–822. DOI 10.1007/s00374-011-0589-6.

Sinha, B. and Bhattacharyya, K. (2014a). Arsenic accumulation and speciation in transplanted autumn rice as influenced by source of irrigation and organic manure. *Intern. J. Bioresour. Stress Managmt.*, **5** (3), 336-368.

Sinha, B. and Bhattacharyya, K. (2014b). Arsenic toxicity in rice with special reference to speciation in Indian grain and its implication on human health. *J. Sci. Food Agric.*, **95** (7), 1435-1444. (wileyonlinelibrary.com) DOI 10.1002/jsfa.6839.

Sinha, B., Bhattacharyya, K., Giri, Pradip K. and Sarkar, S. (2011). Arsenic contamination in sesame and possible mitigation through organic interventions in the lower Gangetic Plain of West Bengal, India. *J. Sci. Food Agric.*, **91**, 2762–2767. (wileyonlinelibrary.com) DOI 10.1002/jsfa.4519.

Tandon, H. L. S. (1997). Organic Resources: An Assessment of potential supplies their contribution to Agricultural productivity and Policy Issues for Indian agriculture from 2000 to 2025. **In**: *Plant Nutrient Needs, Supply, Efficiency and Policy Issues: 2000-2025.* (J. S. Kanwar and J. C. Katyal, Eds.), National Academy of Agricultural Sciences, New Delhi New Delhi, pp.15-28.

Tiwana, N. S., Panesar, R. S. and Kansal, B. D. (1987). Characterization of wastewater of a highly industrialized city of Punjab. **In:** *National Seminar on Impact of Environmental Protection for Future Development of India*, Volume 1, Nainital, pp. 119-126.

Venkata Sridhar, T., Jeevan Rao, K. and Bhopal Raj. G. (2006). Risk assessment of metals in sewage sludge of Hyderabad. *J. Res.*, ANGRAU, **34**, 82-86.

WHO. (1971). International standards for drinking water, Third edition.

WHO. (1993). Guidelines for Drinking-Water Quality, Second edition, Volume 1.

WHO. (1996). Anon., World Health Organization guidelines for drinking water quality,Vol. II, Second Edition, WHO, Geneva.

WHO. (2001). http://www.who.int/inf-fs/en/fact210.html

Zayed, A., Lytle, C.M., Qian, J-H. and Terry, N. (1998). Chromium accumulation, translocation and chemical speciation in vegetable crops. Planta, 206, 293. Doi:10.1007/s004250050403.

Index

T